Trace Metals and Metal-Organic Interactions in Natural Waters

Philip C. Singer

Department of Civil Engineering
University of Notre Dame
Notre Dame, Indiana

ann arbor science PUBLISHERS INC.
POST OFFICE BOX 1425 ● ANN ARBOR, MICHIGAN 48106

Second Printing, 1974

Copyright © 1973 by Ann Arbor Science Publishers, Inc.
P.O. Box 1425, Ann Arbor, Michigan 48106

Library of Congress Catalog Card No. 72-96909
ISBN 0-250-40014-6

PREFACE

    Trace metals in natural waters have become a
significant topic of concern for scientists and
engineers in various fields associated with water
quality, as well as a concern of the general public.
Direct toxicity to man and aquatic life and indirect
toxicity through accumulations of metals in the
aquatic food chain are the focus of this concern.
Mercury, cadmium, and lead are but a few dramatic
examples about which much has been written.
    The presence of trace metals in natural aquatic
systems originates from the natural interactions
between the water and the sediments and atmosphere
with which the water is in contact. The concentrations fluctuate as a result of natural hydrodynamic,
chemical, and biological forces. Man, through
industrialization and technology, has developed the
capacity, perhaps unknowingly, to alter these natural
interactions to the extent that his very use of these
waters and the aquatic life therein is threatened.
    The activity of trace metals in aquatic systems
and their impact on aquatic life vary depending upon
the metal species which are formed. Of major importance in this regard is the ability of metals to
associate with other dissolved and suspended
components of the system. Most significant among
these associations is the interaction between metals
and organic compounds in the water. Many of these
organic species, whether they originate from natural
processes such as vegetative decay or result from
pollution through organic discharges from municipal
and industrial sources, have a remarkable affinity
and capacity to bind metals, a phenomenon which
would naturally alter the reactivity of the metals
in the aquatic system. It is to this association
between metal ions and organic compounds that this
volume is devoted.
    This book developed from a recent symposium,
sponsored by the Division of Air, Water and Waste

Chemistry, held at the 163rd National Meeting of the American Chemical Society in the Spring of 1972. The papers contained herein represent a cross section of the current areas of research related to trace metals and metal-organic interactions. These include analytical, thermodynamic, biochemical, and kinetic considerations presented as review papers, reports of experimental research, and theoretical models summarizing available information from the literature, with papers having been contributed by analytical, physical, and organic chemists, soil scientists and geologists, aquatic biologists and limnologists, and environmental and chemical engineers. As such, the papers should be of great interest to scientists and engineers of similar disciplines with interests in water quality.

The cooperation of the authors and their excellent contributions, the assistance of the Department of Civil Engineering of the University of Notre Dame, and the support of the Division of Air, Water and Waste Chemistry of the American Chemical Society for sponsoring the initial symposium is appreciated.

Notre Dame, Indiana   Philip C. Singer
April, 1973

TABLE OF CONTENTS

Chapter

vii

LIST OF CONTRIBUTORS

Julian B. Andelman. Graduate School of Public Health,
    University of Pittsburgh, Pittsburgh, Pennsylvania.
F. A. J. Armstrong. Fisheries Research Board of Canada,
    Freshwater Institute, Winnipeg, Manitoba, Canada.
Richard T. Barber. Duke University Marine Laboratory,
    Beaufort, North Carolina.
C. W. Childs. Soil Bureau, Department of Scientific and
    Industrial Research, Lower Hutt, New Zealand.
Russell F. Christman. College of Engineering, Department of
    Civil Engineering, University of Washington, Seattle
    Washington.
Howard J. Davis. Celanese Research Company, Summit, New
    Jersey.
Donald S. Gamble. Soil Research Institute, Research Branch,
    Canada Agriculture, Ottawa, Ontario, Canada.
A. L. Hamilton. Fisheries Research Board of Canada,
    Freshwater Institute, Winnipeg, Manitoba, Canada.
Marc A. Horton. Department of Civil Engineering, College of
    Engineering, University of Washington, Seattle, Washington.
Bjorn F. Hrutfiord. College of Forest Resources, University
    of Washington, Seattle, Washington.
Leonard A. Lee. Celanese Research Company, Summit, New
    Jersey.
Harry V. Leland. University of Illinois, Champaign-Urbana,
    Illinois.
A. Lerman. Department of Geological Sciences, Northwestern
    University, Evanston, Illinois.
E. J. Malec. Monsanto Company, St. Louis, Missouri.
Barbara B. Martin. Department of Chemistry, University of
    South Florida, Tampa, Florida.
Dean F. Martin. Department of Chemistry, University of
    South Florida, Tampa, Florida.
Russell E. McDuff. W. M. Keck Laboratory of Environmental
    Engineering Science, California Institute of Technology,
    Pasadena, California.
Roger A. Minear. Department of Environmental Engineering,
    Illinois Institute of Technology, Chicago, Illinois.

Francois Morel. W. M. Keck Laboratory of Environmental Engineering Science, California Institute of Technology, Pasadena, California.

James J. Morgan. W. M. Keck Laboratory of Environmental Engineering Science, California Institute of Technology, Pasadena, California.

Bruce B. Murray. Department of Chemistry, Wisconsin State University--River Falls, River Falls, Wisconsin.

Morris Schnitzer. Soil Research Institute, Research Branch, Canada Agriculture, Ottawa, Ontario, Canada.

Neil F. Shimp. Illinois State Geological Survey, Urbana, Illinois.

Surendra S. Shukla. University of Illinois, Champaign-Urbana, Illinois.

Philip C. Singer. Department of Civil Engineering, University of Notre Dame, Notre Dame, Indiana.

R. D. Swisher. Monsanto Company, St. Louis, Missouri.

T. A. Taulli. Monsanto Company, St. Louis, Missouri.

Thomas L. Theis. Department of Civil Engineering, University of Notre Dame, Notre Dame, Indiana.

1. METHODS OF TRACE METALS ANALYSIS IN AQUATIC SYSTEMS

Roger A. Minear. Department of Environmental
Engineering, Illinois Institute of Technology,
Chicago, Illinois.

Bruce B. Murray. Department of Chemistry,
Wisconsin State University—River Falls, River
Falls, Wisconsin.

INTRODUCTION

The role that trace metals play in natural
biological life cycles is certainly of importance
in defining the behavior and well-being of individual
biological systems directly and, therefore, in defining the overall character of a water ecosystem.
Understanding of this role is intimately connected
to the exact forms and mechanisms of transformation
of the trace metals in the system. These considerations relate both to those metals that are essential
to biological function and those metals that may be
inhibitory to organisms in the aquatic system.
Regarding the latter point, biological transformations may convert inorganic metal forms to organic
compounds of much greater toxicities.[1]
The analysis of trace metals is intimately
related to the setting and enforcement of industrial
and domestic waste discharge standards. The reliability and detectability limits of analytical
methods may be the limiting factor in defining
standards and maintaining environmental surveillance.
Methods employed must be reliable and provide an
unambiguous estimate of the species under consideration. Furthermore, in the case of effective
surveillance programs, methods that are rapid and
which can be accomplished within some reasonable
framework of cost, both of equipment and man power,
are needed.

1

GENERAL

The analysis of water in its natural state for trace metals can be simple or very complex. If one is required to analyze for only one metal species in only a few water samples, the methods are in general very adequate and available. However, it is very common nowadays to study a dozen or more trace metal parameters in samples numbering from a few dozen to several hundred. The samples often are not laboratory samples but field samples taken from rivers, streams, lakes, wastewater, etc. These added factors complicate the whole analytical system and present a challenge to the analyst.

For metals in aqueous systems, trace metals analysis is translated roughly into concentrations of tens of mg/l and extends down to sub µg/l. Table 1 presents a survey of the methods available for trace metals analyses and their approximate lower levels of use. It should be emphasized that

Table 1

Survey of Methods of Analysis and Their
Approximate Concentration Ranges

Analytical Technique	*Approximate Levels of Use, M*
Optical Methods	
Molecular absorption spectrophotometry	$10^{-5} - 10^{-6}$
Molecular fluorescence spectrophotometry	$10^{-7} - 10^{-8}$
Flame emission photometry	$10^{-6} - 10^{-7}$
Spark and arc emission spectrometry	$10^{-5} - 10^{-6}$
Atomic absorption spectrophotometry	$10^{-6} - 10^{-7}$
Atomic fluorescence spectrophotometry	$10^{-6} - 10^{-7}$
Electrical Methods	
Classical polarography	$10^{-5} - 10^{-6}$
Special voltammetric methods	$10^{-7} - 10^{-8}$
Anodic stripping voltammetry	
Hanging drop Hg	$10^{-8} - 10^{-9}$
Thin film Hg or solid electrode	$10^{-9} - 10^{-10}$
Specific ion electrodes	$10^{-5} - 10^{-6}$
Other Methods	
X-ray emission	$10^{-5} - 10^{-6}$
Neutron activation analysis	$10^{-9} - 10^{-10}$
Mass spectrometry	$\sim 10^{-6}$
Gas chromatography (also GC-MS couple)	$10^{-6} - 10^{-8}$

no single method is universally applicable to all
elements or even a particular element in different
chemical forms. Furthermore, the limits are general
ranges and a given element analyzed by a given
method may not necessarily be detected at those
values.

The choice of a particular analytical method
is most often dictated by the available equipment
and facilities. However, when considering a trace
metal method and its adequacy, one should carefully
consider various criteria, among which are sensitivity,
selectivity, accuracy and precision, practicability
(complexity of the method) and economy.

In any trace analysis method the first consideration is the sensitivity of the method. This is
generally stated in parts per million (mg/l) or
parts per billion (μ g/l) but just as often it is
more realistic to state sensitivity at the micro
or nanogram levels. With the very low levels of
some trace metals in natural waters, methods are
needed having sensitivities of a nanogram or less.
Such methods may in many instances involve concentration of the sample but if possible this should
be avoided. The determination of sensitivity or
the stated sensitivity value for a particular method
is generally not an *exact* figure and will vary in
definition from author to author. Sensitivity is
sometimes defined as that concentration or amount
of analyte that yields a 1% of full scale reading[2]
and is used in this manner in atomic absorption.[3]
Another definition[4] for sensitivity in atomic
absorption spectrometry is that concentration of an
element that will produce an absorption of 1%
generally expressed as μg/ml/1%. The detection
limits are usually defined as twice the background.
Still another definition of sensitivity is found in
Elliott.[5] The analyst must realize that stated
values for sensitivity and detection limits can
be largely instrument and operator dependent and in
all cases should be determined experimentally and
carefully defined.

The selectivity of analytical methods is the
degree to which the method analyzes one element
with no interferences from other elements in the
matrix. That is, neutron activation analysis is
very selective but molecular absorption methods
using a general complexing colormetric reagent,
such as 8-hydroquinone, may provide a nonspecific
response from several elements simultaneously.
Ideally one would like methods that are specific

for each element to be analyzed with little or no interferences.

Accuracy and precision of the trace metal procedures are important but data will be less accurate and precise as the concentrations analyzed reach into the µg/l region. Each procedure should be checked for precision on *real* samples and the data reported with respect to the 2σ standard deviation. Determination of the accuracy of a method is also a *must* but in the absence of a wide variety of standard materials, the analyst has to use pseudo synthetic samples. Interlaboratory analyses are used to check standard samples and procedures but one should not expect too much as the results of cooperative interlaboratory analysis[2] have not always been outstanding. A reasonable approach is to spike real samples with known aliquots of the elements to be analyzed and let the data speak for themselves.

Strong emphasis is to be placed on the practicability of the proposed method particularly for routine application. Elegant procedures that are very technique-oriented may work well in the hands of an experienced laboratory worker but fail utterly when used by relatively inexperienced personnel. Published work should provide comments explaining the complexity of the analytical procedure. Also, factors such as interferences, standards, sampling, the analysis of real samples, etc. should be discussed in any proposed method. In general, there seems to be a reluctance to discuss openly the negative aspects of any procedure or method.

In the analysis of water in its natural state, sampling protocol must be evaluated and standardized. There are no universally accepted set rules for water sampling and in any situation the investigator should rely on the experience of others, himself and common sense. General water sampling methods and sample treatment are discussed in the literature.[6-8] The most important factors are obtaining a sufficient representative sample and preserving the sample to prevent loss (plating on container walls) or transformation of chemical and physical form.

The methods and techniques to be considered in greater detail in the following sections are molecular absorption spectrometry (colorimetric methods), molecular fluorescence spectrometry, atomic absorption spectrometry, atomic fluorescence spectrometry, emission and flame spectrometry, x-ray emission spectrometry, neutron activation analysis,

electro-chemical techniques including ion selective
electrodes, mass spectrometry and gas chromatography.
Where appropriate, consideration will be given to
sample concentration and separation as these techniques frequently extend the applicability of a
particular technique.

There will be no attempt to give a complete
review of each technique on all methods of trace
metals analysis. The annual reviews of *Analytical
Chemistry* offer excellent current reference material
on these techniques as applied to trace metals
analysis. The emphasis will be on the methods presently available and in use and those in development
which hold promise for application to water analysis
in the near future. Our focus will be on the current
state of the art but in order to accommodate an
anticipated broad audience, the basic principle
behind each method will be delineated, primarily to
emphasize the advantages and disadvantages of a
particular technique.

METHODS INVOLVING ELECTRONIC
TRANSITIONS

For purposes of quickly and clearly differentiating the principle behind molecular absorption,
fluorescence and emission, and atomic absorption,
fluorescence and emission (including x-ray techniques), such a broad categorization is appropriate.
Each of these methods involves the absorption of
energy and the precise measurement of the energy
absorbed or of the energy released when the excited
species returns to the normal or ground state.

Remembering that each element has a unique
build up of electrons which is a sequential filling
of the lowest energy orbitals, any other distribution
of these electrons constitutes an excited state. To
promote an electron to an excited state takes an
exact amount of energy characteristic of the particular transition and the overall nuclear and
electronic structure of that element. Light
absorbed to promote excitation of outer electrons
or light emitted by electrons returning to lower
excited states or the ground state occurs in a
single narrow range of wavelengths. It is this
narrow line specificity that makes atomic methods
preferable to molecular methods for metals analysis.
(See Figures 1 and 2 for further understanding.)

Molecular absorption in the ultraviolet and
visible spectral regions is usually the result of

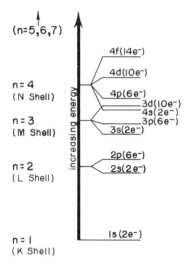

(n=5,6,7)

n = 4
(N Shell)

4f(14e⁻)

4d(10e⁻)

4p(6e⁻)
3d(10e⁻)
4s(2e⁻)
3p(6e⁻)
3s(2e⁻)

n = 3
(M Shell)

n = 2
(L Shell)

2p(6e⁻)
2s(2e⁻)

n = 1
(K Shell)

1s(2e⁻)

increasing energy

Atomic Energy Levels: extent to which levels are filled sequentially defines the chemical element under consideration.

Processes:

(I) Excitation - Input of energy promotes electron in lower energy level to next higher level. Extreme is complete removal or ionization. Energy may be thermal or electro-magnetic (EM).

(2) Atomic Absorption - Outermost electrons absorb (EM) radiation required for transition from ground state (normally last level occupied) to higher levels. Incident light energy diminishes in proportion to number of atoms in path

(3) Emission - Excited electrons in atom or ion return to lower energy levels (ground or lower excited levels) emitting EM radiation corresponding to the energy of transition.

   (a) Flame Emission - usually outermost electrons only, therefore only few characteristic lines per element.

   (b) Emission Spectroscopy - Excitation energy much greater (spark or arc) and the inner electrons (middle shells) are excited yielding more emission lines.

   (c) Atomic Fluorescence - excitation initiated by incident radiant energy rather than thermal means.

(4) X-ray Emission - Transitions involve inner most electrons of the K and L shells for elements containing electrons in 3p and higher energy levels. Excitation is by X-rays which remove inner shell electrons from the K and L shells and when these vacant shells are filled by electrons from the L or M shells the characteristic K and L X-rays are emitted.

Figure 1. Processes of energy emission and absorption involving electronic transitions in elements in the free atomic state.

Processes:

(1) Absorption - Incident radiation corresponding to transition from particular combination of ground electronic and given vibrational and rotational state to particular combination of E_1 or E_2 and attendant vibrational and rotational states. Within each electronic transition, more molecules will be in a particular vibrational and rotational combination; this corresponds to λ max.

(2) Deactivation – Return to Ground State –

(a) Thermal or Collisional - Most common in solution. In absence of rapid deactivation, absorption would demonstrate a time dependent decay.

(b) Radiant (Fluorescence) - If partial vibrational deactivation occurs by thermal and other processes to the ground vibrational state of the lowest excited electronic state, radiant energy release to the ground electronic state may occur in preference to thermal deactivation. The range of λ emitted will generally be the same and independent of the excitation λ.

(c) Radiant (Phosphorescence) - When internal conversion occurs between an excited singlet state and a triplet state, radiant deactivation is much slower in rate and of still higher λ than fluorescence.

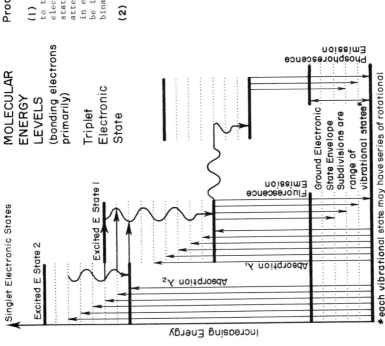

Figure 2. Processes of energy absorption and emission involving electronic transitions in molecular species.

bonding electrons being excited. The actual energy
of transition between electronic states depends upon
the vibrational and rotational levels of each
molecular species before and after excitation. The
vibrational and rotational energy levels are relatively close together and, in solution, molecular
electronic transitions appear as a continuous band
over the excitation energy scale simply because
instruments cannot resolve the individual lines.
When more than one absorbing species is present,
overlapping of the individual spectra represents a
lack of response specificity which is dependent upon
the degree of overlap and the relative efficiency
of energy absorption (and emission for fluorescence).
        Transitions involving inner electrons (K and L
Shell, Figure 1) correspond to absorption and release
of greater energy in the form of x-rays and reflect
the elemental nature rather than the compound containing the element. X-ray methods have the
specificity of atomic methods but can be applied
to elemental analysis in a molecular matrix.

Molecular Methods

Molecular absorption spectrometry

        Molecular absorption methods, more commonly
referred to as colorimetric analyses, are perhaps
the most commonly used for general trace metals
analysis in water with the notable exception of the
alkali and, to a lesser extent, alkaline earth
metals. Many books have been written on colorimetric
analysis[9-13] and literally hundreds of new methods
or modifications of existing methods have been
introduced in recent years.[14,15] The great
popularity of such methods can be attributed to
the relative ease with which determinations can be
made and the low cost of routine instrumentation.
With the technological advances in instrument
stability and wavelength resolution, the precision
of these routine analyses has greatly improved with
respect to instrumental variables. It has been
noted, however,[14] that the amount of attention given
to molecular absorption in the United States has
decreased considerably. This has been attributed
to greater focus on development of atomic absorption,
electrochemical and gas chromatographic methods.
        The primary criteria for molecular absorption
analyses are high selectivity, high sensitivity,
relative insensitivity of color intensity to pH,

temperature and solution composition fluctuations, stability of color intensity with time, and simplicity of analysis. The introduction of automated analysis systems[16],[17] has afforded control over the time, temperature, and pH variations by providing highly reproducible conditions. However, such systems, which may be continuous or incremental in their operation, become economical only for a large number of samples taken on a routine basis. With recent concern directed to on site analysis and the development of relatively reliable and compact battery-operated filter photometers (*e.g.*, Hach Chemical Co.), field analysis units are returning to use. Such units could be called second generation color comparators. For certain metals (*i.e.*, Cu, Cr) measurements are claimed to be $\pm$ 20% on a direct concentration readout. For preliminary field data, such results may be more than adequate.

Specificity and sensitivity can present more formidable problems. Even though some aquo metal ions and oxy anions absorb visible radiation in solution, within the range of realistic sample path lengths, detection is possible only at the 0.1 to 1 g/l levels. Therefore, most metal analysis procedures require conversion of the metal ion to species with a greater efficiency of energy absorption or absorptivity. Usually, this entails forming a complex with an organic ligand. It is desirable that only the metal of interest form a light-absorbing species with the organic compound and thus achieve specificity of response. Alternately it is desirable that those other compounds formed have absorption bands sufficiently displaced from and/or absorptivities sufficiently less than the metal of interest to accomplish specificity.

Much of the continuing work in colorimetric methods for trace metals analysis involves synthesizing compounds with a high specificity for a particular metal form. The other avenue is to find a mechanism for preventing other species from complexing. A case in point is the analysis of Cd, Zn and Pb with dithiozone.[7] Each metal ion forms a complex which absorbs strongly in the visible spectrum. Although different in spectral characteristics, each complex overlaps sufficiently to greatly interfere with the others. However, reaction specificity is accomplished by pH control and the resultant product may be partitioned into an immiscible organic phase. Selective precipitation of the interfering metal ions provides another avenue but with attendant increase in sample manipulation, particularly if filtration is mandatory.

A much more suitable approach to removing interference is by masking or selectively complexing the interfering ions with other organic or inorganic ligands. The following conditions must necessarily be met: the masking agent has a stronger affinity for the interfering ion than the color reagent, lesser or none for the analyte ion, and either does not absorb in the visible region or does not overlap with the analyte complex. As West and West[18] point out, this technique is essentially the standard tool for enhancing specificity. Thousands of such agents are reported in the literature and the general subject is discussed by West and West[18,19] who indicate the literature is so extensive as to be impossible to review. Some of the more common masking agents are citrate (iron), cyanide, fluoride, ethylenediamine-tetraacetic acid (EDTA) and tartarate.

Molecular fluorescence spectrometry

When a molecular species, excited by the absorption of ultraviolet or visible radiation, returns to the ground electronic state via energy released as light, the process is designated as fluorescence (overlooking the special case of phosphorescence). Fluorescence is a much more specific phenomenon than simple absorption and consequently affords a greater specificity of response, especially for molecular species in solution since the fluorescence radiation is at longer wavelengths than the excitation radiation.[20-24] Frequently, more than one excitation wavelength can be used, further enhancing specificity.

Since aqueous metal ions do not fluoresce in solution, with the exception of weak fluorescence by lanthanides,[25] organic chelates must be formed. Fifty-four metals form "luminescent" complexes most commonly with multidentate ligands containing the imido, hydroxy, and carbonyl functional groups.[25] However, application of fluorescence methods for analysis of metals in solution is not nearly as extensive as absorption in spite of the frequently much greater sensitivity and detectability (Table 2). Prior to the improvement and development of other techniques such as atomic absorption, fluorescence methods were superior to other methods for a number

Table 2

Summary of Selected Fluorescence Methods for Metal Ion Analysis
[Reagents and individual source references
listed in Winefordner et al.[24]]

Metal	Number of Fluorescence Reagents Listed	Limit of Detection in mg/l
Ag	3	0.01 - 0.004
Al	5	0.0008 - 0.0002
As	2	50.0 - 7.0
Au	1	0.5
Be	4	0.2 - 0.00004
Bi	1	0.5
Ca	2	0.02 - 0.01
Cd	1	0.02
Co	2	0.06 - 0.0001
Cu	6	3.0 - 0.0002
Fe	2	0.001 - 0.0008
Hg	1	0.002
In	6	5.0 - 0.04
Li	2	1.0 - 0.2
Mg	3	0.004 - 0.00001
Mn	1	0.002
Mo	1	0.1
Ni	1	0.00006
Pb	1	5.0
Ru	1	1.0
Sb	2	0.1 - 0.05
Se	2	0.01 - 0.005
Sn	1	0.1
Th	3	0.5 - 0.02
Cl	2	0.5 - 0.01
V	1	2.0
W	1	0.04
Zn	6	0.03 - 0.002
Zr	1	0.02

of inorganic ions, notably Al, Be, Cd, Li, Zn and
Zr.[26]   As with molecular absorption methods,
specificity can be enhanced further by pH control
as with the Al - 2,4-bis [N,N'-di(carboxymethyl)
aminomethyl] fluorescein complex at pH 2.[20]   Selective partitioning into an organic phase can provide
specificity for trace metals analysis[21],[22],[26] while
also serving as a concentration step by suitable
volume ratio selection. Masking techniques used
in absorption spectrometry are applicable to
fluorescence spectrometry.

The much greater sensitivity of fluorescence
methods over absorption lies, in part, with the
difference in response measurement and its relationship to analyte concentration. Absorption methods
rely upon measuring the ratio of light energy passing
through the solvent and that passing through the
analyte solution. At low concentration the small
signal is actually a small difference between two
large numbers. Fluorescence, however, involves a
small signal above background noise, which can be
amplified. Since emission energy is proportional
to incident energy, increasing the source output
will enhance sensitivity within the limits of
solution heating or photo decomposition.

Problems in fluorescence analysis are
luminescence background and solvent Raman spectrum.[27]
The Raman problem is usually readily remedied by
proper excitation - emission wavelength resolution.
The background fluorescence, however, is not so
simple, particularly in natural waters where many
organic species may exhibit strong fluorescence.[28]
A high background reduces precision and detectability. This latter problem may well be responsible
for the reduced popularity of fluorescence analysis
in natural waters. Sample ashing alleviates such
problems with attendant increased sample handling
and processing requirements.

Photo-decomposition and quenching due to dissolved salts impose further limitations on fluorescence measurements and certainly must be evaluated
prior to method application and development.

Although a sizeable amount continues to be
published regarding fluorescence analysis of metal
ions in solution,[29],[30] Fishman and Erdman[27]
reported only five applications to water analysis
in their comprehensive review.

Atomic Methods

Since quantitation of the metal species is the fundamental goal, measurements involving electronic transitions of the elemental or atomic form directly affords a specificity not attainable with molecular species. Thus, atomic methods are characterized by their high degree of specificity. Atomic absorption, atomic fluorescence, atomic emission, and x-ray emission spectroscopy are all atomic methods and are discussed in this section.

Atomic absorption spectroscopy

Atomic absorption spectroscopy (AAS) is an important technique for analyzing metals in water. The general method involves atomization of samples by flames or thermal sources and the absorption of radiation of a specific wavelength by the atomic specie as it is excited. The radiation source used is a hollow cathode lamp containing, as its cathode, the same element under analysis. The quantity of radiation absorbed by the atomic vapor is proportional to the concentration of the atoms in the ground state. If large numbers of atoms can be atomized without excitation to an excited state, a high sensitivity of $10^{-6} - 10^{-12}$ grams and high selectivity can be realized.

Interferences in AAS are caused by nonatomic absorption due to the flame continuum, the presence of organic matter and high salt levels. These can be corrected by double beam background correction techniques,[31-33] by determination of the background near the analyte line using a nonabsorbing line, or, in some cases, by a matrix blank or standard addition techniques.

AAS is generally done by direct aspiration of liquid samples into a flame. However, advances in flame and nonflame atomization techniques have greatly improved the sensitivity of AAS. Such devices as the tantalum boat,[34] Delves cup,[35,36] carbon rod atomizer,[37,39] graphite furnaces,[40,43] and electrothermal atomizer[43,44] should all be considered when applying AAS to metal analysis in aqueous systems. Table 3 compares some of these atomizing techniques to conventional flame AAS.[31,39,45,46] These systems allow the analyst to present the total sample to the analyzing radiation

Table 3

Comparison of Stated Sensitivities[a]
of Various Atomization Devices

Element	Conventional Atomic Absorbtion µg/ml[45]	Carbon Rod in Picograms[39]	Ta Ribbon/ Inert Gas in Picograms[45]	Carbon Furnace in Picograms[46]
Ag	0.030	1.2	50	
Al	0.70	63	900	150
As	0.96	92	9000	160
Au	0.16	21	100	
Be	0.022	1.1	50	3.4
Bi	0.23	10		280
Ca	0.10	0.65		3.1
Cd	0.011	0.64	8	0.8
Co	0.06	12	500	120
Cr	0.065	9.2	200	18
Cs	0.17	34		71
Cu	0.036	20	60	45
Eu	0.33	61		
Fe	0.061	4.4		120
Ga	1.0	29		120
Hg	2.4	340	2000	15000
K	0.01	2.0		
Li	0.015	6.1		
Mg	0.004	0.43	2	
Mn	0.024	0.67	30	7
Mo	0.60	45		
Na	0.005	0.14		
Ni	0.073	28	2000	330
Pb	0.13	6.8	300	23
Pd	0.10	150		250
Pt	1.05	220		740
Rb	0.06	4.3		41
Sb	0.50	53	4000	510
Se	0.6	72		
Si	2.0			24
Sn	0.91	96	500	5500
Ti	1.25			280
Tl	0.27	11		90
V	1.0	92	2000	320
Zn	0.0095	0.35	20	2.1

[a]Sensitivity defined as the concentration necessary to give
1% absorption change in References 31, 39, and 47, but as
4.4 milliabsorbance in Reference 46.

in a fraction of a second and affords sensitivities in the range of $10^{-9} - 10^{-14}$ grams.

In the analysis of water samples, many elements can be analyzed with little or no sample preparation. Fernandez and Manning[47] reported on the analysis of 12 trace metals in natural water. They used a heated graphite atomizer and reported data for Al, As, Cd, Co, Cr, Ca, Fe, Mn, Pb, Sn, and Zn. The best sensitivity was for Cd at 1.3 picograms and the worst was for Sn at 1500 picograms. Background absorption effects were corrected with a deuterium background corrector and matrix interferences were studied by standard addition techniques. Matrix effects were observed for Mn, Pb, Cr, Sn, and As and the interference of NaCl in the analysis of Pb was studied. Segar and Gonzales[48] in an analysis of sea water, reported that the high salt content of sea water made the direct atomization of samples impossible. Cu, Fe, Mn, Co, Ni, and V could be determined by selective volatilization techniques but Zn, Pb, Cd, and Ag could not be separated from the volatilization of the major salts. Fe had good sensitivity but Cu, Mn, Co and Ni had reduced sensitivity. They recommended a separation of the trace metals before analysis.

Analysis of Hg by flameless AAS[49] continues to be the major analytical method for Hg. The sensitivity of the method is usually in the range of $10^{-9} - 10^{-10}$ grams. The main interference is organic vapors and this can be corrected for by using $PdCl_2$[50] or Ag wool to remove Hg from the vapor phase and thus obtain a true background. Hg can be concentrated from natural water by electrodeposition or amalgamation.[51,52] Sample concentration for conventional AAS can be accomplished by several methods including simple evaporation or freeze drying with dissolution in acid, partitioning of metal chelates in smaller volumes of organic solvent[53] or collection of metal ions on ion exchange or chelating resins.[54,55] Concentration factors of up to 2,000 have been claimed by the latter.[55]

AAS has become a standard method for the analysis of trace metals in water, although only one element can be analyzed for each sample atomized. Multielement AAS has not been successful.

Atomic fluorescence spectroscopy

Atomic fluorescence spectroscopy (AFS) and its applications to trace metal analysis in water is not

as well-developed as AAS. The method involves a
measurement of the fluorescence emission of atomic
species that have been excited by either a continuum
or line source. The concentration of the atomic
species in the flame is proportional to the number
of atoms that can be excited and their resulting
fluorescence. The fluorescence radiation is
generally observed at 90° to the exciting source.
Both discharge tubes and high intensity line sources
have been used to excite AF. In general, inter-
element interference is a problem and the S/N ratios
are poor.

The most intriguing application of this method
is its multielement capabilities.[56],[57] Aqueous
samples have been analyzed for four to eight elements
sequentially with good precision and an excellent
turn around time of one minute per sample. Detection
limits are 0.01 - 0.05 µg/ml. The use of nonflame
excitation such as the carbon rod provides increased
sensitivity and less interference from other elements.
The carbon rod atomizer operated in Ar/H_2 or N_2/H_2
atmosphere[58] is reported to show less interference
and good sensitivity. Table 4 gives comparative
detection limits for AFS and AAS using the carbon
rod atomizer.[59]

Table 4

*Comparison of Atomic Absorption Spectroscopy to Atomic
Fluorescence Spectroscopy -- Detection Levels in Picograms*

Element	AAS	AFS
Ag	1	0.5
As	1000	500
Au	40	10
Be	400	30
Ca	1	0.1
Cd	2	0.03
Cu	20	5
Fe	10	2
Mg	0.4	0.03
Pb	20	3
Zn	1	0.1

On examination of references cited in Reference 60, one wonders why AFS has not been more successful in its application to water analysis for metals, in spite of comments by other investigators who do not believe that AFS can compete with AAS.

Atomic emission spectroscopy

Atomic emission spectroscopy (AES) and flame photometry offer complementary analytical methods to AAS. Both AAS and AES can be conducted with the same instrumentation, and AES affords good sensitivity and selectivity for most metal elements. Sensitivities in the $10^{-8}-10^{-10}$ grams range are common.[5,61] Atomic emission involves the measurement of the radiation emitted when an excited atom or ion loses energy. Generally, samples are excited in flames but plasmas can be used. Normally, flame photometry is carried out in colder flames such as air/propane, but hotter flames such as O_2/C_2H_2 and NO_2/C_2H_2 can be used to reduce interferences. Many elements can be excited simultaneously and excellent linearity over 4-5 decades is observed. The major problem in AES is the separation of the desired signal from the flame continuum, and a high resolution monochromator is necessary. Background correction and matrix effects can be severe.

Applications of AES to metal analysis in water are not numerous. Elliott[5] discusses the use of an argon plasma to analyze 21 elements. The sensitivities obtained were between 0.05-35 nanograms with an aspiration rate of 0.2 ml/min and a 15 second integration time. A comparison study of detection limits for AAS and AES has been made by Christain and Feldman.[62] Using a NO_2/C_2H_2 flame, 68 elements were examined and, of these, 42 could be determined as well by AES as by AAS. Searle and Kennedy[63] discuss the determination of Ca in rainwater, using a high temperature NO_2/C_2H_2 flame and KCl as an ion suppressant. No interferences were noted and the working data was linear from 0-2 mg/l.

Flame emission spectroscopy (FES) could lend itself to multielement determinations if a wide band detector system were used. An optical multichannel analyzer (OMA)[64] coupled to a good monochromator would make possible multielement FES but no commercial instruments have been developed. One could expect several elements to be analyzed simultaneously with good signal to noise ratios and good sensitivity.

Emission spectroscopy

The application of emission spectroscopy methods to trace metals analysis of water offers several advantages: sensitivity, multielement analysis, quantitative as well as qualitative data, and large sample throughput. The disadvantages are matrix effects, high instrumentation costs, and relatively low precision. Also, in analysis of pure water, large samples of 1-10 liters must be concentrated.

The basic approach in emission spectroscopy is to present the sample to a dc arc or spark excitation source between two electrodes. The electrodes are usually high purity graphite but metal electrodes have been used. Plasma torches have also been used for excitation. The atomic species are excited to extremely high energy states and when they de-excite the emitted radiation is separated by a grating monochromator and detected by photographic emulsion or photomultiplier tubes. The extremely high temperature of excitation creates a high background and large variations in the excitation which results in a varying background and signal levels. Books by Slavin[65] and Grove[66] cover most of the aspects of instrumentation and methods.

Kershner *et al.*[67] give an excellent account of the analysis of impurities in mineral acids and their methodology could be applied to water analysis. Thirty-three elements were analyzed with detection limits of 0.1-50 µg/l based on 100 gram samples. Recovery of spiked samples was determined for 18 elements. Niedermeir *et al.*[68] reported on the analysis of biological fluids and this method could also be adapted to water analysis. Fourteen elements were analyzed and the minimum detectable quantities for all elements was less than 0.1 µg/l based on a 2.00 milliliter sample. Kopp and Kroner[69] used a direct reading spectrograph for the measurement of 14 minor elements in water. They used preconcentration by evaporation and excitation by a high voltage spark using a rotating electrode. The analytical range was 0.01 to 100 mg/l with recoveries of 80-113% at the 90% confidence level. Other reports on water analysis by emission spectroscopy are found in References 70-74.

X-ray emission spectroscopy

X-ray emission spectroscopy (XES) is one of the most specific elemental techniques in metals

analysis and is independent of the chemical form
of the element. Characteristic x-rays are excited
by electrons, x-rays or radioisotope sources and
the x-rays emitted measured by dispersive or non-
dispersive methods. X-ray energies from 1-100 Kev
can be dispersed by various diffraction crystals and
detected by gas proportional or NaI(Tl) x-ray detectors or the energy can be resolved and detected
by Si(Li) solid state detectors. With either method,
the energy of the x-rays can be identified with a
specific element and the integrated x-ray intensity
related to the concentration of the element. The
method is matrix dependent and does not afford sensitivities below a few mg/l without concentration.
Optimum results for water analysis have been reported
for sample concentrations in the 5-500 mg/l range.
Chemical separation has been employed to reduce
interferences and improve recovery of trace
metals.[75]

Tackett and Brocious[76] using Cr as an internal
standard, analyzed for iron in polluted streams and
mine waters by direct pipeting of samples onto
chromatographic paper with a precision of +7% at
114 mg/l and cited surface characteristics of the
paper as their principal source of error. Luke[75]
used prior chemical separation to reduce saturation
of exchange capacity and concentrated metals on ion
exchange paper by batch equilibration. Campbell
et al.[57] used repeated filtration through ion
exchange paper but found seven filtrations necessary
for 100% recovery of Ca, Co, Cr, Cu, Fe, Mn, Ni,
and Zn (all analyzed as a single sample) while only
60% of Cs could be collected. Determinations were
made directly on the unpelletized filter papers for
21 metals with detection limits ranging from 5.0
micrograms (Al) to 0.05 micrograms (Co).

To overcome the multiple pass or long equilibration time problems of earlier works, Leyden *et
al.*[78] employed Chelex-100 chelating resin in batch
configuration to concentrate chromium (III) ion and
produced a pellet under high pressure. The problems
of ion distribution in filter discs and matrix
irregularities are reduced with this technique.

Solvent extraction techniques also provide
concentration plus some specific isolation of
transition metals from alkali and alkaline earth
metals. Morris[79] analyzed sea water for V, Cr, Mn,
Fe, Co, Ni, Cu, and Zn by adding ammonium pyrollidine
dithiocarbamate to 800 milliliters of acidified
sample and partitioning the chelated metals into

methylisobutyl ketone with a continuous liquid-
liquid extractor. The organic solvent is distilled
off, residue ashed, dried and formed into pellets
with cellulose.

Park[80] determined lead and zinc in the 2.5 to
100 μg/l range by extracting the dithiozone complexes
into benzene using pulsed column extraction. The
isolated dithionates were oxidized, re-extracted
into aqueous solution, and the metals precipitated
as the sulfides which were analyzed directly.
Relative standard deviations were of the order of
9 to 10%.

Rhodes *et al.*,[81] using a Si(Li) detector and
radioisotope sources, recently reported on the
analysis of 17 metals in air filter samples. The
sensitivities were in the range of 0.01 to 0.05
nanograms for each element and the results correspond favorably with AAS. This approach could be
applied to the analysis of resin-loaded filter
papers used to separate trace metals from water
samples. Major element concentrations and matrix
effects could cause problems.

General references to the application of XES
to water analysis can be found in References 82 and
83. Although lacking in sensitivity, when compared
to other methods, XES does have the utility of
automatic sample analysis and routine multielement
analysis via computer techniques.[81]

NEUTRON ACTIVATION ANALYSIS

Neutron activation analysis (NAA) is not a
general method for the analysis of trace metals in
natural waters. Many problems are associated with
the application of this method to water analysis.
Liquids are not generally irradiated in nuclear
reactors unless they are frozen or the irradiation
time is short. Therefore, water samples must be
concentrated in most instances. Low temperature
evaporation of large water volumes, freeze drying,
ion exchange, semipermeable membrane ion exchange
and solvent extraction are the best approaches.
Access to a nuclear reactor and a sophisticated
counting and radioisotope laboratory increase the
cost of NAA. The advantages that can be realized
are high selectivity, high sensitivity, low matrix
effects, multielement analysis, and good precision
and accuracy.

Sample size for water analysis should be in
the range of 1 to 10 liters and should be filtered,

acidified and then concentrated. The simplest
concentration method is low temperature evaporation
in Teflon. However, for field work, water can be
pumped through a prefilter and then through an ion
exchange bed to collect the trace metal ions. The
ion exchange bed can be eluted later. Samples and
standards should be irradiated in polyethylene or
polypropylene for short irradiation times (minutes
to hours) and in quartz tubes for longer irradia-
tions. Standards should be freshly prepared and
added to real duplicate samples as well as pure
standards. Irradiation should be performed in the
reactor at the lowest ratio of fast/slow neutrons.
For direct counting, the samples should be counted
on Ge(Li) detectors with at least 6-10% efficiencies
(with respect to 3" x 3" NaI), resolution of 1.5 -
2.0 Kev at 1.332 Mev, and peak/compton ratios of
30/1. The gamma photopeak counting rates of the
standards are compared directly to the unknown
samples by peak area measurements.[84] This approach
will allow 5-10 elements or more to be determined
in preconcentrated water samples. Na, Mn, K, V,
Al, Cu, Br, Fe, Zn, Co, etc. can be measured by
varying irradiation and counting periods. Sensitivities will depend on the preconcentration,
irradiation time, the nuclear cross section and
neutron flux. Typical sensitivities for an integrated neutron flux of 10^{16} neutrons/cm^2 will be
approximately: Mn, 0.050 ng; Cu, 0.50 ng; Na, 1.0
ng; Br, 5.0 ng; Hg, 100 ng; and Co, 100 ng.[85]

General methods for NAA analysis can be found
in References 86-90. The best selection of NAA
papers are currently being published in Reference
89. Specific references on the applications of
NAA to water analysis are few. Piper and Goles[92]
reported on the analysis of nine elements in sea
water. One-liter samples were filtered and freeze
dried. Co, Cr, Fe, Zn, Sr, Rb, Cs, Sc, and Sb were
determined.

Weiss and Crozier[93] have analyzed for Hg in
sea water by precipitation of HgS with a carrier
precipitate, and irradiation of the filtered solids
on Millipore filters. Recoveries were 81 ± 9% and
the sensitivity at 3 standard deviations was 4 ng
for 1.0 liter samples. Becknell, Marsh, and Allie[94]
have used NAA and an ion exchange resin paper to
determine Hg in water. Hg levels from 0.05 to 250
micrograms could be determined using 0.50 liters
water samples. Hg recoveries were 100% for samples
containing less than 250 micrograms of Hg. Weiss

reported on the determination of Se in glacial ice.[95] Chemical separation yields were about 50% and sensitivity of about 1 nanogram for 1-4 liter samples. Nozki *et al.*[96] reported on the analysis of U in drinking water by NAA. Their method could determine 5×10^{-11} grams and they found 9×10^{-9} µg U/l in drinking water. A method for the low level determination of U in water is reported by Dorshel and Stolz.[97] Evaporated water samples are positioned between nuclear track plates and the system irradiated with thermal neutrons. Fission tracks are counted and calibrated with standard U solutions.

A major problem with instrumental NAA is the presence of elements having high yields for neutron activation and prominent gamma ray spectra. Their gamma spectra cover the gamma spectra of less sensitive elements. Examples are ^{24}Na, ^{42}K, ^{82}Br, ^{46}Sc, etc. ^{24}Na can be removed by adsorption of Na on hydrated antimony pentoxide[98,99] or by isotopic exchange with excess NaCl solid. The resulting sample allows the gamma counting of several radioisotopes not detectable in the original samples.

Other approaches involve chemical separation of elements into chemical groups or individual element fractions. Chemical recoveries must be measured with another radioisotope of the same element, by reirradiation of separated standards, or by recoveries based on standards in the same matrices. Chemical separation schemes can involve ion exchange, solvent extraction, precipitation, etc. and can vary from a simple to a very complex scheme. Typical examples are found in References 100-102 and involve the determination of Cd in biological matter and multielement analysis in biological, "environmental," and rock samples.

The NAA method is an excellent approach for the analysis of trace elements in water, but requires specialized, expensive instrumentation. It should be noted, also, that NAA is an excellent reference method with which to check other analytical procedures.

ELECTROCHEMICAL TECHNIQUES

Of all the electrochemical techniques, only variations of voltammetric methods and ion selective electrodes appear adequate in sensitivity to analyze metals in water to the mg/l levels and below. The latter category is nearing the limits of its

application when applied to trace metals analysis,
while special variations of voltammetry are capable
of measuring extremely low metal concentrations.

Voltammetric Methods

The methods that will be discussed are pulse
polarography, differential pulse polarography,
anodic stripping, differential anodic stripping
and differential pulse anodic stripping. These
specific methods allow the analysis of several
metals in the range of 1 milligram to 1 nanogram
per liter with adequate precision.

Classical dc polarography relies upon a
polarized electrode whose current is independent
of the voltage applied to it. The working polarographic electrode behaves in this fashion until the
potential is great enough to allow a reaction to
occur at the electrode surface. Over a short range
of voltage increases, the current becomes a function
of the applied voltage because the rate of current
transfer is dependent upon the rate of reaction at
the electrode surface. Above some specific voltage,
the electrode reaction will be faster than the
rate of transfer of reactant to the electrode
surface and the current will again be independent
of the applied voltage. Called the diffusion
current, its magnitude is proportional to the
concentration gradient between bulk solution and
the electrode surface. If the gradient width can
be maintained constant and the reaction rate is
such that the surface concentration is essentially
zero, the familiar Ilkovic equation results:

$$i_d = K \cdot \text{concentration}, \quad K = \text{constant}$$

The classical dropping mercury electrode system
approximates the requisite conditions and generally
is the basis for dc polarography. The representation of current versus voltage or polarogram will
display characteristic voltages at which electrode
reactions occur. These voltages are dependent upon
the chemical nature of the ion, solvent characteristics, and the presence of complexing agents.
Since the diffusion current is additive, several
ions can be determined from a single polarogram if
the individual increases or waves can be resolved.
However, below a certain ion concentration, i_d is
no longer reliably measureable without concentrating
the sample or amplifying the signal.

Pulse polarography[103-105] uses the dropping
mercury electrode but instead of applying a slowly
increasing dc voltage, a series of increasing
voltage pulses of millisecond duration are applied
to the electrode. One pulse is applied during
each drop cycle, say in the last 50-100 milliseconds of a 2-8 second drop cycle. The pulse
generates a charging current which decays rapidly
and the faradic current (due to the reduced species)
is measured in the last 10-20 milliseconds of the
pulse life. The cycle starts over with each new
mercury drop but with a regularly increasing pulse
amplitude for each cycle. Actually, when the
faradic current is measured there is still a
current contribution due to the charging current
but its major portion has decayed. Pulsed polarography can have sensitivities better than 10^{-6} M.
Differential pulse polarography uses a slowly increasing dc voltage ramp, and pulses of constant
amplitude and duration are applied to the ramp
voltage. An integrated current is read for a fixed
time before the pulse is applied and for the same
time just before the end of the pulse duration.
Times used in commercial instruments are 16.7
milliseconds integrated current, 56.7 milliseconds
pulse with the current again integrated over the
last 16.7 milliseconds of the pulse. The difference
of the two integrated currents is plotted versus the
dc voltage ramp. The output is a differential of
the normal polarographic curve. Trace metals can
be determined at the 10^{-8} M level, with good resolution of peaks and at low electrolyte concentrations.
Pulse polarography methods have been applied
to the analysis of natural water by Abdullah and
Royle.[106] Cu, Pb, Cd, Zn, Ni, and Co were determined after preconcentration from fresh or sea
water using the Ca form of Chelex-100 resins.
Sample sizes up to 1 liter were used and were
analyzed in the 0.10 to 100 µg/l concentration
range with good accuracy and precision.
The sensitivity of differential pulse polarography is shown by Osteryoung *et al.*[104,107] for the
determination of Cd. The detection level for Cd
by differential pulsed polarography is about 1 µg/l
and the concentration response is linear to 2% over
the concentration range of 0.02 to 0.56 mg/l. Both
of these techniques appear ideally suited for water
analysis particularly if preconcentration is used.
Anodic stripping techniques involve the concentration of trace metals by electrodeposition of

the metals on a suitable electrode and then stripping the metals from the electrode with a linear
voltage ramp. [108-110]

Anodic stripping can be performed with a
standard polarograph that has the capability of
scanning in both the positive and negative direc-
tions. Both two and three electrode systems can
be used and readout is on a strip chart recorder.
The major consideration in anodic stripping is the
test electrode. The most common are the hanging
mercury drop electrode (HMDE) of which there are
several variations, the wax-impregnated graphite
electrode (WIG), glassy carbon electrode (GCE),
the pyrolytic graphite electrode, carbon paste
electrodes, mercury coated wire electrode, etc.
The most popular of these electrode systems are
the HMDE and the WIG.

The applications of the anodic stripping
technique which are described below illustrate
both its utility and some of the problems encountered. Eisner and Mark[111] used a wax-impregnated
pyrolytic graphite electrode polished to a very
smooth surface on one end for analysis of Ag in
rain and snow samples. Concentration of Ag above
3×10^{-9} grams could be determined directly by
linear voltage anodic stripping, but for lower
concentrations the following preconcentration step
was employed. One-liter water samples were passed
through Dowex 50W-X7 cation exchange column and
the Ag eluted with 0.25 M NH_4SCN. Sample concentrations as low as 4×10^{-11} M could be determined.
Precision and accuracy were checked at Ag concentrations of 4×10^{-9} M with reproducibility of
peaks better than $\pm$ 5%. Standard addition methods
at 6×10^{-9} M Ag provided agreement between data
of $\pm$ 7%.

Allen *et al.*[112] used a WIG electrode to measure
the concentration of cadmium, lead and copper in
natural waters. Their work clearly demonstrates
the extent to which metal ions can exist in the
complexed state in natural waters. Acidified
samples frequently assayed at much higher levels
(100 to 200% increase). Unless precautions are
taken to make the necessary pH adjustments, anodic
stripping will not respond to the total soluble
metal concentration.

Anodic stripping using a GCE where Hg^{+2} was
added with the sample and supporting electrolyte
has been reported by Florence.[113] Pb was analyzed
at 5×10^{-9} M Pb^{+2} and a relative standard deviation

of + 7.5% was obtained. The resolution of the
stripping peaks was significantly better than reported for other stripping methods and probably was
due to the thin Hg film coating. The GCE was
rotated at 2000 rpm in this method. The Hg film
was removed from the electrode after each run and
a new film plated with the next sample. Excellent
stripping peaks were obtained for mixtures of Cu,
Pb, In, and Cd at 2 x 10^{-7} M in 0.10 M KNO_3.

Anodic stripping has certain disadvantages:
the large capacitance current limits the measurement
of the faradic current, and long electro-deposition
times are required for good sensitivity. In order
to circumvent these problems, differential, pulse
differential, and pulse anodic stripping have been
developed. Christian[114] has reported on a pulse
anodic stripping method using a hanging drop mercury
electrode. Cd was analyzed at a 10^{-8} M concentration
with good precision.

Another method for improving anodic stripping
has been developed by Zirino and Healey.[115] Zn, Cd,
Pb, and Cu were measured in natural waters by differential anodic stripping analysis. Both fresh
and sea waters were studied. Two hanging mercury
drop electrodes (HMDE) were used in the same cell
versus a standard calomel electrode (S.C.E.) reference. One HMDE electrode was used as the test
electrode and the other as the differential electrode.
The latter electrode was connected to a variable gain
inverting operational amplifier with a gain of (-1).
The output from the two identical HMD electrodes
was fed to the summing amplifier in the polarographic
analyzer. Calibrations were made at varying pH's
and the pH was adjusted by using $CO_2 - N_2$ mixtures as
the purging gas. Studies of interferences in the
analysis of sea water were also made. Samples were
routinely analyzed at the μg/l level with adequate
precision.

Ion Selective Electrodes

In the strict sense, the term ion selective
electrodes encompasses both metal-based electrodes
and membrane electrodes.[116] The latter category
contains the glass, doped crystal, and ion exchange
(solid and liquid matrix) electrodes which comprise
the commercial products sold as ion selective or
specific ion electrodes. Consideration of these
electrodes should be qualified at the outset by
saying that direct application to trace metals

analysis is limited. Only Ca, Mg, K, Na, Ag, Cd, Cu, and Pb electrodes are commercially available.[117] Even though these electrodes can be applied to the measurement of other metals indirectly,[117,118] the advantage of *in situ* measurements is lost.

In any potentiometric method the response measured is a difference in potential between the working electrode and a reference electrode of constant potential. In the ideal case, the variation in potential can be related to the activity of the solution ion of interest by the Nernst equation:

$$E \text{ measured} = \left(\begin{array}{cc} E^{\circ\prime} - E \\ \text{working} \quad \text{reference} \\ \text{electrode} \quad \text{electrode} \end{array} \right) + \frac{RT}{F\, n_{ion}} \ln A_{ion}$$

where the $E^{\circ\prime}$ term is the standard electrode potential at unit activity plus all interface potentials and strain potentials in the working electrode system (R = gas constant, F = Faraday, T = absolute temperature and n = charge on the ion).

The phenomenon generating the potential across the electrode membrane (glass, crystal, liquid or solid ion exchanger) is considered to be essentially an ion exchange process[116,119,120] at each of the membrane surfaces. By maintaining a constant ion activity on one side (interior) of the membrane, the potential drop across the membrane forms a classical concentration cell and the $RT \ln A_{ion}$ term for the inner membrane surface is included in the $E^{\circ\prime}$ term.

The potential measured is an equilibrium value and the time required to achieve this state will determine response time for a particular electrode configuration. The specificity of response relates directly to the nature of the ion exchange process. If more than one species can undergo ion exchange at the external membrane surface, the response obtained will be a mixture of potentials given by:

$$E \text{ measured} = \left(\begin{array}{cc} E^{\circ\prime} - E \\ \text{working} \quad \text{reference} \\ \text{electrode} \quad \text{electrode} \end{array} \right) + \frac{RT}{Fn_1} \ln \left(a_1 + \Sigma k_i a_i^{n_1/n_i} \right)$$

where k_i is the relative selectivity of the ion exchange process for the ith ion compared to the ion of interest. The ideal situation obtains when k = 0 for all other ions. As k_i increases for other ions, the interference increases. Rechnitz[120] points out that the selectivity also can have a strong impact on the operational sensitivity of an electrode. This occurs because deviation from Nernst behavior is manifested at higher analyte concentration as the interference level increases. When this occurs, fluctuations in the total signal obscure the analyte signal.

The reproducibility or reliability of the measured value mentioned above imposes limitations on ion selective electrode use. In order to achieve 1 to 2% precision, measurements must be reproduced to 0.1 to 0.2 millivolts. Frequently, it is not instrument limitation that prevents realization of this goal but rather failure of the reference electrode to remain stable due to variable junction potentials.[121]

The electrodes listed above are reported to be capable of determining metal ion activities to sub mg/l levels.[117,122] Weber[123] has reported the following limits in mg/l: Na (0.02), K (2.0), Ca (0.4), Hardness (0.0001), Cd (0.01), Cu (0.006), Ag (0.01) and Pb (0.02). Muller *et al.*[124] found a linear response for the silver ion electrode in the 10 to 130 μg/l range and claimed precision of $\pm$ 0.2%. However, ionic strength was held constant at 0.1 M for laboratory systems.

Lal and Christian[125] found the lead electrode functional between 2.1 x 10^{-5} and 10^{-2} M with an optimum above 10^{-4} M and optimum pH values much less than found in natural waters. Further problems were related to high response to monovalent cations. Frant[126] in discussing industrial applications cites similar experiences and claims direct measurement precisions are low.

As is apparent from the equation describing the electrode system response, the ion activity not its concentration is providing the measured value. Thus, when measurements are taken in natural waters, variations in ionic strength between samples and standards must be considered. Taking samples and standards to uniform ionic strength overcomes this problem but obviates *in situ* measurements and may reduce the sensitivity. When low ionic strength waters are involved such as alpine lake waters[123] little or no problem is encountered. The magnitude of ionic strength effects has been clearly demonstrated.[127]

Also, because only the activity of the free metal ion is measured, any complexed metal will go unmeasured.[128] This may be appreciable as indicated

by Williams[9] who reports up to 28% of the soluble
copper in sea water may be bound organically.
Others[129] have indicated higher values of up to 50%.
At the same time the advantage of ion selective
electrodes in studying complexes between metals and
natural water species should be apparent. Orion[117]
has reported that activities as low as 10^{-10} to
10^{-17} M can be measured for some ions as long as
the total metal present in solution is 10^{-7} M or
greater.

Although used widely for anions, ion selective
electrode applications for metals do not currently
offer the wide spread advantage of atomic absorption
for single species analysis of the previously
troublesome species, Na, K, Ca, and Mg. Under
special conditions of interference in other methods
(*i.e.*, without sample filtration) as pointed out by
Reynolds,[122] potentiometry may be more practical.

GAS CHROMATOGRAPHY

The basic principle of gas chromatography is
the differential partitioning of a volatile substance
between the stationary liquid or solid phase and the
mobile carrier gas in a column. Response specificity
depends upon completeness of separation of injected
components, the nature of the sensing device at the
column exit, or some combination of the two.

Traditionally thought of as an organic analysis
tool, gas chromatography has been finding increasing
application for metals in analytical chemistry. The
incorporation of specific sections pertaining to
metals in biennial reviews of *Analytical Chemistry*[130,131]
is clear support of this increase. With the exception
of high temperature separations of metal vapors or
certain metal chlorides, both of which generally require special modifications of routine gas chromatographic systems,[130,132] these analyses involve
organometalic compounds. Most naturally occurring
organometallic compounds are either not of sufficient
volatility and thermal stability for gas chromatography
or are simply not found in aqueous solution. Gas
chromatography has been applied to the analysis of
several organomercury compounds in biological
matter.[133,134]

Developments in the formation of volatile metal
chelates and their susceptibility to gas chromatography[132,139] led to the evolution of analytical
techniques for ultratrace amounts of metals. These
methods, which represent the ultimate sensitivity,

accuracy and precision for some metals, appear likely
to find routine application in several disciplines.
Water and wastewater analysis would certainly be
included.

Many metals form volatile derivatives with
various β-diketones, fluorinated β-diketones and
their sulfur derivatives. Table 5 lists several of
those compounds in common use. Not all of these

Table 5

Some Chelating Agents Forming Volatile Metal Chelates

Chelating Agent	*Metals Reacting*	*References*
Trifluoroacetylacetone	Be, Cr, Rh, Co, Cu, Fe, Al, Ga, In, Li, K, Na, Sc	132, 140-154
Hexafluoroacetylacetone	Li, K, Na	154
Monothiohexafluoroacetylacetone	Cd, Cu, Fe, Ni, Pb, Pd, Pt, Na	158
Heptane-3,5-dione	Fe, Al, Ca, Ni, Zn, Cd, Mn, Co, Pd	156
2,6-Dimethylheptane-3,5-dione	Fe, Al, Ca, Ni, Zn, Cd, Mn, Co, Pd	156
2,2-Dimethylheptane-3,5-dione	Fe, Al, Ca, Zn, Ni, Cd, Mn, Co, Pd	156
2,2,6,6-Tetramethylheptane-3,5-dione	Li, Na, K, Ca, Mg, Si, Fe, Al, Ni, Zn, Cd, Mn, Co, Pd	154-156
Pentafluorodimethylheptane-3,5-dione	Li, K, Na	154
Heptafluorodimethylheptane 3,5-dione	Li, K, Na	154
Trifluoroacetyl-2,2,6,6-tetramethylheptane-3,5-dione	Li, K, Na	154
1,1,2,2,3,3 heptafluoro-7,7-dimethyloctane-4,6-dione	Cr, Cu, Ni, Co, Pd, Be, Al, Rh	140

chelates have been separated by gas chromatography
nor do they all react quantitatively. A number of
the chelates have been formed from aqueous solutions
resulting from the digestion of solid material
(minerals, metals and biological substances) which
demonstrates the potential application to water
analysis. Notably the trifluoroacetylacetonates of
beryllium,[140-147] chromium[140,144,148-150] aluminum,[140,
151-153] and copper[140,152] have been utilized in the
analytical determination of the respective metals
from aqueous solutions.

Formation of the chelates from aqueous solution
and thermal stability of the chelate does not guarantee successful gas chromatographic separation.
The alkali metal chelates of trifluoroacetylacetone,
hexafluoroacetylacetone, 2,2,6,6-tetramethylheptane-
3,5-dione, and trifluoroacetyl-2,2,6,6-tetramethyl-
heptane-3,5-dione,[154] and the alkaline earth metal
chelates of 2,2,6,6-tetramethylheptane-3,5-dione,[155]
while capable of being gas chromatographed individually in pure form, apparently exist as polymers in
the gas phase and undergo metal exchange on the
column to form mixed metal chelates (*i.e.*, $[M_1L]_2$ +
$[M_2L]_2 \rightleftharpoons 2 M_1M_2 [L]_2$) with the complexity increasing
with increasing number of metal ion species. This
problem does seem to have been overcome for the
alkali metals with the use of penta- and hepta-
fluorodimethylheptane-3,5-dione.[134]

Formation of chelates from aqueous solutions
does not guarantee that a successful analytical
method can be formulated. Frequently, less than
100% reaction will occur[140,148,156] based upon
extraction of the metal chelate from the aqueous
phase into the organic phase. This has been
attributed primarily to the formation of the
diketonate hydrate form[148,157] for hexafluoroacetylacetone. Formation of the monothio derivative of
hexafluoroacetylacetone seems to overcome this
limitation.[158]

Similarly, the improvement in gas chromatographic
detection systems has promoted the utility of this
method. Earlier investigations[152,153] using thermal
conductivity detectors required high concentrations
of metals, and frequently this led to overlapping
of peaks. Use of flame ionization detectors
(FID)[137,152] reduced detection of Be, Al, Cr, and Rh
trifluoroacetylacetonates to $10^{-8} - 10^{-6}$ M. Electron
capture detectors (ECD)[141-143,145,146,148,149] have
allowed detection of as low as 0.01 µg/l of Be and
Cr in the organic phase.

Such an increase in sensitivity is not without attendant problems, most notably interference from the unreacted β-diketone which also extracts into the organic phase. With trifluoroacetylacetone, back extraction with aqueous NaOH selectively removes the β-diketone.[142,143,145,146,148,149] However, not all chelating agents appear to be so easily removed and result in greatly reduced sensitivity of the electron capture detector.[159]

Another failure of the electron capture detector (similarly for FID) is its lack of specificity. This may prove to be a serious problem with natural waters due to the presence of potentially interfering organic matter and metal ions. Masking with other chelating agents such as EDTA seems to overcome some metal interferences.[141,146,147] Clean-up procedures similar to those used in pesticide analysis could also be implemented. Work with Cr and Be acetyl-acetonates in serum and plasma[144] has demonstrated the use of gas chromatography - mass spectrometry interfacing to provide absolute specificity to the response while sensitivities comparable to those of electron capture detectors are obtained. The attractive feature of this technique is that the gas chromatographic peaks need not be resolved at all since observation of one mass is not interfered with by the other metal species. Obviously, cost of instrumentation would be a major drawback.

MASS SPECTROMETRY

A mass spectrometer measures the mass to charge ratio (m/e) of ions which have been vaporized. In conventional mass spectrometry, the sample is introduced at slow rate into a high vacuum system. Gases and liquids of sufficient vapor pressure can be introduced directly. Other liquids and solids may be heated to vaporize them. The ions are then produced by bombardment with an electron beam, accelerated, and then focused by electrical or magnetic fields or both upon a detector. The m/e of an ion focused on the detector is related to the accelerating potential, the electric field strength, and the magnetic field strength. Variation of the field strengths provides selective measurement of m/e while the detector current provides quantitative response for a given m/e at fixed field strengths.

Since metal salts generally are not sufficiently volatile, conventional vaporization techniques are not applicable. However, vaporization can be achieved as in emission spectroscopy by making the sample part of a set of electrodes. The ions thus produced in a vacuum chamber by a spark or arc source are accelerated and focused as with conventional mass spectrometry. An obvious restriction for water analysis is that only evaporated samples can be employed.

Spark source mass spectrometry (SSMS) analysis for trace metals has the advantage, as do emission spectrographic, neutron activation and x-ray fluorescence methods, of simultaneous multielement analysis which may involve a wide variety of sample matrices. Of these four methods, only emission spectroscopy could be considered to be of lower cost in a relative sense but it also suffers the most from matrix effects and lower sensitivity. The primary reason for the high cost involved with mass spectrometry ($40,000 - $125,000 dependent upon detection system and computer interfacing) is the necessity for double focusing (Mattauch-Herzog) geometry required for the spark source ionization system.[160-163] The ion beam produced by a radio frequency discharge of 50-150 KV through electrodes constructed from the sample material has a large energy spread making the above restrictions essential.[160,162,163]

Advances such as automatic spark gapping[164] enhance the method but photographic detection, though advantageous in compensating for the erratic total ion current, limit the method to semiquantitative status.[160,162,165] For homogeneous samples with careful standardization of techniques, precision and accuracy of $\pm$ 5 to 20% relative standard deviation can be achieved, similar to emission spectroscopy.[160] Recent advances in electrical detection involving monitoring of the total ion current[160,162-164] allow greater survey analysis versatility and can result in single element analysis precision and accuracy of $\pm$ 3 to 5%.

Morrison,[160] who also outlines the numerous practical problems associated with SSMS, reports absolute detection limits of 0.1 to 0.01 nanograms for most metals. These values compare favorably to neutron activation analysis and in some cases represent significant improvements in detectability, notably Cr (0.05 vs 100),* Fe (0.05 vs 5,000), Mo (0.3 vs 10), Ni (0.07 vs 5), Pb (0.3 vs 1,000), Se (0.1 vs 500) and Zn (0.1 vs 10).

---

*
Values are quantities in nanograms as the metal.

Examples of SSMS application to natural water samples include Crocker and Merritt[164] who analyzed freeze-dried, low temperature ashed samples simultaneously by NAA and SSMS and found 32 elements in concentrations ranging from 1.5×10^{-8} to 1.5×10^{-2} g/l. Quick scan (10 min) yielded $\pm$ 30% precision of 10^{-4} mg/l but peak switching for single elements provided $\pm$ 5% precision. Wahlgren *et al.*[166] analyzed 21 elements in filtered water samples at concentrations of 20 to 0.04 μg/l with relative precision of $\pm$ 20%. Similar results have been reported for airborne particulates, whose analogous component in water could be treated in the same manner.[167] A 2 milligram sample provided $\pm$ 30% relative standard deviation for 28 elements of which 23 were metals.

Removal of sample matrix interferences can be achieved by direct computer interfacing which identifies possible interferences in the sample matrix. For water samples, detectabilities of ng/ml without preconcentration and pg/ml for preconcentrated samples have been claimed.[162]

The cost of equipment involved, the necessity of having a solid sample which is relatively free of organic matter,[168] and formation of electrodes from the sample, all reduce the likelihood of routine application of spark source mass spectrometry in the immediate future. One time-accounting study has estimated that 1.5 man days per sample are required for SSMS analysis of water samples.[166] Use of electronic detection instead of a photoplate system might reduce this time considerably.

A more limited application of mass spectrometry, in the sense of being restricted to those metals forming chelates under the proper conditions, would be direct probe insertion or even simple volatilization of the volatile metal chelates discussed earlier (under G.C.). Conventional mass spectrometers of variable geometry and resolution could be readily employed. Such an application has been reported for the analysis of 14 metal chelates of 1,1,1,2,2,3,3-heptafluoro-7,7-dimethyloctane-4,6-dione.[169] The metals were Al, Cr, Fe, Ni, Cu, Y, Pd, Nd, Sm, Dy, Ho, Tm, Yb, and Pb. Quantitative results were obtained at the 10^{-12} gram level with a precision of $\pm$ 5% and the precision limit was attributed to the microliter syringe used to introduce the sample to the direct insertion microprobe. A linear relationship was reported between picograms of aluminum and peak area of the time resolved scane. Ultimate detectabilities were of the order of 10^{-14} grams.

Before such methods can be of general survey
use with respect to water analysis, chelates that
can be formed directly from aqueous solutions or
salts upon evaporation must become available. In
some cases, however, when other methods do not provide the analysis capabilities needed, this method
could prove of value. A case in point is the
analysis of chromium as reported by Booker *et al.*[170]
They claim other methods have high interferences and
neutron activation analysis is not sufficiently
sensitive. The trifluoroacetylacetonate of chromium
forms directly from aqueous solution and provides a
linear response from 0.5 to 20 nanograms as Cr. The
standard deviation is $\pm$ 2.2% at 100 nanograms and
+ 6.5% at 2 nanograms. At 0.05 nanograms, noise
Is 25% of the signal.

SUMMARY

In conclusion, it can be seen that a large
number of analytical methods are available for trace
metals determinations. Not all methods provide the
sensitivity, selectivity, species differentiation
or precision that may be desired for a particular
metal. Cost, time or available sample quantity can
frequently impose overriding consideration upon a
particular application forcing a compromise.
A primary consideration fundamental to all
analytical situations is that the more sample manipulation involved in the method, the greater the
chances for contamination of the original sample.
Consequently, trends in trace metals analysis have
demonstrated a move towards methods requiring
smaller sample sizes and minimum concentration
procedures while retaining extremely high
sensitivities.
Response specificity is another area of
paramount concern. Methods involving atomic
transitions, notably atomic absorption, have, in
their development, greatly furthered response
specificity for metal analysis in water systems.
Similarly, mass spectrometry affords extreme
specificity but has a number of notable drawbacks.
There is a continuing need in the study of water
systems for development of better techniques for
metals analysis at the trace level, which provide
the desired specificity but also provide the
capacity to analyze large numbers of samples. The
subject matter of the following papers is testimony
to this continuing need.

REFERENCES

1.   McBride, B. C. and R. F. Wolf. *Biochem 10,* 4312 (1971).
2.   Hume, D. *Analysis of Water for Trace Metals,* Advances in Chemistry Series, #67 (1967).
3.   Hwang, J. Y., P. A. Ullucci, S. B. Smith, Jr., and A. L. Malenfant. *Anal. Chem. 43,* 1319 (1971).
4.   *Perkin-Elmer Manual 303 AA,*
5.   Elliott, W. *American Laboratory,* 45 (August 1971).
6.   Pinta, M. *Detection and Determination of Trace Elements.* (Ann Arbor: Humphrey Science Publishers, 1970).
7.   *Standard Methods,* 13th ed. (American Public Health Association, 1971).
8.   *ASTM, Annual Stds., Water; Atmospheric Analysis,* Part 23 (1970).
9.   Williams, P. M. *Limnol. Oceanogr. 14,* 158 (1969).
10.   Sandell, E. B. *Colorimetric Determination of Traces of Metals.* (New York: Interscience, 1959).
11.   Charlot, G. *Colorimetric Determination of Elements, Principles and Methods.* (Elsevier Publ. Co., 1964).
12.   Calder, A. B. *Photometric Methods of Analysis,* (American Elsevier Publ. Co., 1969).
13.   Stearns, E. I. *The Practice of Absorption Spectrophotometry.* (New York: Wiley-Interscience, 1969).
14.   Blotz, D. F. and M. G. Mellon. *Anal. Chem. 42,* 152R (1970).
15.   Blotz, D. F. and M. G. Mellon. *Anal. Chem. 44,* 300R (1972).
16.   Mancy, K. H. *Instrumental Analysis for Pollution Control.* (Ann Arbor: Ann Arbor Science Publishers, 1971).
17.   *Methods for Chemical Analysis of Water and Wastes.* Environmental Protection Agency 16020-7/71 (1971).
18.   West, P. W. and F. K. West. *Anal. Chem. 42,* 99R (1971).
19.   West, P. W. and F. K. West. *Anal. Chem. 44,* 251R (1972).
20.   Udenfriend, S. *Fluorescence Assay in Biology and Medicine,* Volume II. (New York: Academic Press, 1969).
21.   Udenfriend, S. *Fluorescence Assay in Biology and Medicine,* Volume I. (New York: Academic Press, 1962).
22.   Hercules, D., ed. *Fluorescence and Phosphorescence Analysis* (New York: Wiley-Interscience, 1966).
23.   White, C. E. and R. J. Argauer. *Fluorescence Analysis - A Practical Approach.* (New York: Marcel Dekker, 1970).
24.   Winefordner, J. D., S. G. Schulman, and T. C. O'Haver. *Luminescence Spectrometry in Analytical Chemistry.* (New York: Wiley-Interscience, 1972).
25.   Lytle, F. E. *Appl. Spec. 24,* 319 (1970).
26.   White, C. E. "Fluorometry," In *Trace Analysis,* Yoe, J. H. and J. R. Koch, eds. (New York: Academic Press, 1957).

27. Fishman, M. J. and D. E. Erdmann. *Anal. Chem. 43*, 356R (1971).
28. Ghassemi, M. and R. F. Christman. *Limnol. Oceanogr. 13*, 583 (1968).
29. White, C. E. and A. Weissler. *Anal. Chem. 42*, 57R (1970).
30. White, C. E. and A. Weissler. *Anal. Chem. 44*, 182R (1972).
31. *Manual for Varian Model 1000 Atomic Absorption Spectrometer*, Varian Associates, 1970.
32. Kahn, H. *Atomic Absorption Newsletter 7*, 40 (1968).
33. Kahn, H. and D. Manning. *American Laboratory*, 51 (August, 1972).
34. Kahn, H., G. Peterson, and J. Schallis. *Atomic Absorption Newsletter 7(2)*, 35 (March-April, 1968).
35. Delves, H. *Analyst 95*, 431 (1970).
36. Fernandez, F. and H. Kahn. *Atomic Absorption Newsletter 10(1)*, 1 (1971).
37. West, T. and X. Williams. *Anal. Chem. Acta 45*, 27 (1969).
38. Amos, M. *American Laboratory*, 33 (August 1970).
39. Matousek, J. *American Laboratory*, 45 (June 1971).
40. L'Vor, B. *Spectrochem. Acta 17*, 761 (1961).
41. Massman, H. *Z. Anal. Chem. 225*, 203 (1967).
42. Kahn, H. *American Laboratory*, 35 (August 1971).
43. Woodriff, R., R. Stone, and A. Held. *Appl. Spectrosc. 22*, 408 (1968).
44. Electro Thermal Analyzer, Barnes Engineering, ETA - 1 and 2, May 1972.
45. Hwang, J., P. Ullucci, and S. Smith. *American Laboratory*, 41 (August 1971).
46. Manning, D. and F. Fernandez. *Atomic Absorption Newsletter 9(3)*, 65 (May-June, 1970).
47. Fernandez, F. and D. Manning. *Atomic Absorption Newsletter 10(3)*, 65 (1971).
48. Segar, D, and J. Gonzales. *Anal. Chim. Acta 58*, 7 (1972).
49. Hatch, W. and W. Ott. *Anal. Chem. 40*, 2085 (1968).
50. Windham, R. *Anal. Chem. 44*, 1334 (1972).
51. Fishman, M. *Anal. Chem. 42*, 1462 (1970).
52. Doherty, P. and R. Dorsett. *Anal. Chem. 43*, 1887 (1971).
53. Brooks, R., B. Presley, and I. Kaplan. *Talanta 14*, 809 (1967).
54. *Bulletin #114*, Bio-Rad Labs, Richmond, California (1969).
55. Riley, P. and D. Taylor. *Anal. Chem. Acta 40*, 497 (1968).
56. Malmstadt, H. and E. Cordos. *American Laboratory*, 35 (August 1972).
57. Demers, D. and D. Mitchell. *Advances in Automated Analysis, Technicon*, Vol. II, 507 (1970).
58. Fihn, S., S. Subben, and C. West. Paper 134, Pittsburgh Conference, March, 1972.

59. Amos, M., D. Bennett, K. Brodie, P. Lung, and J. Matousek. *Anal. Chem. 43*, 211 (1971).
60. *Analytical Reviews, 1972, Anal. Chem. 44(5)*, (1972).
61. Pickett, E. and S. Koirtyohann. *Anal. Chem. 41*, 28A (1969).
62. Christain, G. and F. Feldman. *Appl. Spect. 25(6)*, 660 (1971).
63. Searle, P. and G. Kennedy. *Analyst 97*, 457 (1972).
64. Zatzick, M. and G. Olson. Paper 23, Chicago Spectroscopy Forum, June 1972.
65. Slavin, M. *Emission Spectrochemical Analysis*. (New York: Wiley-Interscience, 1971).
66. Grove, E. L. *Analytical Spectroscopy Series*, Vol. I, Part II (New York: Marcel Dekker, Inc., 1972).
67. Kershner, N., E. Joy, and A. Barnard. *Appl. Spec. 25*, 542 (1971).
68. Niedermeier, N., J. Griggs, and R. Johnson. *Appl. Spec. 25(1)*, 53 (1971).
69. Kopp, J. and R. Kroner. *A-pl. Spect. 19*, 155 (1965).
70. Hughes, R., P. Miiran, and G. Gunderson. *Anal. Chem. 43*, 691 (1971).
71. Mallory, E. *Advances in Chemistry Series, No. 73* (American Chemical Society, 1968).
72. Steiner, R., J. Austin, and D. Lander. *Env. Sci. Tech. 3*, 1192 (1969).
73. Hitchcock, R. and W. Starr. *Appl. Spec. 8*, 5 (1954).
74. Baer, W. and E. Hodge. *Appl. Spect. 14*, 141 (1960).
75. Luke, C. L. *Anal. Chem. 36*, 318 (1964).
76. Tackett, S. L. and M. A. Brocious. *Anal. Let. 2*, 649 (1969).
77. Cambell, W. J., E. F. Spans, and T. E. Green. *Anal. Chem. 38*, 987 (1968).
78. Leyden, D. E., R. E. Channell, and C. W. Blount. *Anal. Chem. 44*, 607 (1972).
79. Morris, A. W. *Anal. Chim. Acta 42*, 397 (1968).
80. Park, Y. K. *Doehan Hawhak Hwojce 13*, 45 (1969); *Chem. Abstr. 72*, 24460n (1970).
81. Rhodes, J., A. Pradzynski, C. Hunter, J. Payne, and J. Lingren. *J. Env. Sci. Tech. 6(10)*, 922 (1972).
82. Rose, H. J., Jr. and F. Cuttitta. *Adv. X-Ray Anal. 11*, 23 (1968).
83. Birks, L. S. *Anal. Chem. 44*, 557R (1972).
84. Quittner, P. *Anal. Chem. 41*, 1504 (1969).
85. Cooper, R., D. Linekin, and G. Brownell. *AEC Symposium Series #13*, Conf. 671111, (1967).
86. Adams, F., P. Vander Winkel, R. Gijbels, D. DeSoete, J. Hoste, and J. OpdeBeek. "Activation Analysis," *Critical Reviews of Analytical Chemistry 1(4)*, (January 1971).

87. Devor, J., and P. LaFleur, eds. *Modern Trends in Neutron Activation Analysis*, Special Publication 312, Vol. I and II, USNBS (1969).

88. Rahovic, N. *Activation Analysis*. (Cleveland, Ohio: CRC Press, 1970).

89. Kruger, P. *Principles of Activation Analysis*. (New York: Wiley-Interscience, 1971).

90. Barier. *Induced Radioactivity*. (Amsterdam: North-Holland Publ. Co., 1969).

91. *J. Radio. Anal. Chemistry 1-8* (1967-1972).

92. Piper, D. and G. Goles. *Anal. Chim. Acta. 47*, 563 (1969).

93. Weiss, H. and T. Crozier. *Anal. Chim. Acta 58*, 231 (1972).

94. Becknell, D., R. Marsh, and W. Allie. *Anal. Chem. 43*, 1230 (1971).

95. Weiss, H. *Anal. Chim. Acta 56*, 139 (1971).

96. Nozacki, T., M. Ichikawa, T. Sasaga, and M. Inarida. *J. Radio. Anal. Chem. 6*, 33 (1970).

97. Dorschell, E. and W. Stolz. *Radiochem. Radio. Anal. Letters 4*, 277 (1970).

98. Gills, T., W. Marlow, and B. Thompson. *Anal. Chem. 42*, 1831 (1970).

99. Ralston, H. and E. Sata. *Anal. Chem. 43*, 129 (1971).

100. Lieberman, K. and H. Kramer. *Anal. Chem. 42*, 266 (1970).

101. Edgington, D. and H. Lucas. *J. Radio. Anal. Chem. 5*, 233 (1970).

102. Morrison, G., J. Gerard, A. Travesi, R. Currie, S. Peterson, and N. Potter. *Anal. Chem. 41*, 1633 (1969).

103. Barker, G. and A. Gardner. *Z. Anal. Chem. 173*, 79 (1960).

104. Osteryoung, J. and R. Osteryoung. *American Laboratory 4*, 8 (July 1972).

105. Burge, D. *J. Chem. Edc. 47*, A81 (1970).

106. Abdullah, M. and L. Royle. *Anal. Chim. Acta 58*, 283 (1972).

107. Osteryoung, R. and E. Parry. *Rev. Polarography* (Japan), *14*, 134 (1967).

108. Barendrecht. *Electroanalytical Chemistry*. Vol. 2 (New York: Marcel Dekker, 1971).

109. Kemula, W. *Pure Appl. Chem. 21*, 449 (1970).

110. Brainina, K. *Talanta 18*, 513 (1971).

111. Eisner, U. and H. Mark. *J. Electro Anal. Chem. 24*, 345 (1970).

112. Allan, H., W. Matson, and K. Mancy. *J. Wat. Poll. Control Fed. 42*, 573 (1970).

113. Florence, T. *J. Electroanal. Chem. 27*, 233 (1970).

114. Christian, G. *J. Electroanal. Chem. 23*, 1 (1969).

115. Zirino, A. and M. Healey. *Env. Sci. Tech. 6*, 243 (1972).

116. Whitfield, M. *Ion Selective Electrodes for the Analysis of Natural Waters.* (Sydney: Australian Marine Science Association, 1971).

117. Orion Research Incorporated. *Analytical Methods Guide.* (Cambridge, Mass., 1971).

118. Seigerman, H. and O'Dom. *American Laboratory 6,* 59 (1972).

119. Durst, R. A. *Am. Sci. 59,* 353 (1971).

120. Rechnitz, G. A. *Chem. Eng. News,* 146 (June 12, 1967).

121. *Orion SIE Technology Newsletter 1(4),* 21 (1969).

122. Reynolds, R. C. *Water Resources Res. 7(5),* 1333 (1971).

123. Weber, S. J. *Amer. Labs,* 1 (July 5 1970).

124. Müller, D. C., P. W. West, and R. H. Müller. *Anal. Chem. 41,* 2038 (1969).

125. Lal, S., and G. P. Christian. *Anal. Chim. Acta 52,* 41 (1970).

126. Frant, M. S. *Plating 58,* 686 (1971).

127. Bagg, J., and R. Vinen. *Anal. Chem. 44,* 1773 (1972).

128. Riseman, J. M. *Anal. Inst. 8,* V-3 (1970).

129. Slowey, J. F., L. M. Jeffrey, and D. W. Hood. *Nature 214,* 377 (1967).

130. Juvet and Cram. *Anal. Chem. 42,* 1R (1970).

131. Cram and Juvet. *Anal. Chem. 44,* 213R (1972).

132. Moshier, R. W. and R. E. Sievers. *G. C. of Metal Chelates.* (Pergamon, 1965).

133. Westöö, G. *Acta Chem. Scand. 20,* 2131 (1966).

134. Westöö, G. *Acta Chem. Scand. 21,* 1790 (1967).

135. Westöö, G. *Acta Chem. Scand. 22,* 2277 (1968).

136. Tatton, S. O'G., and P. J. Wagstaffe. *J. Chromatogr. 44,* 284 (1969).

137. Jernelöv, A. *Limnol. Oceanogr. 15,* 958 (1970).

138. Bache, C. A. and D. J. Lisk. *Anal. Chem. 43,* 950 (1971).

139. Hill, R. D. and H. Cesser, *J. Gas Chrom.* 11 (October 1963).

140. Ross, W. D., L. C. Hansen, W. G. Scribner. ARLP-710285 Wright-Patterson Air Force Base.

141. Ross, W. D. and R. E. Sievers. *Talanta 15,* 87 (1968).

142. Taylor, M. L. and E. L. Arnold. *Anal. Chem. 43,* 1328 (1971).

143. Eisentraut, K. S., D. G. Griest, and R. E. Sievers. *Anal. Chem. 43,* 2003 (1971).

144. Wolf, W. R., M. L. Taylor, B. M. Hughes, T. O. Tiernan, and R. E. Sievers. *Anal. Chem. 44,* 616 (1972).

145. Noweir, M. H. and J. Cholak, *Env. Sci. Technol. 3,* 927 (1969).

146. Ross, W. P. and R. E. Sievers. *Env. Sci. Technol. 6,* 155 (1972).

147. Foreman, J. K., T. A. Gough, and E. A. Walker. *Analyst 95,* 797 (1970).

148. Hansen, L. C., W. G. Scribner, T. W. Gilbert, and R. E. Sievers. *Anal. Chem. 43*, 349 (1971).

149. Savory, S., P. Mushak, F. W. Sunderman, Jr., R. H. Ested, and N. O. Roszel. *Anal. Chem. 42*, 294 (1970).

150. Booth, G. H., Jr. and W. J. Dorby. *Anal. Chem. 43*, 813 (1971).

151. Ross, W. D., R. E. Sievers, and C. Wheeler, Jr. *Anal. Chem. 37*, 598 (1965).

152. Moshier, R. W. and J. E. Schworberg. *Talanta 13*, 445 (1966).

153. Morie, G. P. and T. R. Sweet. *Anal. Chem. 37*, 1552 (1965).

154. Belcher, R., J. R. Majer, R. Perry, and W. I. Stephan. *Anal. Chim. Acta 45*, 305 (1969).

155. Schworberg, J. E., R. E. Sievers, and R. W. Moshier. *Anal. Chem. 42*, 1878 (1970).

156. Koshimura, H., and T. Okubo. *Anal. Chim. Acta 49*, 67 (1970).

157. Schultz, B. G. and E. M. Larsen. *JACS 71* (1949).

158. Bayer, E., H. P. Muller, and R. E. Sievers. *Anal. Chem. 43*, 2012 (1971).

159. Minear, R. A. and C. Palesh. Unpublished research, 1972.

160. Morrison, G. H. *Annal. New York Acad. Sci. 199*, 162 (1972).

161. Smith, D. S. "Mass Spectrometry," In *Guide to Modern Methods of Instrumental Analysis,* Gouw, I. H., ed. (New York: Wiley Interscience, 1972).

162. Brown, R., P. Powers, and W. A. Wolstenholne. *Anal. Chem. 43*, 1079 (1971).

163. Ahearn, A. J. *Mass Spectrometric Analysis of Solids* (Elsevier, 1966).

164. Crocker, I. H., and W. F. Merritt. *Wat. Res. 6*, 285 (1972).

165. Chastagner, P. *Anal. Chem. 41*, 796 (1969).

166. Wahlgren, M. A., D. N. Edgington, and F. F. Rawlings. ANL. 7860 III. Radiological Physics Division. *Annual Report* (1971), page 55.

167. Brown, R. and P. G. T. Vossen. *Anal. Chem. 42*, 1820 (1970).

168. Evans, C. A. and G. H. Morrison. *Anal. Chem. 40*, 869 (1968).

169. Kowalski, B. R., T. L. Isenhour, and R. E. Sievers. *Anal. Chem. 41*, 998 (1969).

170. Booker, J. L., T. L. Isenhour, and R. E. Sievers. *Anal. Chem. 41*, 1705 (1969).

2. ADVANCED INSTRUMENTAL TECHNIQUES FOR THE IDENTIFICATION OF ORGANIC MATERIALS IN AQUEOUS SOLUTIONS

Bjorn F. Hrutfiord. College of Forest Resources, University of Washington, Seattle, Washington.

Russell F. Christman, Marc A. Horton. College of Engineering, Department of Civil Engineering, University of Washington, Seattle, Washington.

INTRODUCTION

A large number and variety of organic compounds are present in the receiving waters of the world. Urban runoff, industrial effluents, and domestic wastes all contain organic compounds of both natural and synthetic origin. A recent review of the literature[1] has shown the classes of compounds listed in Table 6 to be present in the nation's water resources. Each class of compounds listed

Table 6

Classes of Organic Compounds Reported in Waste Water

Humic Substances	Sulfur Compounds
Polyphenolics	Phenolic Compounds
Amino Acids	Plasticizers & Esters
Peptides	Organic Acids
Purines	Nitrogen Compounds
Carbohydrates	Chlorinated Compounds
Detergents	Resin Acids
Aliphatic Hydrocarbons	Terpenes
Aromatic Hydrocarbons	Alcohol & Ketones
Polynuclear Hydrocarbons	

can be expanded to include compounds within the
class (Tables 7 and 8). The compounds vary in
their persistence in the environment and some of
the more stable have led to water pollution problems such as color,[2] taste and odor,[3] or toxicity.[4]

Table 7

Chlorinated Compounds Reported in Waste Water

Penta Chlorophenol	Aldrin
0-chloronitrobenzene	Dieldrin
Chlorobenzene	Eldrin
1,2 & 1,4-Dichlorobenzene	Lindane
1,2,4-Trichlorobenzene	Heptachlor
5-chloro-2-methylbenzofuran	DDT
Hexachlorophene	DDE
Polychlorobiphenyls	2,4-D

Table 8

Organic Acids Reported in Waste Water

Formic	Citric
Acetic	Palmitic
Propionic	Stearic
Isobutyric	Oleic
Butyric	Benzoic
Isovaleric	4-hydroxyphenyl acetic
Valeric	Phenyl lactic
Caproic	Gallic
Glycolic	0-methoxybenzoic
Lactic	Alkyl benzoic

In nearly all cases, very minute amounts of
organics are sufficient to cause pollution problems,
and current isolation and identification procedures
are often pressed to their limits of sensitivity.
The purpose of this paper is to review and assess
the current status of various isolation and
identification procedures.

The first step in almost any analytical scheme
for the identification of organics in water is a

concentration process. There are a variety of
methods currently in use, including adsorption and
solvent extraction. Following concentration,
separation and identification are generally carried
out utilizing one of the many analytical tools
available. In organic compound identification
carried out to date, various spectrometric and
chromatographic methods have been employed.
Christman and Ghassemi[5] employed freeze concentration, oxidative degradation and chromatographic
separation in identifying components of color-
causing macromolecules in natural water supplies.
Sugar and Conway[6] have presented operating characteristics for application of gas-liquid chromatography
to wastewater analysis. Liquid chromatography in
the form of high resolution ion exchange chromatography is now being applied to the determination
of organics in sewage effluent. Hunter and
Heukelekian[7] have also examined domestic sewage
effluents using elaborate organic extraction
schemes. Middleton and Lichtenberg[8] have used
carbon columns for concentration followed by solvent
extraction and then IR spectroscopy for identification. The application of UV spectroscopy in the
examination of organic substances in water has been
demonstrated by Hanya and Ogura.[9] Garrison[10,11]
has utilized almost all the above analytical tools,
including Nuclear Magnetic Resonance (NMR) and a
coupled Gas Chromatograph/Mass Spectrometer (GC/MS)
system to characterize industrial organic pollutants
in water. Examples of instrumental application are
numerous in the literature, but emphasis will be
placed here on the most important advances which
have occurred through the coupling of elaborate
instrumentation and systematic separation schemes.

ISOLATION OF ORGANIC COMPOUNDS FROM WATER

Carbon Adsorption Method

Today, the isolation and concentration steps
are still necessary and are perhaps still the most
difficult part of the scheme. For the past two
decades, carbon adsorption has been one of the most
widely used concentration methods, dating back to
the work of Middleton and his co-workers.[12] In
this method the aqueous sample is passed through
multiple carbon adsorption columns where the
organic material is retained. The carbon is then
removed from the columns, dried under controlled

conditions and the organic material is solvent-extracted, almost always with chloroform and ethanol. The carbon adsorption method has the advantage of applicability to large sample volumes, allowing detection of trace amounts of organics. A large volume of water can be processed with a relatively small volume of carbon; the amounts can be adjusted to meet almost any sensitivity requirement. Samples of 5000 gallons of water have been used and detections in the order of mg/l or less can be achieved if required. Also, carbon is capable of adsorbing a wide variety of organic compounds from aqueous solution, making the method quite versatile. Generally nonpolar compounds are adsorbed more strongly than polar compounds.

While this procedure is very useful and often the only effective procedure of organic compound isolation, there are enough problems involved with it to encourage researchers to look for improved methods. Actually, the carbon adsorption method is relatively slow and cumbersome when compared with other methods to be mentioned. Careful attention must also be paid to details such as drying the carbon. Oxidation, hydrolysis, reduction or isomeration may occur on the surface of the carbon. Also, with carbon, some compounds are very difficult to desorb quantitatively, making accurate quantification somewhat difficult.[13] These are again generally the more polar compounds.

Solvent Extraction Methods

Solvent extraction methods have also been used for many years and the majority of the identified compounds referred to earlier have been isolated by this procedure.[14-16] This method works quite well in some cases. Solvent extraction essentially involves: (a) the removal of organic material from water by solvent extraction, usually with chloroform, (b) concentration and drying of the chloroform extract, and (c) analytical identification utilizing, for instance, a GC/MS system.

It is frequently possible to complete a simple solvent extraction with chloroform, remove the solvent and go directly to the GC/MS system for identification, thus producing identification of organic compounds very quickly. Recent applications of this technique have been reported by Garrison and his co-workers,[11] who have presented quantitative data showing relative changes in organic content of pulp mill effluents across stabilization ponds. Samples as small as one gallon can be processed for identification of compounds in the few mg/l range by this method.

The solvent extraction method is simple and relatively rapid to perform, and has the advantage of being applicable to class fractionation techniques. Extraction of an alkaline solution with hexane, for example, removes only nonpolar compounds. Chloroform extraction would then remove more polar organics. Acidification, perhaps in two steps, would allow separation of weak and strong acids. In some cases, the isolated organics can be analyzed directly by GC/MS application without further separations. The major disadvantage of this method is its limited applicability. With dilute solutions and the necessary large volumes to be handled, some limitation of detection may occur unless extractions are mechanized to a high degree.[17] Counter-current flow is generally applied in these circumstances.[18] There are also many compounds that cannot be extracted successfully from aqueous solutions with organic solvents, *i.e.*, very polar materials such as sugars.

Perhaps the area of most active research today is in the use of high surface area macroreticular adsorption resins such as the XAD series manufactured by Rohm and Haas. These resins are similar to carbon in that they will adsorb a wide spectrum of organics from very dilute solutions.[19,20] They are easily applicable to large sample volumes. In addition, they have the distinct advantage that adsorbed organics may be selectively desorbed into classes of compounds in an organic solvent.[21] Briefly, the method involves adsorption of organic material on the resin by passing aqueous samples at neutral pH through resin columns followed by selective elution of the classes of compounds by schemes utilizing various pH solutions and organic solvents. By repeating the adsorption on the resin, final elutions of classes of compounds can be made with methanol, accomplishing concentration, class separation and elimination of most of the water from the sample. Figure 3 presents an extraction scheme utilized by the Ames Laboratory.[21] It appears as if this method may well be the method of choice in the recovery and isolation of organics from water, both in the analytical as well as the practical sense.

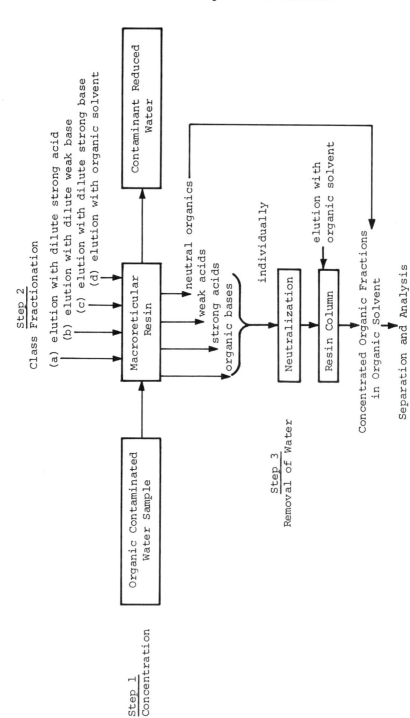

Figure 3.   *Separation and analysis scheme using macroreticular resins.*

The methods just mentioned for isolation of organics from water are not the only ones, but they appear to be the most widely used and useful. They are all preparative methods, resulting in organics in a form suitable for further analysis and identification.

LIQUID CHROMATOGRAPHY

The use of liquid chromatography in the organics field is generally not an identification procedure as such, but it is and will be the method of choice in the routine monitoring of organics in water. A general schematic of one instrument type is shown in Figure 4.[22] The general method has been in use for years and recently commercial instrument packages have been made available.

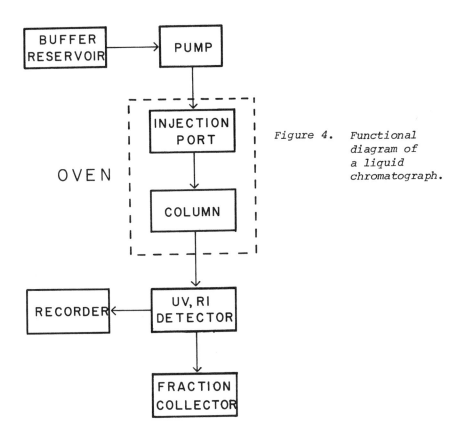

Figure 4. *Functional diagram of a liquid chromatograph.*

Two important developments have led to the
increased popularity of liquid chromatography.
First, there has been the development and general
availability of excellent continuous monitoring
detectors utilizing the UV absorbance properties
or the refractive index of organic materials. This
has given liquid chromatography a wider range of
applicability. Secondly, the development of ad-
sorbents or ion exchange materials capable of high
pressure operation have greatly increased the speed
of chromatographic operations.

Only brief mention of the application of liquid
chromatography will be made. Figure 5 shows a
typical urine analysis completed on a liquid chromatograph utilizing a UV detector.[23] Application of
this method to primary and secondary sewage treatment facilities has been made and the results
clearly show the effectiveness of organic removal
in these treatments. This method is relatively
simple, requiring only sample concentration, usually
by evaporation, and application of the concentrate
to a liquid chromatograph.

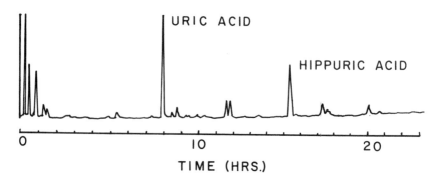

Figure 5.   UV-absorbing constituents of urine.

IDENTIFICATION TECHNIQUES

In the actual instrumental identification of
organic compounds, the best single tool available
today is the GC/MS combination, preferably coupled
with a computer. This combination exceeds the advantages of each of the individual components.

Gas chromatography as a separation tool is
highly developed at present.[6] Its basic requirement, that compounds have some volatility before
GC analysis can be performed, has been overcome

for many compounds with the development of high
temperature liquid phases and through the use of a
wide spectrum of useful derivatives, which have
the effect of increasing volatility and decreasing
polarity of compounds. Sensitivities on the order
of 10^{-12} to 10^{-14} g are not unusual in GC work.[24],[25]
Recently the gas chromatograph has been coupled
with the mass spectrometer. The main types of
interfaces, as described by Karasek[25] and Junk,[26]
are shown in Figure 6. The Watson-Biemann separator
is undoubtedly used most widely at present and it
works well for many compounds. The porous tube

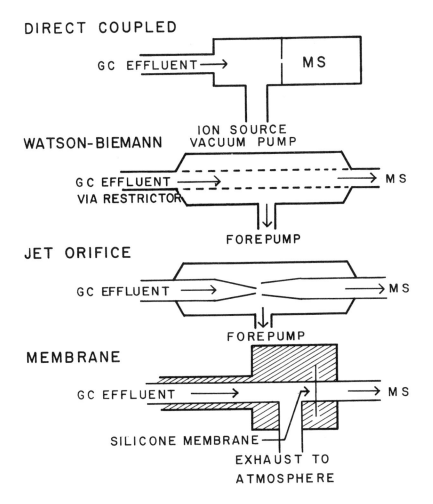

Figure 6.   GC/MS interfaces.

(Watson-Biemann) separator operates on the selective
diffusion of the lighter carrier gas through a
fritted glass section of tubing while heavier components pass through. The driving force for the
diffusion is supplied by placing a vacuum on the
glass section. This type of separator is most
common because it is easy to manufacture, relatively
inert to surface adsorption, and consequently supplied by instrument manufacturers. However, yields
tend to be somewhat lower than the jet type, especially with polar and higher molecular weight
compounds. The membrane type also appears to work
well in many cases. Membrane enrichment devices
contain a semipermeable membrane selective to the
organic compounds of the GC effluent. Diffusional
driving force is supplied in this case by the vacuum
of the MS inlet. Consequently, these separators
require no additional vacuum connections or pumps,
leading to their compactness. Membrane separators
seem to give high yield. However, their performance
does have strong temperature dependence and inertness
problems with the membrane can develop. The jet
type essentially works on the principle that a high
velocity stream of GC effluent, passing through a
drift section under vacuum, and aimed at the inlet
of the MS will lose the lighter carrier gas to the
vacuum while passing the sample to the inlet. This
type of separator may be in more than one stage.
Generally, the jet separator has a high yield and
good enrichment. Other interfaces continue to be
developed, and the ultimate answer to interfacing
problems has yet to appear. However, at present,
the jet type seems to possess the greatest
applicability.

A wide variety of mass spectrometer instruments
are available. It is important to point out that
the instruments in a GC/MS/Computer system require
only a resolving power in the 600 to 1000 range.
The molecular weight range does not have to be much
over 750. Essentially this means that the time-of-
flight and quadrupole instruments as well as the
magnetic sector instruments can all be used.

Obviously, compound identification cannot
always be made on the basis of mass spectra alone.
In working with specific organic contaminants in
water, one is almost always dealing with a known
compound, and identification is completed by
comparison with known spectra. This is where the
computer becomes an essential segment of the system.
Besides eliminating the mechanical data reduction

necessary with MS data, computer searches can readily be made assuming a data bank of known spectra is available. Such a data bank is now being established by EPA at Battelle Columbus and should prove to be a valuable asset.

Junk[26] has compiled an extensive work on GC/MS combinations and their application. He has given detailed and complete descriptions of system components, interfacing problems, and system parameters. This paper should be consulted for more detailed information.

A serious drawback to the GC/MS/computer systems is their relatively high cost. The GC/MS instrumentation alone may cost $30-40,000. The computer can cost an additional $50,000, bringing the total system cost to over $80,000. It should be mentioned that the equipment, regardless of cost, must work well and some systems do not.

SUMMARY

In conclusion, it appears as if the GC/MS/ computer combination will continue to be the foremost analytical tool in the identification of organic compounds from water in the near future. The choice of interfacing devices will be left to the state of the technology and the individual researcher. At present, the most efficient and universal interface is the jet type as described by Karasek[25] and Junk.[26] In the matter of isolation of organic compounds and class separation, use of the new Rohm and Haas XAD series of resins appears to be the best method available, having distinct advantages over the carbon adsorption columns. As pressure is increased on waste treatment operations to closely monitor organic removal efficiencies, with particular organics in mind, high pressure liquid chromatographs should be more widely used. These chromatographs will obviously be utilized in some identification capacity also.

REFERENCES

1. Department of the Army, U.S. Army Corps of Engineers. *Assessment of the Effectiveness and Effects of Land Disposal Methodologies of Wastewater Management.* Wastewater Management Report 72-1. Appendix (1972).
2. Ghassemi, M., and R. F. Christman. "Properties of the Yellow Organic Acids of Natural Waters," *Limnol. Oceanogr. 13,* 538 (1968).

3. Buchhoft, C. C., F. M. Middleton, H. Braus, and A. A. Rosen. "Taste- and Odor-Producing Components in Refinery Gravity Oil Separator Effluents," *Ind. and Eng. Chem. 46,* 284 (1954).

4. Howe, R. H. "Toxic Wastes Degradation and Disposal," *Process Biochem. 4,* 25 (1969).

5. Christman, R. F., and M. Ghassemi. "Chemical Nature of Organic Color in Water," *JAWWA,* 723 (1966).

6. Sugar, J. W., and R. A. Conway. "Development of Gas-Liquid Chromatography Technology for Wastewater Analysis," Presented at ACS Meeting, Division Air Water Waste, April, 1967.

7. Hunter, J. V., and H. Heukelekian. "The Composition of Domestic Sewage Fractions," *JWPCF 37,* 1142 (1965).

8. Middleton, F. M., and J. J. Lichtenberg. "Measurements of Organic Contaminants in the Nation's Rivers," *Ind. and Eng. Chem. 52,* 99A (1960).

9. Hanya, T., and N. Ogura. "Application of UV Spectroscopy to the Examination of Dissolved Organic Substances in Water," *Adv. Org. Geochem.* (Proc.) (New York: Pergamon Press, 1964).

10. Garrison, A. W. "Analytical Studies of Textile Wastes," Presented at ACS Meeting, Division Air, Water & Waste. September 1969.

11. Garrison, A. W., L. H. Kieth, and M. M. Walker. "The Use of Mass Spectrometry in the Identification of Organic Contaminants in Water from Kraft Paper Mill Industry," Presented at Amer. Soc. Mass Spec. Meeting, San Francisco (1970).

12. Rosen, A. A., and F. M. Middleton. "Identification of Petroleum Refinery Wastes in Surface Waters," *Anal. Chem. 27,* 790 (1955).

13. Allen, S. C., R. H. Pahl, and K. G. Maghan. "Organic Desorption from Carbon-1," *Water Research 5,* 3 (1971).

14. Lamar, W. L., and D. F. Goerlitz. *Organic Acids in Naturally Colored Surface Waters.* Geo. Survey Water Supply Paper 1817-A. (Washington, D.C.: U.S. Government Printing Office, 1966).

15. Rebhun, M., and J. Manka. "Classification of Organics in Secondary Effluents," *Environ. Sci. and Tech. 5,* 606 (1971).

16. Hoak, R. D. "Recovery and Identification of Organics in Water," *Proc. Int. Conf. on Water Pollution Res.* (London, 1962).

17. Kawahara, F. H., J. W. Eichelberger, B. H. Reid, and H. Stierli. "Semiautomatic Extraction of Organic Materials from Water," *JWPCF 39,* 572 (1967).

18. Department Health, Education and Welfare. *Organic Chemicals in Waters--Course Manual--Water Supply and Water Pollution Training.* R. A. Taft Sanitory Engineering Center (1957).

19. Gustafson, R. L., and J. Paleos. "Interactions Responsible for the Selective Adsorption of Organics on Organic Surfaces," *Organic Compounds in Aquatic Environments,* S. D. Faust and J. V. Hunter (eds.) (New York: Marcel Dekker, Inc., 1971).

20. Paleos, J. "Adsorption from Aqueous and Nonaqueous Solution on Hydrophobic and Hydrophylic High Surface-Area Copolymers," *J. Coll. and Inter. Science 31,* 7 (1969).

21. "Find New Method for Isolating Minute Amounts of Organic Compounds Found in Water," USAEC Ames Laboratory News Release. December, 1970.

22. Burtis, C. A., and D. R. Gere. *Nucleic Acid Constituents by Liquid Chromatography.* Varian Aerograph Publication A-101 (1970).

23. Burtis, C. A., and R. L. Stevenson. *The Separation of the UV-Adsorbing Constituents of Physiological Fluids by High Resolution Liquid Chromatography.* Varian Aerograph Publication A-1012 (1971).

24. Lamar, W. L., D. F. Goerlitz, and L. M. Law. *Detection and Measurement of Chlorinated Organic Pesticides in Water by Electron Capture Gas Chromatography.* U.S. Department of Interior, Geo. Survey (1964).

25. Karasek, F. W., and W. H. McFadden. "Flow and Vacuum: Keys to GC/MS Interfacing," *Res./Devel. 22,* 52 (1971).

26. Junk, G. A. "Gas Chromatograph-Mass Spectrometer Combinations," *Int. J. Mass Spectrom. Ion Phys. 8,* 1 (1972).

3. INCIDENCE, VARIABILITY AND CONTROLLING FACTORS FOR TRACE ELEMENTS IN NATURAL, FRESH WATERS

Julian B. Andelman. Graduate School of Public Health, University of Pittsburgh, Pittsburgh, Pennsylvania

INTRODUCTION

There is a great deal of current interest in the incidence of trace elements in the environment and the quantities to which man is exposed. Although health effects have been related to high exposures to specific trace elements such as cadmium and mercury, of equal concern is the possible chronic disease relationships to the uptake of more typical quantities from food, water and air. A growing number of reports over the past dozen years show statistical correlations between drinking water quality and cardiovascular disease, such as the work of Schroeder in the United States,[1,2] and Crawford and co-workers in Great Britain.[3,4] In many of these studies, reviewed by Winton and McCabe,[5] water hardness has been shown to correlate negatively with cardiovascular mortality rates. More recently similar negative correlations have been found between infant mortality and municipal water hardness and calcium concentration.[6] Schroeder also found negative correlations between cardiovascular mortality and trace element concentrations for vanadium, barium, strontium, and lithium, and a positive one for copper. Masironi has shown that there is a high positive correlation between hardness and the concentration of several trace elements in surface water in the United States.[7] Berg and Burbank have shown statistically significant positive correlations between mortality from various types of cancer and concentrations of several trace elements in water supplies, including arsenic, beryllium, cadmium, lead and nickel.[8] Voors found statistically significant negative correlations for lithium and vanadium in municipal water with atherosclerotic heart disease.[9]

Such epidemiological investigations indicate the importance of and the need to understand the nature, incidence, variability and controlling factors for trace elements in natural waters. Aside from man's direct exposure to trace elements from drinking water obtained from natural fresh waters, both surface and subsurface, and with various degrees of treatment, water can also be a source of such trace elements for aquatic organisms and terrestrial edible plants, ultimately reaching man and foraging animals through their food chains.

This paper will outline and discuss the various processes that influence and control the trace element composition of natural waters, the emphasis being on fresh waters. For this purpose trace elements are defined as those which generally occur at concentrations in natural fresh waters below 1 mg/l. The paper will consider the various chemical and physical forms and species in which trace elements can occur in natural waters and which can influence their behavior; the variability of these elements with time, river flow, water depth and geographic location; the influence of flora and fauna in water and the lithologic environment; and the contributions of such sources as precipitation and fallout from air, as well as man's urban, industrial and agricultural activities.

CHEMICAL AND PHYSICAL STATES

The chemical and physical states of trace elements are of considerable interest in understanding controls on their concentrations in natural waters. Whether an element exists in a particulate, colloidal, or soluble state may influence the result of its chemical analysis. This can be shown in a comparison of analyses of suspended and dissolved trace elements in surface waters of the United States by Kopp and Kroner.[10] They distinguished operationally between these two fractions by filtration with 0.45 micron pore size membrane filters, the results for several of the elements studied being shown in Table 9.
It is apparent from these data that the frequency of detection and mean concentrations can vary considerably, depending on the trace elements, so that the greater concentration can occur either in the

Table 9

A Comparison of Incidence and Concentration of
Several Suspended and Dissolved Trace
Elements in U.S. Surface Waters[a]

	Suspended		Dissolved	
Element	Frequency Found (%)	Mean ($\mu g/l$)	Frequency Found (%)	Mean ($\mu g/l$)
Zn	64	62	77	64
Al	97	3860	31	74
Pb	2	120	19	23
Ba	95	38	99	43
Sr	10	58	100	217

[a]Data taken from Reference 10.

suspended or dissolved fraction. Similar differences
have been obtained for trace elements in sea water.[11]
In a more detailed study of Fe, Mn, and Al, Kopp and
Kroner also showed that the range of concentrations,
as well as the mean, were different for the soluble
and particulate fractions.[10]

Another indication of the presence of trace
elements in particles of different size may be seen
in the data of Feder, shown in Table 10.[12] Surface
waters were analyzed unfiltered and after filtration
with 0.45 and 0.1 micron pore size filters. For some
elements, such as barium, strontium, and copper, the
differences between the analyses of the filtered and
unfiltered samples were not large, while for other
elements, such as aluminum and iron, the results of
the analyses depended not only on whether the sample
was filtered but also on the pore size of the filter.
Although membrane filtration might be expected to
provide evidence of particle size distribution
based on pore size, colloidal particles (those less
than approximately one micron in diameter) are known
to readily sorb onto a wide variety of materials,
so that caution must be used in the interpretation
of such data.

The association of trace elements with particulate and colloidal species can be attributed to a
variety of mechanisms. The larger sand and silt

Table 10

Comparison of Effects of Filter Pore Size on
Concentrations (µg/l) of Selected Elements
in the Filtrate of Surface Water Samples
From a Missouri Creek[a]

Element	Unfiltered	Filtered 0.45µ pore size	Filtered 0.1µ pore size
Aluminum	310	77	57
Barium	90	87	80
Boron	31	16	17
Copper	2	1	2
Iron	740	240	49
Lithium	< 10	< 10	< 10
Manganese	200	170	160
Nickel	8	6	< 5
Strontium	67	73	70
Titanium	7	2	< 2

[a]From Reference 12.

particles consist of a variety of minerals, within
which are incorporated many trace elements. Clay
minerals, which are major constituents of the
smaller suspended particles and colloidal matter
in streams, also contribute large quantities of
trace elements. The mechanism by which these particles carry and contribute trace elements to a
stream can be varied and complex. There can be
sorption and ion exchange onto the mineral surface,
as well as into its structure, with uptake and
release being influenced by a variety of solution
variables. Turekian and co-workers have made
preliminary calculations of soluble Co, Ag, and Se
which can be desorbed from suspended clay minerals
in streams, compared to those present in an already
dissolved state.[13] Although there were variations
among the rivers analyzed, in general the desorbable
Ag fraction compared to the total was smallest, Se
intermediate, and Co the highest, some of their data

being shown in Table 11. The calculations showed
surprising uniformity among the rivers studied.
For comparison, the data of Kopp and Kroner indicate
that Ag and Co were not detected in any suspended
samples,[10] indicating agreement with these calculations for Ag but not for Co.

Table 11

*Total Soluble and Adsorbed Trace Elements
in Several Rivers.*[a]

	Co		Ag		Se	
River	Sorbed %	Total g/l	Sorbed %	Total g/l	Sorbed %	Total g/l
Mississippi	79	0.52	8	0.26	21	0.14
Susquehanna	67	1.06	5	0.39	15	0.39
Rhone	79	0.47	7	0.41	17	0.18
Amazon	59	0.29	4	0.24	16	0.25

[a]Values calculated by Turekian *et al.*[13]

A further complication of such phenomena is the
effect of particle size of the sorbing mineral.
Since smaller particles have larger surface areas,
they would be expected to sorb larger quantities of
trace elements for a given mass of sorbent. This
was shown to apply for a range of particle sizes in
a calcium carbonate ooze in sea water, although for
submicron size particles a reversal of this effect
was noted.[14]
The sorption of trace elements onto such mineral
surfaces can occur without ion exchange as a result
of coulombic interactions with the surface in its
electric double-layer. Most particles in water are
negatively charged, and trace cations can be held
at the negatively charged surface of such a colloidal
or particulate mineral or other material. Similarly,
colloidal trace elements may be held as well. For
such interactions, variations in fresh water calcium
concentrations would be expected to have a significant
effect, because of the well-known ability of multi-
charged ions to compress the electric double-layer
at solid-solution interfaces.

Ion exchange can also exert a considerable effect on trace element uptake by suspended and sediment minerals. It is considered to be the most important mechanism for chemical control at the sediment-water interface in the oceans,[14] and is probably of similar importance in fresh waters as well. Thus, clay minerals exposed to sea waters for about five years have been shown to lose considerable quantities of Ca and Na while gaining Mg.[14] Such ion exchange involving Ca in fresh water is likely to exert an effect on the exchange of trace cations with minerals. Thus one might expect that waters with larger Ca concentrations would have higher concentrations of *dissolved* trace elements because of the inhibition of their exchange from the solution phase onto the minerals. This effect is in agreement with the positive correlations between trace element and Ca concentrations found by Masironi,[7] provided that the concentrations are attributed to soluble species only. Alkali metal ions, such as Na and K, would have a similar but smaller effect because they are univalent.

In some cases one might expect solubility to act as a limit on trace element concentrations in natural waters. One of the difficulties in making predictions on this basis is the uncertainty of the crystallographic form of the precipitate, or even its chemical composition. Such elements as Fe, Al, Mn, and Ti, which form quite insoluble hydroxides, are likely to exceed solubility limits. However, these limits are based on equilibrium considerations, and natural waters are often acknowledged to be out of equilibrium. In considering the extent to which the concentrations of several trace elements in sea water might be limited by their solubilities, it has been shown that Sr and Ba might, but other trace elements considered probably were not.[15] These conclusions probably apply to many trace elements in fresh waters as well. Thus, for example, it has been shown that, although there seemed to be a decrease in zinc concentrations with increasing pH in tributaries of the Chesapeake Bay, the measured concentrations, usually less than 100 μg/l, were considerably less than the solubility limits, which were generally above 1 mg/l in the pH range encountered.[16] Even these considerations are, however, misleading because of the possibility of coprecipitation.[15] Thus, the solubility of calcium carbonate may be exceeded in water, resulting in its precipitation and trace elements being "brought down" at the same time, such as by inclusion or entrapment in the solid calcium carbonate particles.

Complexation of trace elements by both organic and inorganic ligands plays a potentially important role in their effect on and movement through natural waters, influencing their solubility, their sorption on suspended and bottom sediments, and their uptake by biota. Organic compounds in both fresh and sea water are known to interact with trace elements, especially through the mechanisms of complex and chelate formation.[17],[18]   A large percentage of the copper in sea water has been shown to be associated with organic compounds in nonlabile combinations, as has up to 99% of the copper present in soil solutions. Other trace elements, such as cobalt and boron, have been similarly associated with organic materials, and it has been concluded that the control and limitation of phytoplankton productivity in sea water may be the most vital effect of metal-organic interactions in the environment.[18]

Naturally occurring organic species like fulvic acid, which can complex these metals and sorb onto such clay minerals as montmorillonite, are important in this regard,[19] but organic pollutants may play a role as well. The latter can concentrate at the air-water interface, where they have been shown to enrich the concentrations of such trace elements as lead, iron, copper and nickel.[20]   Once concentrated in the surface layer, these organics and their associated trace metals can enter the food chain and be concentrated in higher trophic members.

Organic compounds in water and soil-water systems can affect the oxidation state of trace elements, thereby exerting a strong influence on their behavior, such as their solubility and ability to be absorbed and ion-exchanged.[17]   Thus, for example, the reducing environment created by organic pollution may cause the reduction of chromium-VI to chromium-III, thereby affecting its transport in water, since sediments and the biomass have different capacities for sorbing these two species.[21]   Complexation of trace elements by organics can also influence both the rate and equilibria of their redox reactions.[17]

Some indication of possible complexation of copper, manganese, and zinc in the Gulf of Mexico may be seen in the data of Slowey and Hood,[22] shown in Table 12. These data indicate that measureable fractions of the elements occur in the colloidal and larger particle sizes, perhaps 10 to 20%, depending on the element. In addition, however, the extractable

Table 12

*Average Concentrations (μg/l) of Various Forms of Copper,
Manganese and Zinc in Open Sea in the Gulf of Mexico*[a]

	Cu	Mn	Zn
Total	1.3	0.31	3.5
Extractable	0.90	0.26	2.6
Nondialyzable	0.10	< 0.02	1.5
Particulate (>0.45μ)	0.25	0.03	0.3
Particulate (10 mμ - 0.45μ)	0.15	< 0.01	0.1

[a]Data of Slowey and Hood.[22]

(by an organic chelate in chloroform) portions
constituted the greater fraction and were judged to
contain both ionic and organically complexed metal.
Some measure of the latter in the samples was the
"nondialyzable" fraction, which for zinc constituted
perhaps 40% of the total.

In a study of the various forms of copper in
sewage effluents and river water, Stiff found that
much of the copper in river waters was associated
with suspended solids, but also that most of the
soluble portion was complexed.[23]   He concluded that
the most likely soluble forms are the free cupric
ion and its complexes with carbonate, cyanide,
amino acids and polypeptides, and humic substances.

Studies such as these, in which the speciation
of both the soluble and insoluble forms of trace
elements in natural waters is determined, have been
relatively few. Yet, in order to understand the
role and influence of trace elements in the natural
environment, these kinds of information are necessary.

GEOGRAPHIC VARIABILITY

The possible variability in concentration of
trace elements among geographical locations, both
on a macro and micro scale, is of interest in
attempting to characterize their behavior in natural
waters. Turekian and co-workers found that the
concentrations of several dissolved trace elements
in streams of the United States and elsewhere showed

two types of distributions.[13] Silver, cobalt, selenium, rubidium, and cesium have a relatively narrow range of concentration among the rivers. In contrast, molybdenum, chromium and antimony vary considerably. For this latter group, it was noted that "there is no immediate correlation of the concentrations of any of these elements with the superficially evident properties of the stream, its sediment, or its drainage basin,"[13] a conclusion similar to that of Silvey for trace elements in California surface waters.[24]

An example of the variability encountered in river basins of three areas of the United States, using the data of Kopp and Kroner,[10] is shown in Table 13, representing the mean positive concentrations encountered in each area. The underlined

Table 13

Mean Positive Trace Element Concentrations in U.S. River Basins (µg/l)[a]

Element	All U.S.	Northeast	Lake Erie	Alaska
Zn	64	96	205	28
As	64	34	308	34
B	101	32	210	28
Cu	15	15	11	9
Ni	19	8	56	5
Pb	23	17	39	12
V	40	9	54	32

[a]From Kopp and Kroner, 1962-1967.[10]

values are above the U.S. averages. On this macro-geographic scale it is apparent that, except for copper, concentrations in Lake Erie are significantly higher than the U.S. average, in Alaska they are significantly lower, while in the Northeast there is a more mixed behavior.

However, even on a micro-geographic scale one can encounter significant differences. Thus, for example, in Lake Michigan in one relatively short period zinc concentrations varied from 3 to 80 µg/l, depending on location within the lake, with the comparable concentrations for chromium ranging from 0.55 to 1.1 μg/1.[25]

In a study of samples collected from 25 points along the 150 mile Neuse River in North Carolina, a river reasonably free of industrial and municipal pollution, an attempt was made to determine if there was any correlation with regional ground water lithology, which was varied and well-defined.[26] Results for four of the trace elements at six of the sample points are shown in Table 14.

Table 14

Variations Along the Neuse River, North Carolina[a]

Ground Water Lithology	Ca mg/l	Sr	Ba	Co	Ag
			μg/l		
Slate	6.2	53	17	0.044	0.52
Shale	8.8	45	7.3	0.110	0.56
Granite	7.6	33	6.8	0.051	0.25
Slate & Schist	7.6	56	9.7	0.085	0.86
Cretaceous Sand	6.9	33	13	0.097	0.30
Tertiary Lime	8.0	61	22	0.078	----
Average[b]	7.7	50	12	0.078	0.37
% Stand. Deviation[b]	14	24	33	56	49

[a] Data of Turekian et al.[26]

[b] Based on samples at 25 points.

There doesn't seem to be any systematic change as one proceeds downstream, and the authors concluded that there were no correlations along the river between the concentrations of the major elements sodium, chloride, calcium and silica with the trace elements shown. Similarly, they could not make correlations with the prevailing ground water lithology, also shown for the sample points of Table 14. The cobalt and silver seems to be the most highly variable, as shown by their standard deviations. It was concluded that the changes in concentrations were probably due to ground water variations and removal by biological and inorganic processes.

A similar variability can be seen in Table 15 for the length of the Ohio River, which is heavily industrialized, with the data being taken from Kopp and Kroner.[10]   Boron, barium and strontium were

Table 15

Variation in Some Trace Element Concentrations During 1962-1967 in Ohio River, U.S.A.[a]

	Concentration					
	B		*Ba*		*Sr*	
Sampling Point	*Mean (μg/l)*	*Max/ Min*	*Mean (μg/l)*	*Max/ Min*	*Mean (μg/l)*	*Max/ Min*
Allegheny River[b]	50	15	37	14	95	5
Monongahela River[b]	89	34	30	11	160	10
Toronto, O.	84	7	28	7	105	5
Addison, O.	106	42	42	11	131	4
Huntingdon, W.Va.	61[c]	164	37	25	123	4
Cincinnati, O.	83	8	41	8	155	3
Louisville, Ky.	64	4	58	8	154	2
Evansville, Ind.	78	5	58	7	157	3
Cairo, Ill.	43[c]	6	41	5	103	4

[a]Data taken from Reference 10.

[b]These rivers join to form the Ohio River at Pittsburgh, Pa.

[c]Detected in at least 95 percent of the samples. All other elements were detected in all samples.

tabulated because positive values were obtained in essentially all the samples, the mean values being shown for the period from 1962 through 1967. Just as in the much shorter, relatively uncontaminated Neuse River, there are relatively large variances for these three metals as one proceeds down the Ohio River.

It is of interest to compare mean values for several trace elements in the two major tributaries of the Ohio River, the Allegheny and Monongahela, just before they join at Pittsburgh to form the Ohio. Some data in this regard are shown in Table 15, but Table 16 presents additional data at these sampling points for six trace elements studied over a one-year period in 1964-65.[27]   As shown in this

Table 16

Comparison of Macro and Trace Element Content and Variation in Two Tributaries of Ohio River at Pittsburgh in 1964-65[a]

	Allegheny River			Monongahela River		
Element	Mean	Range	Max/Min	Mean	Range	Max/Min
Trace (µg/l)						
Ba	34	8-70	9	20	10-34	3.4
B	100	60-155	2.5	127	74-210	3
Cu	30	14-97	7	25	6-50	8
Ni	< 26	<8-44	>5	<13	<7-15	> 2
Sr	74	28-130	4.6	92	35-175	5
Zn	89	38-180	5	98	26-180	7
Macro (mg/l)						
Ca	35	14-66	5	47	18-90	5
Cl	20	7-46	7	16	3-47	16
Na	20	5-52	10	36	8-90	11
K	2.6	1-5	5	3.3	1.2-6.5	5
SO_4	143	43-350	8	221	72-450	6

[a]From Reference 27.

table, the mean values for the two rivers at sampling points less than a mile apart were significantly different in every instance. As a point of comparison, the concentrations of the macro elements in the two rivers also were markedly different, as shown in Table 16.

It should also be noted that variations in concentration occur with depth. An example of

such variability is shown in Table 17, which tabu-
lates the concentrations of several trace elements
at three depths in Williams Lake in the state of
Washington.[28]   The highest concentration for each
element is underlined, and, except for mercury,
always occurred in the hypolimnion. This is
probably due to interactions with bottom sediments.
For some elements there were also differences
between samples taken at and one meter below the
surface.

Table 17

*Depth Variations in Trace Elements
in Williams Lake, Washington*[a]

Element	Surface	1.0 Meter	Hypolimnion
Cr	2.43	2.25	5.46
Sc	0.04	0.02	0.37
Zn	13.3	38.2	125
Sb	0.18	0.15	0.37
U	2.81	1.62	23.7
Au	0.09	0.08	0.15
Ba	16.5	18.9	144
Hg	5.13	3.43	3.70

[a]Data of Funk, *et al.*[15] Concentrations in µg/l.

All of these studies indicate that there can
be large differences in trace element concentrations, on both a macro and micro geographic scale,
and that such variations often occur in an
unsystematic and nonpredictable fashion.

TIME VARIABILITY

Another important consideration in assessing
trace elements in natural waters is their variability in concentration with time. If this
variability is large, at the very least

it has implications for the number and frequency of samples required to determine average concentrations or loads. Some indication of this time variability may be found in Table 15 for several sampling points along the Ohio River, where the ratio of the maximum to minimum concentration values obtained over a 5-year time span are tabulated at each sampling station. Although this may not be an ideal measure of time variability, it is a useful measure of the range of values encountered. Depending on the sampling point and element, there is a large variability in this maximum/minimum ratio. It is certainly greater, for example, than the ratios of the comparable values for the range of the mean concentrations among the sampling points.

Similarly Table 16 shows values for the ranges and the ratio of the extreme values for the Allegheny and Monongahela Rivers at Pittsburgh. This time variability is similar to that encountered along the Ohio. It is also useful to note that the time variability of the trace elements may be of the same magnitude as that encountered for the macro elements, as also shown in Table 16. The average values obtained for alternate monthly periods for these same two sampling points are shown in Table 18. These

Table 18

Time Variation of Concentration (µg/l) of Several
Trace Elements in the Allegheny (A) and
Monongahela (M) Rivers at Pittsburgh[a]

Element	1964 Nov A	1964 Nov M	1965 Jan A	1965 Jan M	1965 March A	1965 March M	1965 May A	1965 May M	1965 July A	1965 July M	1965 Sept A	1965 Sept M
Barium	27	34	59	15	40	19	30	25	22	12	30	24
Boron	100	129	118	74	111	130	100	105	155	205	69	85
Copper	21	29	97	6	28	27	27	50	27	25	24	32
Nickel	20	<8	<8	<8	<13	<10	25	15	40	20	44	20
Strontium	118	143	45	53	28	47	50	65	100	90	120	130
Zinc		74	118	50	26	81	44	140	120	180	160	120	150

[a]Data of Shapiro, Andelman, and Morgan.[27]

provide additional indications of the differences
obtained for the two rivers just before they join
to form the Ohio, and they also give a more detailed
impression of the time variabilities. It is also
of interest to note that there is no readily
apparent systematic relationship between concentration and season.

Another indication of time variability is
shown in Table 19, which presents analyses for
several trace elements at one sample station in
Lake Michigan at three different times.[25] Just as

Table 19

*Time Variation of Trace Element Content (µg/l)
At One Sampling Point in Lake Michigan*[a]

Element	Oct. 1969	April 1970	June 1970
Al	15	10	21
As	0.38	2.6	1.1
Ba	29	19	35
Ce	0.30	0.79	0.22
Cr	1.6	1.6	1.6
Hg	0.025	0.020	0.020
La	0.11	0.12	0.12
Sr	56	85	96
Zn	5	33	15

[a]From Reference 25.

with the data in the river systems discussed above,
these data show that time variability may be encountered in a large lake and that its magnitude
depends on the particular element. Thus, for
example, chromium and lanthanum concentrations were
essentially constant, while zinc and arsenic showed
considerable variability.

EFFECT OF FLOW

In flowing waters one might anticipate that
variations in trace element concentrations could be

correlated with stream flow, but this is often not
the case, as pointed out by Durum and Haffty.[29]
Predictable effects of flow that might influence
trace element content are scouring and the variation
in suspended sediment; perhaps even more important
is the load of trace elements associated with
varying quantities of water from tributaries with
varying characteristics.

In considering the concentrations of Al, Ba,
Cu, Ni, and Sr at low and high flow in four major
rivers in the United States, Cambodia, Canada, and
South Africa, Durum and Haffty showed that, depending
on the river, there could be either a large increase
or decrease in trace element concentration with
flow.[29] It was also noted that such metals as Fe,
Al, Mn, and Ti, which readily precipitate in natural
waters to form hydroxides as particles or colloids,
may be dissolved from suspended minerals under the
turbulent conditions of high flow, perhaps aided by
complex formation with tannic acid or similar materials
leached from decaying vegetation. In such a situation
one could obtain concentrations of these metals higher
than solubility and redox relationships might predict, and which would increase with stream flow.

Often, however, there is not even any systematic
effect of flow, aside from a predictable relationship.
Such a situation is shown in Figure 7, in which the
mean monthly concentrations of copper are plotted
versus flow at two sampling stations on the Allegheny
River in western Pennsylvania.[27]   The monthly

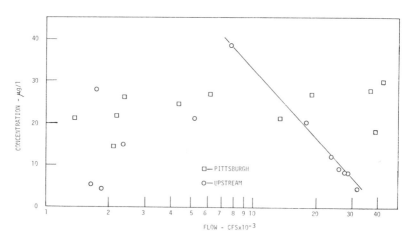

Figure 7.   *Effect of monthly flow rate on mean copper
concentrations in two stretches of Allegheny
River. Data of Shapiro, Andelman and Morgan.*[27]

concentrations were obtained from several samples, composited by flow, and the data represent a one-year period. As is apparent from Figure 7, at least at the downstream (Pittsburgh) station, there is no systematic correlation with flow. At the upstream station, the same can be said at the lower flows, although at higher flows there appears to be a fairly regular decrease in concentration with increase in flow.

In a study of trace element concentrations in the Columbia River and its tributaries, some approximate correlations of concentration with water flow were found, and in some instances these could be associated with contributions from tributaries.[30] Thus, it was found that concentrations of uranium and phosphorus tended to decrease with flow; copper, manganese, arsenic and lanthanum increased to a maximum in April, although not directly with flow; and zinc and cobalt closely followed river flow. The concentrations of other trace elements, such as iron, were highly variable.

In attempting to develop input-output models for trace elements in natural waters, it is necessary to determine the effect of flow, since this is the critical variable determining load. If, for example, the flow is fairly independent of load, as is the case for the copper at the Pittsburgh station in Figure 7, this implies a stream load essentially proportional to flow, and perhaps a continuous leaching process as the ground and surface waters pass over soils and minerals. If, on the other hand, the concentration decreases with flow, as at the upstream station at higher flows, this may indicate a relatively constant source or load of trace element, which is merely diluted as flow increases.

INTERACTION WITH BIOTA

Interactions with aquatic animals and plants may exert a major influence on trace element concentrations in water. The ability of a wide variety of aquatic organisms to concentrate trace elements is well known. Goldberg has presented data indicating that their uptake increases with their ability to be complexed by ligands.[31] Thus, the alkaline earth metals Ba, Sr, Mg, and Ca were found to concentrate considerably less than the transition metals Cu, Ni, Co, Zn, and Mn. At least five mechanisms of concentration have been suggested by various investigators for sea water biota, and summarized by Martin.[32] These are: (a) particulate ingestion of aqueous suspended matter, (b) ingestion in food material, (c) complexation by biological chelating agents, (d) incorporation into physiological systems, and (e) ion exchange and sorption on tissue or membrane surfaces. This important phenomenon of biological uptake has yet to be understood to the point of predicting its effect on trace element concentrations in fresh surface waters and is certainly a confounding factor in attempts to relate them to the lithologic environment. Nevertheless, it should be pointed out that calcium levels in water have been shown to reduce trace element uptake by some plants, and this relationship may provide an important control on trace elements.

In a study of various surface and subsurface waters in California, Silvey concluded that aquatic plants and fresh water biota may exert the most important influence on the concentrations of trace elements.[24] He found differences among their concentrations in various types of water, examples from some of his data being shown in Figure 8. These represent mean concentrations of trace elements for those analyses above the sensitivity limit. It is apparent that there are some differences among these three types of water. From these and other data, he concluded that the greater the biological activity in the water, the lower were the concentrations of trace elements.

A study of several trace elements in tropical, oceanic zooplankton showed concentration factors as high as 200,000 (the latter for lead).[33] Most of the trace element concentrations were higher in the deep samples, and it was suggested that a critical factor in the uptake was likely to be time. Because of the large quantities of such zooplankton and their high surface to volume ratio, which provides a large absorbing surface in both the living animal and in their exoskeleton, it was concluded that such organisms will prove to be a major factor in the biological transport of trace metals in the world's oceans.

CORRELATIONS WITH LITHOLOGY AND
OTHER ELEMENTS

An important consideration in evaluating trace elements in natural waters is the possible influence

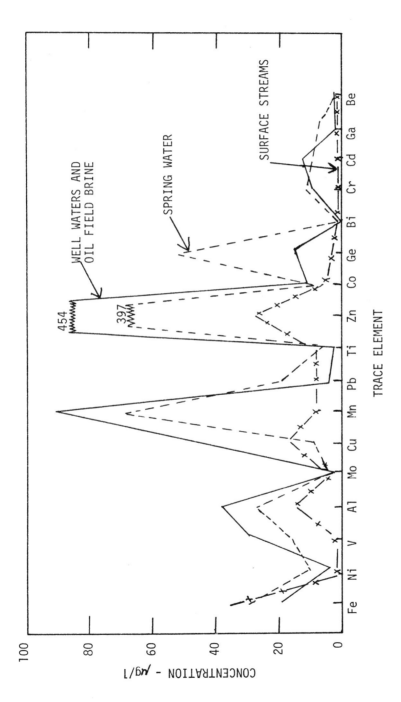

Figure 8.   *Mean Positive Concentrations of Trace Elements in California Waters. Data of Silvey.*[40]

of the lithologic environment as well as the inter-relationships among the concentrations of the trace elements and between them and specific macro elements. If such relationships exist and can be elaborated, then one would have a predictive tool for determining the levels of trace elements that might be encountered.

Examples of such relationships can be found in the discussion by Durum and Haffty,[29] in which they note in a comparison of large North American rivers that the ratio of the mean concentrations for Ni/Cr varied from 1.1 to 2.5, and for Ni/Cu from 1.4 to 3.1. Similarly they found that Ba/Cl varied from 3.8×10^{-3} to 14×10^{-3}, and Sr/Cl from 5.0×10^{-3} to 20×10^{-3}. However, in a study of the Neuse River in North Carolina, as noted elsewhere in this discussion, no correlation along the river could be found between the concentration of the major element and any of the trace elements studied.[26]

One interesting study in this regard is that of Masironi,[7] who found that the concentrations of 19 trace elements in river waters of the United States show positive correlations with water hardness. The correlation coefficients and levels of significance are given in Table 20. The higher

Table 20

Correlations of Hardness with Various Trace Elements
in River Waters of the U.S. 1961-1962[a]

Element	r	Element	r
Bi	0.95	Zn	0.90
Sb	0.95	Pb	0.87
Sn	0.94	B	0.85
Co	0.94	Mn	0.65
Cd	0.93	F	0.57
Cr	0.93	Be	0.56
V	0.92	Ag	0.45
Ni	0.92	Ba	0.44
Mo	0.91	Fe	0.30
		Cu	0.21

Levels of Significance

r	Probability -%
0.24	1
0.3	0.1

[a] Data of Masironi (116 samples).[7]

the correlation coefficient r, the smaller the probability that the correlation is spurious. The correlation is a positive one; that is, the greater the hardness, the higher the concentration of trace element. As is apparent from this table, the correlation varied considerably for the different trace elements. To some extent such correlations are not unexpected, since limestone and other calcium (and magnesium) bearing minerals, exposed to ground and surface waters, contain variable quantities of trace elements which can be leached.[34,14]

As noted previously, several investigators were not able to establish any relationship between lithologic environment and trace element content of the contacting waters.[13,24,26] One might expect, however, that ground water in well-defined and distinctly different geohydrologic formations would reflect this in their trace element composition. The results of one such study to ascertain this are given in Table 21, the data being that of Feder.[35]

Table 21

Influence of Geohydrologic Environment on Trace Elements in Missouri Ground Waters[a]

Geohydrologic Unit	Mean Concentration µg/l			
	Al	Ba	B	Sr
Quaternary alluvium	11	540	49	240
Glacial drift	15	200	70	460
Strata of Cretaceous and Tertiary age	12	100	34	130
Strata of Pennsylvanian age	23	140	200	670
Strata of Mississippian age	11	53	11	71
Strata of Cambrian and Ordovician age (SW)	10	51	26	130
Strata of Cambrian and Ordovician age (SE)	11	43	11	46
% Variance due to geohydrologic units	23	43	50	50

[a]Data of Feder, 1971.[35]

He found differences among the mean concentrations
of these trace elements, depending on the geohydrologic unit.   In addition, a statistical analysis
showed that a large portion of the variance, as
indicated in Table 21, can be attributed to differences among the geohydrologic units, in contrast to
differences among samples at given locations or
within these units.
     Such correlations among elements and between
them and the lithologic environment are useful, both
as predictive tools and for understanding the controls on trace elements.   However, such relationships
and studies in the scientific literature are few
indeed.

CONTRIBUTIONS FROM AIR

     It is important to consider precipitation,
washout, and fallout from air as possible sources
of trace elements to terrestrial waters.   In this
regard questions of interest are: (a) does spray
from sea water control or have an influence? (b) is
air a major source? and (c) do terrestrial dust or
air pollutants contribute?
     Table 22 is useful as a starting point in that
it indicates, in summary fashion, mean concentrations
of typical trace metals in surface water and precipitation in three river basins of the United States.

Table 22

*Comparison of Mean Concentrations (μg/l) of Several Trace
Elements in Some U.S. Surface Waters[a] and Precipitation[b]*

Basin	Pb		Zn		Mn		Cu		Ni	
	Pptn.	*Sur-face*	*Pptn.*	*Sur-face*	*Pptn.*	*Sur-face*	*Pptn.*	*Sur-face*	*Pptn.*	*Sur-face*
Northeast	50-90	17	40-70	96	2-19	4	16-55	15	0-5	8
Western Gulf	0-4	4	0-30	92	2-4	10	5-15	11	3-5	10
Pacific Northwest	0-4	15	0-30	40	2-19	3	5-15	9	0-8	10

[a]Data of Kopp and Kroner, 1970.[10]

[b]Data of Lazrus *et al.*[38]

The sampling points for precipitation and surface
waters are not identical, but the data do give some
indication of the concentration levels encountered.
The striking point is that the concentrations in
precipitation in general are not very different
from those in surface water. Depending on the
river basin, the surface water concentrations may
be higher, or vice versa. Since the concentrations
are comparable, however, one should properly consider that either the precipitation essentially
establishes the concentration in surface waters or
that both the latter and the precipitation are
influenced by the same source, such as soils and
underlying minerals, with the soil dust going into
the atmosphere and subsequently being incorporated
into precipitation.

One useful approach in this regard is to first
establish the contribution from sea spray to the
trace element levels in precipitation. It is
generally accepted that sea spray is the major
source of chloride in precipitation.[36] Thus,
although its concentration in precipitation generally
decreases going from the sea inland, its ratio to
that of trace elements is a useful guide as to
whether their presence in precipitation is due to
the sea as a source. There is some fractionation
(change in concentration) of macro elements in the
process of sea spray formation. Thus magnesium in
some cases is up to 20% higher in ocean clouds than
would be expected from its ratio to that of chloride
in the ocean. However, if one neglects such fractionation effects, the ratios for chloride versus
trace elements in precipitation compared to those
in sea water should indicate the significance of the
latter as a source.

Table 23 makes such comparisons, using data
from various sources.[14,37,38] As indicated, the
average concentrations of trace elements in United
States precipitation are somewhat higher than in
sea water, while for chloride the opposite is true
by a factor of about 10^5. By comparing the ratio
of M/Cl (metal to chloride concentration) in precipitation to that in sea water, as shown in the
last column, it is thus apparent that if one assumes
that there is no significant fractionation in the
formation of sea spray and that the chloride in
precipitation comes primarily from the sea, there
must be large contributions of trace elements from
terrestrial sources, such as air-borne soil dust or
air pollution. One can also conclude that, although

Table 23

The Influence of Sea Water on Trace Element
Concentrations in Precipitation

Element	Avg. Conc. in U.S. Pptn.[a] μg/l	Avg. Conc. in Sea Water[b] μg/l	$\frac{(M/Cl)pptn.}{(M/Cl)sea}$
Pb	34	0.03	7.1×10^7
Mn	12	2	3.8×10^5
Cu	21	3	4.4×10^5
Zn	107	10	6.8×10^5
Ni	4.3	2	1.4×10^5
Cl[c]	300 (200-6000)	1.9×10^7	

[a]From Lazrus *et al.*[38]

[b]From Horne, 1969.[37]

[c]From Junge, 1963.[14]

the precipitation may contribute large quantities
of these trace elements to the surface waters, the
latter may at the same time be leaching these elements from these same soils and underlying minerals,
both processes being capable of contributing to the
similarity between the concentrations in surface
waters and precipitation.

That particulates in air can contribute trace
elements has been demonstrated.[39]   In that study of
12 trace elements in particulates in the air of six
cities of the United States, the average annual
metal concentrations ranged from 0.02 to 3.2 $\mu g/m^3$.
There were striking contrasts among the cities, as
well as time variations. It was concluded that the
concentration levels of the trace metals in these
particulates were influenced by emissions of lead
from gasoline, vanadium from fuel oil, with coal
combustion at power plants, municipal incineration,
and industrial processes contributing other elements.
Thus, one can conclude that both dust from soil, and
industrial and urban pollution contribute significant
quantities of trace elements to surface waters via
the air route.

POLLUTION SOURCES AND EFFECTS

Trace elements in natural waters are known to be contributed from specific waste outfalls, as well as associated with urban and industrial activity in general. For example, in a survey of selected trace elements in surface waters of the United States,[40] it was found that a sample in a creek downstream from an industrial complex in North Carolina contained 1100 µg/l of arsenic, the highest of any sample in the study. Similarly, it was noted that the higher concentrations of cadmium generally occurred in areas of high population density. The probably somewhat cautious conclusion was that the survey showed that for the heavy metals "there is some evidence that the concentration levels are related to man's activities in certain instances."

Such correlations are, however, not always easily made. Thus, in a study of trace elements in the Ohio River and its two major tributaries at its source, there was sufficient variability with time to make it difficult to judge the possible contributions of zinc and copper to the Ohio River from the Pittsburgh urban-industrial center.[27] The monthly mean concentrations of copper and zinc in the source tributaries, the Allegheny and Monongahela, are plotted versus the similar values for the Ohio at a point 25 miles downstream from Pittsburgh in Figure 9. Depending on the month, the mean concentrations downstream in the Ohio were either greater than or less than those in the source tributaries, although the yearly means, weighted by flow, tended to be equal. Thus, there were times of the year when there were effective losses of copper and zinc along this section of the Ohio River, while at other times there were additions due to the urban-industrial activity.

An example of increased concentrations of a particular trace metal associated with man's activities is that of uranium in the rivers draining into the Gulf of Mexico.[41] It was found in this study, reported in 1972, that the uranium concentrations in samples taken from 15 such rivers had considerably higher concentrations than those found in North American rivers in 1952. Several of these rivers flow through uranium bearing strata which are mined in this region, and it was estimated that such rivers contained an additional concentration of uranium of about 1 µg/l. However, it was also noted that the commercial phosphate fertilizers used in this region

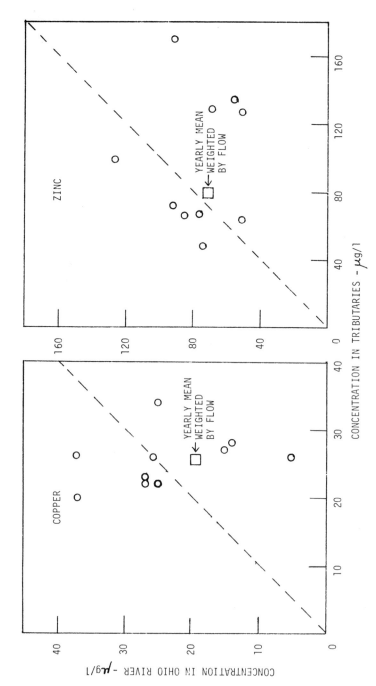

Figure 9.   Comparison of copper and zinc concentrations (composited monthly by flow) in tributaries at source of Ohio River and 25 miles downstream for one year period. Data of Shapiro, Andelman and Morgan, 1967.[27]

contain significant quantities of uranium, up to
180 mg/kg, that this uranium leaches from the soil
to which the fertilizer is applied, and that a rough
estimate is that the difference in concentration of
uranium between the rivers draining the agricultural
and nonagricultural areas is 0.3 µg/l.[41]

Similarly, in a study of chromium and zinc
concentrations at 19 sampling stations in Lake
Michigan, it was found that the highest concentrations of these elements were found in the region
traversed by the plume of the Grand River, which
enters the lake in the southeastern region.[25] The
authors concluded that it appears that both the
chromium and zinc entered the lake primarily from
the Grand River, and, in turn, they probably entered
the river from industrial plating operations.

Another possible effect of pollutants is an
indirect one, namely that of other chemical pollutants influencing the behavior of trace elements in
natural waters. Earlier in this discussion it was
noted that organic pollutants can concentrate at
the air-water interface, and it was found in
Narragansett Bay that such layers concentrate trace
metals.[20] The authors noted that, once concentrated,
they then become accessible to microorganisms and
plankton and are eventually concentrated in the
food chain.

A similar phenomenon is the possible effect of
runoff from road deicing salt, particularly on the
release of mercury from contaminated sediments.[42]
In laboratory studies of sandy and highly organic
sediments it was found that the addition of calcium
and sodium chloride increased the relative amounts
of mercury released from the sediments into the
water by several orders of magnitude. The effect
was probably due to complexation of the mercury by
the chloride, as well as competition of added
sodium or calcium with mercury for exchange sites
on the sediment. It was indicated that other toxic
heavy metals could be similarly affected.

Specific industrial processes are known to
contain trace elements in their waste effluents.
For example, a report compiled by an association
of dye manufacturers considered the sources and
likely concentrations of eight metals in dyehouse
effluent streams from a variety of textile dyeing
and related operations.[43] It was noted that "tramp"
metal gets into the dye from both chemicals and
water used to prepare the dye. In addition, heavy
metals are used intentionally, such as for catalysts

in synthesizing dyes, and also in applications related to dyeing and finishing processes, such as
aluminum and antimony for flame retardant finishes,
and chromium in oxidation processes. It was also
noted that fibers and fabrics entering a mill sometimes contain appreciable quantities of heavy
metals. It was estimated that the tramp heavy
metals alone in the dyes averaged from less than
1 mg/l to 80 mg/l. However, it was concluded that
these are commonly diluted by about 10,000 fold by
process water, and generally occur at concentrations
of a few micrograms per liter or less in untreated
textile mill effluents.

Such heavy metals as lead, cadmium, and mercury
have been widely investigated in relation to specific
pollution sources, their distribution in water and
other parts of the environment, and their possible
effects on and uptake by humans, animals, and
plants.[44-46] The sources of these metals from man's
activities are varied and numerous. Any assessment
of movement of these and other trace elements through
natural waters must consider such human-influenced
inputs in order to accurately estimate their cycling,
behavior, and concentration level.

CONCLUSIONS

The variability of concentrations of trace
elements in natural fresh waters is great, both on
a macro and micro geographic scale, as well as with
time at a given location. In most instances it is
difficult to correlate their concentrations with
such characteristics as stream flow or lithologic
environment.

In order to accurately characterize trace
elements and attempt to develop models for predicting
their behavior and cycling through water and associated environments, it is necessary to distinguish
their various chemical and physical states, such as
particle size and incidence and forms of complexes
and chelates. Such speciation probably plays an
important role in their movement in water, as well
as their availability to and uptake by aquatic
animals and plants. There is some evidence that
such biological interactions influence the concentration levels of these trace elements, as well as
their transport in natural waters.

There are a variety of sources that contribute
trace elements to natural fresh waters, including
fallout and precipitation from air, human activities,

and flow of water over soils and minerals. In some cases specific contributions from these sources can be identified. However, the status of knowledge in this regard in general is quite primitive, and adequate models for the movement of trace elements into and through these waters have yet to be developed.

REFERENCES

1. Schroeder, H. A. "Relation Between Mortality from Cardiovascular Disease and Treated Water Supplies," *J. Amer. Med. Assoc. 172*:1902 (1960).

2. Schroeder, H. A. "Municipal Drinking Water and Cardio-vascular Death Rates," *J. Amer. Med. Assoc. 195*:125 (1966).

3. Crawford, M. D., M. J. Gardner, and J. N. Morris. "Mortality and Hardness of Local Water Supplies," *The Lancet 1*:827 (April 20, 1968).

4. Morris, J. H., M. D. Crawford, and J. A. Heady. "Hardness of Local Water Supplies and Mortality from Cardiovascular Disease," *The Lancet 1*:860 (April 22, 1961).

5. Winton, E. F., and L. J. McCabe. "Studies Relating to Water Mineralization and Health," *J. Amer. Water Works Assoc. 62*:26 (1970).

6. Crawford, M. D., M. J. Gardner, and P. A. Sedgwick. "Infant Mortality and Hardness of Local Water Supplies," *The Lancet 1*:988 (May 6, 1972).

7. Masironi, R. "Cardiovascular Mortality in Relation to Radioactivity and Hardness of Local Water Supplies in the U.S.A.," *Bull. World Health Org. 43*:687 (1970).

8. Berg, J. W. and F. Burbank. "Correlations Between Carcinogenic Trace Metals in Water Supplies and Cancer Mortality," In *Geochemical Environment in Relation to Health and Disease*, Hopps, H. C. and H. L. Cannon, eds. *Annals N.Y. Acad. Sci. 199*:249 (June 28, 1972).

9. Voors, A. W. "Minerals in the Municipal Water and Atherosclerotic Heart Death," *Amer. J. Epid. 93*:259 (1971).

10. Kopp, J. F. and R. C. Kroner. *Trace Metals in Waters of the United States* (Cincinnati: Federal Water Pollution Control Administration, 1970).

11. Piper, D. Z. and G. G. Goles. "Determination of Trace Elements in Seawater by Neutron Activation Analysis," *Anal. Chim. Acta 47*:560 (1969).

12. Feder, G. L. In *Geochemical Survey of Missouri, Open-File Report* (Denver: U.S. Geological Survey, 1972).

13. Kharkar, D. P., K. K. Turekian, and K. K. Bertine. "Stream Supply of Dissolved Silver, Molybdenum, Antimony, Selenium, Chromium, Cobalt, Rubidium and Cesium to the Oceans," *Geochim. Cosmochim. Acta 32*:285 (1968).

14. Horne, R. A. *Marine Chemistry* (New York: Wiley-Interscience, 1969).

15. Krauskopf, K. B. "Factors Controlling the Concentrations of Thirteen Rare Metals in Sea Water," *Geochim. Cosmochim. Acta 9:1* (1956).

16. O'Connor, J. T., C. E. Renn, and I. Wintner. "Zinc Concentrations in Rivers of the Chesapeake Bay Region," *J. Amer. Water Works Assoc. 56:280* (1964).

17. Lee, G. F. and A. W. Hoadley. In *Equilibrium Concepts in Natural Water Systems, Advances in Chemistry Series,* Vol. 67 (Washington, D.C.: American Chemical Society, 1967).

18. Siegal, A. In *Organic Compounds in Aquatic Environments,* Faust, S. D. and J. V. Hunter, eds. (New York: Marcel Dekker, 1971).

19. Schnitzer, M. In *Organic Compounds in Aquatic Environments,* Faust, S. D. and J. V. Hunter, eds. (New York: Marcel Dekker, 1971).

20. Duce, R. A., J. G. Quinn, C. E. Olney, S. R. Piotrowicz, B. J. Ray, and T. L. Wade. "Enrichment of Heavy Metals and Organic Compounds in the Surface Microlayer of Narragansett Bay, Rhode Island," *Science 176:161* (1972).

21. Gloyna, E. F., Y. A. Yousef, and T. J. Padden. In *Non-equilibrium Systems in Natural Water Chemistry, Advances in Chemistry Series,* Vol. 106 (Washington, D.C.: American Chemical Society, 1971).

22. Slowey, J. F. and D. W. Hood. "Copper, Manganese, and Zinc Concentrations in Gulf of Mexico Waters," *Geochim. Cosmochim. Acta 35:121* (1971).

23. Stiff, M. J. "The Chemical States of Copper in Polluted Fresh Water and a Scheme of Analysis to Differentiate Them," *Water Res. 5:585* (1971).

24. Silvey, W. D. *Occurrence of Selected Minor Elements in the Waters of California,* Water Supply Paper 1535-L (Washington, D.C.: U.S. Geological Survey, 1967).

25. Copeland, R. A. and J. C. Ayers. *Trace Element Distributions in Water, Sediment, Phytoplankton, Zooplankton and Benthos of Lake Michigan* (Ann Arbor: Environmental Research Group, Inc., 1972).

26. Turekian, K. K., R. C. Harriss, and D. G. Johnson. "The Variations of Si, Cl. Na, Ca, Sr, Ba, Co, and Ag in the Neuse River, North Carolina," *Limnol. Ocean. 12:702* (1967).

27. Shapiro, M. A., J. B. Andelman, and P. V. Morgan. *Intensive Study of the Water at Critical Points on the Monongahela, Allegheny, and Ohio Rivers in the Pittsburgh, Pennsylvania Area* (Pittsburgh: University of Pittsburgh, 1967).

28. Funk, W. H., S. K. Bhagat, and R. H. Filby. *Trace Element Measurements in the Aquatic Environment* (Pullman: Washington State University, 1969).

29. Durum, W. H. and J. Haffty. "Implications of the Minor Element Content of Some Major Streams of the World," *Geochim. Cosmochim. Acta 27*:1 (1963).

30. Silker, W. B. "Variations in Elemental Concentrations in the Columbia River," *Limnol. Oceanog. 9*:540 (1964).

31. Goldberg, E. D. In *Chemical Oceanography*, Vol. 1, Riley, J. P. and G. Skirrow, eds. (New York: Academic Press, 1965).

32. Martin, D. F. *Marine Chemistry*, Vol. 2 (New York: Marcel Dekker, 1970).

33. Martin, J. H. "The Possible Transport of Trace Metals via Moulted Copepod Exoskeletons," *Limnol. Ocean. 15*: 756 (1970).

34. Boynton, R. S. *Chemistry and Technology of Lime and Limestone* (New York: Interscience, 1966).

35. Feder, G. L. In *Geochemical Survey of Missouri*, Open-File Report (Denver: U.S. Geological Survey, June, 1971).

36. Eriksson, E. "Air Borne Salts and the Chemical Composition of River Waters," *Tellus 7*:243 (1955).

37. Junge, C. E. *Air Chemistry and Radioactivity* (New York: Academic Press, 1963).

38. Lazrus, A. L., E. Lorange, and J. P. Lodge, Jr. "Lead and Other Metal Ions in United States Precipitation," *Environ. Sci. Technol. 4*:55 (1970).

39. Lee, R. E., Jr., S. S. Goranson, R. E. Enrione, and G. B. Morgan. "National Air Surveillance Cascade Impactor Network. II. Size Distribution Measurements of Trace Metal Components," *Environ. Sci. Technol. 6*: 1025 (1972).

40. Durum, W. H., J. D. Hem, and S. G. Heidel. *Reconnaissance of Selected Minor Elements in Surface Waters of the United States, October 1970*, Circular 643 (Washington, D.C.: U.S. Geological Survey, 1971).

41. Spalding, R. F. and W. M. Sackett. "Uranium in Runoff from the Gulf of Mexico Distributive Province: Anomalous Concentrations," *Science 175*:629 (1972).

42. Feick, G., R. A. Horne, and D. Yeaple. "Release of Mercury From Contaminated Freshwater Sediments by the Runoff of Road Deicing Salt," *Science 175*:1142 (1972).

43. Allen, W., E. Altherr, R. H. Horning, J. C. King, J. M. Murphy, W. E. Newby, and M. Saltzman. "The Contribution of Dyes to the Metal Content of Textile Mill Effluents," *Textile Chem. Colorist 4*:275 (1972).

44. Dunlap, L. "Mercury: Anatomy of a Pollution Problem," *Chem. Eng. News 22* (July 5, 1971).

45. Lagerwerff, J. V. "Lead, Mercury and Cadmium as Environmental Contaminants," In *Micronutrients in Agriculture* (Madison: Soil Science Soc. of America, 1972).

46.   Patterson, C. C.   "Contaminated and Natural Lead
       Environments of Man," *Arch. Environ. Health* *11*:344
       (1965).

4. FACTORS AFFECTING DISTRIBUTION OF LEAD AND OTHER TRACE ELEMENTS IN SEDIMENTS OF SOUTHERN LAKE MICHIGAN

Harry V. Leland, Surendra S. Shukla. University of Illinois, Champaign-Urbana, Illinois

Neil F. Shimp. Illinois State Geological Survey, Urbana, Illinois

INTRODUCTION

Sediments are an integral part of the cycling of minerals in surface waters; thus a comprehensive investigation of trace elements in an aquatic eco-system must include characterization of recently deposited materials. Fine-grained sediments in large lakes reflect the nature of overlying waters at the time of deposition because of their capacity to incorporate organic and inorganic constituents during transport and deposition. Lake sediments thus provide both a record of past climatic and geologic events and an indication of the activities of man in the surrounding watershed.

This report considers the distribution of trace elements in sediments of Lake Michigan (Figure 10), the third largest in surface area (58,000 km^2) of the St. Lawrence Great Lakes. The trace element composition of Lake Superior sediments has been discussed previously.[1] The importance of the Great Lakes as a freshwater resource and the continually mounting evidence[2] that recently accelerated ecological changes are occurring in some areas emphasizes the need for baseline data on biological and chemical characteristics of these lakes and a critical assessment of the changes taking place. One area clearly undergoing accelerated change, as evidenced by historical trends in major chemical constituents in the water,[3,4] benthos,[5] and

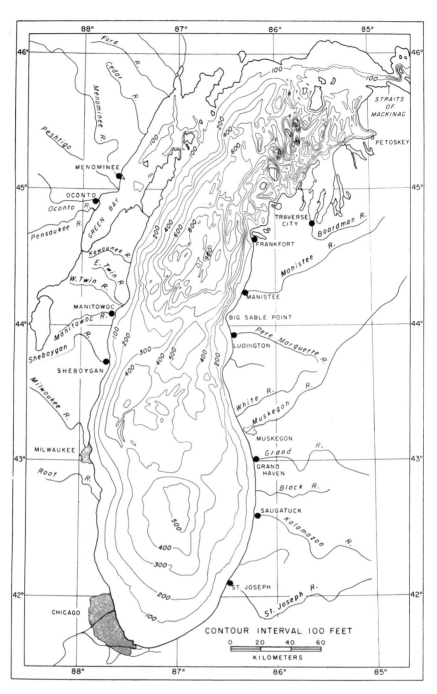

Figure 10. *Lake Michigan (based in part on Kittrell, 1969).*[6]

phytoplankton,[7,8] is the southern basin of Lake
Michigan. The shallower of two basins, its area
spans approximately the southern one-third of the
lake. Most land in the watershed of the southern
basin is agricultural or forested, although extensive
and highly industrialized urban areas exist. Major
cities on the shoreline are Chicago, Illinois; Gary,
Indiana; Milwaukee, Wisconsin; and Muskegon, Michigan.
Sheet erosion of agricultural lands and subsequent transport of eroded soils via tributaries
is the major source of sediments today.[9] Tributary
sediments are also derived from urban areas under
construction, roadsides, mining and streambank
erosion. Shoreline erosion and wind-blown dust may
also be effective sources. The sediments presently
being deposited in the southern lake basin are
derived principally from several small rivers (the
Muskegon, Grand, Black, Kalamazoo, and the St.
Joseph) draining the southern peninsula of Michigan.
Published data on trace elements in Lake
Michigan are more numerous than for the other Great
Lakes.[10,11] Nevertheless, quantitative information
on sources, dispersion, and reservoirs of trace
elements in the lake are so fragmentary that a
comprehensive chemical model cannot be constructed
for any trace element at the present time. Recent
research at the Illinois State Geological Survey
and the University of Illinois, which is summarized
herein, does provide a quantitative description of
the distribution of trace elements in sediments of
southern Lake Michigan. Available information on
geology, mineralogy, major constituents of the
sediments, and properties of the lake water permit
qualitative interpretation of factors influencing
individual trace element concentrations in the
sediments. The distribution of lead and the factors
involved in its transport and deposition in the
southern basin of Lake Michigan are emphasized in
this review. Shukla and Leland[12] discuss the general
distribution of lead in the atmosphere, hydrosphere,
and lithosphere.

SOURCES OF TRACE ELEMENTS IN
LAKE MICHIGAN

Information on sources and the relative amounts
of trace elements in Lake Michigan introduced via
surface drainage and atmospheric transport is ex-
ceedingly fragmentary. For some elements inputs
from the atmosphere are clearly comparable in

magnitude to river inputs. Winchester and Nifong[13]
present a partial inventory of air pollution emissions
for 30 trace elements in the Chicago, Milwaukee, and
northwest Indiana metropolitan areas. The authors
consider the major sources which contribute trace
elements to the atmosphere over southern Lake
Michigan to be (1) coal burned for electrical and
heating uses, (2) emissions from coke ovens in the
manufacture of metallurgical coke, (3) fuel oil
burned for electrical and heating needs, (4) auto-
motive fuel burned for transportation, (5) emissions
from iron and steel manufacturing facilities, and
(6) emissions from cement manufacturing plants.
Comparisons of the inventory of airborne emissions
with actual stream inputs of copper, nickel and
zinc indicate that the atmosphere is the more
significant source of zinc in Lake Michigan and
comparable to watershed drainage in inputs of
copper and nickel.[13]   Comparison of the inventory
with estimates of preindustrial, unpolluted stream
inputs for 28 elements indicates that air pollution
probably exceeds unpolluted stream inputs for many
additional elements in Lake Michigan.
    Robbins *et al.*[14] provide a preliminary estimate
of mean daily loadings of Ca, Mg, Na, K, Fe, Mn, Cu,
Ni, Cr, Mo, Zn, Sr, and Ba to Lake Michigan via
tributaries. The Grand River, the largest tributary
source of 10 of the 13 elements analyzed, is singu-
larly high as a source of Cu, Ni, and Cr discharged
into the southern basin. The St. Joseph River,
the tributary with the second largest trace element
loading to Lake Michigan, also flows into the
southern basin. Based on a comparison of mean
daily tributary inputs to Lake Michigan with
estimates of atmospheric inputs from the Chicago-
Gary and Milwaukee urban and industrial regions,
the authors suggest that of the 13 elements
analyzed, the atmosphere is a significant source
of Fe, Mn, Cu, Ni, and Zn. The estimates are based
on the assumption that at least 20% of the daily
air emissions enter Lake Michigan.
    Order-of-magnitude estimates of lead input
into the southern basin of Lake Michigan by precipitation, boating, municipal and industrial wastewater
discharges, soil erosion, and watershed drainage
indicate the most important (greater than 90%)
contribution to be precipitation.[15]   The importance
of dry fall of lead particulates to Lake Michigan
is unknown. Winchester[16] states that an average
deposition velocity of 0.5 cm per sec for particulates

from clean air over Lake Michigan is not unreasonable
to expect. If this is true, transfer of particles
from air to water by mechanisms other than precipitation may be substantial. Concentrations of lead in
the uppermost few centimeters of water near the air
interface were shown to decrease with distance from
Chicago during a period when prevailing winds were
from the southwest.[17] Lead concentrations ranged
from about 2 μg per liter offshore at Chicago to
approximately 0.4 μg per liter ten miles west of
Grand Haven, Michigan. Concentrations of copper and
cadmium determined on the same samples did not vary
with distance from Chicago.

SEDIMENT DISTRIBUTION OF TRACE ELEMENTS
AND GEOLOGY OF SOUTHERN LAKE MICHIGAN

Studies by Shimp *et al.*,[18,19] Kennedy *et al.*,[20]
and Ruch *et al.*[21] show that certain trace elements,
namely As, Br, Cr, Cu, Hg, Pb, and Zn occur in
substantially (up to 20 times) higher concentrations
near the sediment water interface of fine-grained
sediments than in underlying sediments of southern
Lake Michigan. The variation in concentration of
selected trace elements with sediment depth in
south-central and southeastern Lake Michigan is
demonstrated in Figure 11. The concentrations of
each of the above elements, which are here identified
as "accumulating" trace elements in sediments of
southern Lake Michigan, are significantly related
to the amount of organic carbon present in the
sediments at that depth.[19] Trace elements exhibiting
little or no enrichment near the sediment-water
interface in fine-grained sediments of southern Lake
Michigan are B, Be, Co, La, Ni, Sc, and V. From the
standpoint of concentrations, these trace elements
are "nonaccumulating" and are poorly correlated with
organic carbon content. Trace element concentrations
in surficial sandy sediments do not differ substantially from concentrations deeper in the same core.
The relatively high concentrations of
"accumulating" trace elements in the fine-grained
surficial sediments are apparently not a natural
phenomenon but are a consequence of man's activities.
Frye and Shimp[22] examined the trace element composition of deposits of a glacially fed lake, now called
Lake Saline, that existed beyond the limit of
Wisconsinan glaciation and received sediments while
glaciers occupied Lake Michigan. Dating by radio-
carbon methods and stratigraphic correlation show

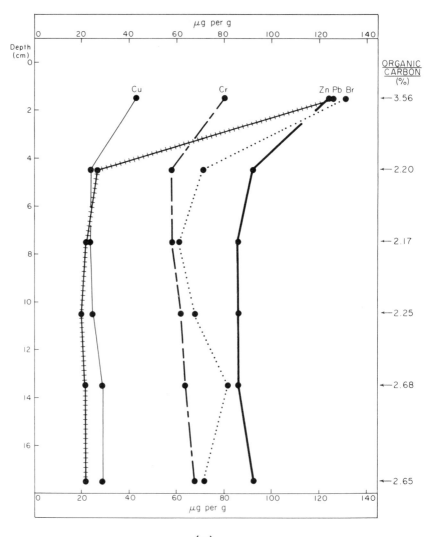

(a)

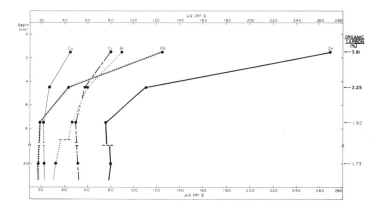

(b)

Figure 11.   Distributions of bromine, chromium, copper, lead,
             zinc, and organic carbon in (a) a sediment core
             from the area of the brown silt facies (42° 2.00'N,
             86° 57.00'W) of the Waukegan Member of the Lake
             Michigan Formation and (b) a core from the area
             of the gray silt facies (42° 9.79'N, 86° 48.09'W).
             Near-surface accumulations of trace elements
             generally extend deeper into the sediments of the
             gray silt facies than of the brown silt facies
             (data from Reference 18).

the youngest deposits of Lake Saline to be 13,000
years before present, which is the same age as the
oldest glacially derived sediments in southern Lake
Michigan. The major, minor, and trace element
composition of underlying (15-100 cm depth interval),
fine-grained sediments of Lake Michigan (Table 24)
is remarkably similar to that of Lake Saline. Such
a favorable comparison supports the conclusions that
the trace element composition of underlying sediments
represent natural (baseline) levels in southern Lake
Michigan and that analysis of the underlying lake
sediments can be used to estimate preindustrial trace
element concentrations. The natural concentrations,
based on the 15-100 cm Lake Michigan and Lake Saline
samples, reported by Frye and Shimp[22] are reproduced
in Table 25. Trace element concentrations in Lake
Saline sediments do not correlate closely with
organic carbon content, as seems to occur in
southern Lake Michigan, and no consistent accumulation of trace elements exists over the 13,000-
35,000 radiocarbon-year time span studied. The
fact that significant recent additions of trace
elements are being made to southern Lake Michigan
is strongly supported by this study.

Lake Michigan has the smallest outflow of the
Great Lakes, 1560 m^3 per sec at the Straits of
Mackinac and a small flow through diversion to the
Chicago River. The ratio of lake volume to outflow
is large and the average flow-through time, an
estimated 100 years, is long.[24] Water-soluble
constituents can be expected to have a mean residence time in Lake Michigan of approximately a
century, whereas constituents which react with
sediments and are deposited may remain in the lake
indefinitely.[16] An examination of the fate of a
constituent added to the lake from the watershed
clearly requires an understanding of the chemical
properties of the constituent and factors affecting
its distribution in the lake environment.

The geographic distribution of "accumulating"
trace elements in southern Lake Michigan is related
to the pattern of sediment deposition.[25,26] Studies
by Lineback *et al.*,[27-29] Gross *et al.*,[30,31] and
Lineback and Gross[26] describe the stratigraphy and
mineralogy of sediments in the southern lake basin.
The following account is adopted from these reports.

Sediments of glacial, glacial-fluvial and
lacustrine origin in the southern basin of Lake
Michigan reduce the bedrock relief to a relatively
smooth lake floor (Figure 12). Glacial till (Wedron

(a)

Age		Formation	Member	Lithology
Pleistocene	Holocene	Lake Michigan	Ravinia	Beach sand
			Waukegan	Silt and clay, dark gray to brown
			Lake Forest	Clay and silty clay, dark gray with black beds
	Wisconsinan		Winnetka	Clay, brownish gray with a few black beds
			Sheboygan	Clay, reddish brown
			South Haven	Clay, reddish gray
		Equality	Carmi	Clay, silt, sand, clay pebble conglomerate, gray to brown
		Wedron	Wadsworth	Glacial till, clayey, gray

(b)

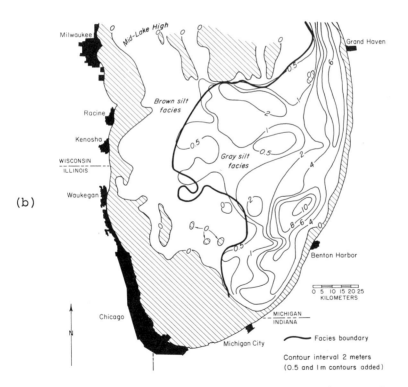

Figure 12. (a) *Late Pleistocene sediments underlying southern Lake Michigan and (b) thickness and facies of the Waukegan Member of the Lake Michigan Formation (adapted from Reference 25).*

Table 24

Chemical Composition of Sediments and Suspended Matter
Near the Sediment Interface in Southern Lake Michigan

Major or Minor Constituent (percent)	Sediments[a,b]			
	Uppermost Sample Interval			
	Mean	Range	Number of Samples	Mean
SiO_2	53.1	40.2 -70.2	24	48.4
TiO_2*	0.36	0.13- 0.55	24	0.49
Al_2O_3*	7.95	4.72-11.70	24	9.54
Fe_2O_3	4.04	2.09- 9.99	24	3.56
MnO	0.072	0.025- 0.212	25	0.050
MgO*	5.26	2.55- 9.18	24	6.65
CaO	7.83	4.37-15.90	24	9.14
Na_2O*	0.62	0.37- 0.88	24	0.63
K_2O*	2.19	0.98- 3.51	24	2.73
P_2O_5	0.15	<0.01- .0.42	56	0.17
S*	0.10	0.02- 0.24	23	0.12
Ignition Loss	16.10	9.15-20.33	24	17.46
Organic Carbon	2.51	0.77-4.73	25	1.40
Trace Element (μg per g)				
As	14.	8-30	15	5.
B	41.	14-71	21	48.
Be	1.8	0.6-3.0	24	1.7
Br	62.	< 20-132	20	35.
Co	13.	7-24	22	14.
Cr	77.	35-165	24	52.
Cu	37.	9-75	22	20.
Hg	0.20	0.06-0.38	21	0.05
Ni	34.	18-58	24	35.
Pb	88.	27-172	24	20.
V	48.	7-83	24	62.
Zn	206.	58-519	22	66.

[a] Data on sediments from Shimp et al.,[18] Ruch et al.,[21] Schleicher and Kuhn,[23] Shimp et al.,[19] and Kennedy et al.[20] The means and ranges for sediment constituents are from Frye and Shimp,[22] with the exception of constituents marked by an asterisk(*).

[b] Top interval sediments selected to contain at least 0.75% organic carbon and 10% less than 2 μ clay-sized material.

| All Sample Intervals>15 cm but<100 cm | | Suspended Matter[c] | | |
Range	Number of Samples	Mean	Range	Number of Samples
41.5--58.9	31	33.1	14.3 -48.7	12
0.41- 0.57	22	0.23	0.17- 0.37	12
8.24-11.60	22	4.90	2.19- 9.78	12
1.74- 5.99	31	4.68	2.47- 7.38	12
0.027- 0.110	24			
4.87- 8.11	22			
3.37-15.00	32	29.2	8.8 -71.1	12
0.55- 0.93	22			
2.05- 3.31	22	1.34	0.83- 2.28	12
<0.01- 0.46	73	0.0033	0.0007-0.0060	12
0.08- 0.20	22	0.18	0.60- 0.36	12
11.09-20.91	32	34.4	19.1 -47.2	12
0.49- 2.65	23			
2-8	24	27.	8-42	12
17-71	24			
0.7-3.2	23			
<20-72	19	9.7	2-19	12
6-22	23			
32-68	24			
8-29	24	39.	17-102	12
0.04-0.08	28			
12-54	23	47.	26-84	12
16-48	23	56.	36-97	12
26-99	23	56.	26-72	12
22-129	23	1420.	500-2100	12

[c]Suspended matter (particles removed by a 0.45 μ membrane filter) selected without regard to organic carbon content or particle size.

Table 25

*First Approximation of Natural (Baseline) Concentrations
of Minor and Trace Elements in Sediments**
of Southern Lake Michigan[22]

Element	Base line (µg per g)	Element	Base line (µg per g)
As	< 8	Hg	0.05-0.10
B	30-50	Ni	15-40
Be	1-2	Pb	15-30
Br	< 40	V	30-60
Co	10-20	Zn	50-100
Cr	20-40	MnO (%)	0.05-0.10
Cu	15-30	P_2O_5 (%)	< 0.3

*Fine-grained bottom sediments only.

Formation) is thin or absent over bedrock highs and in the center of the southern basin but may be as thick as 56 m in bedrock valleys and in nearshore areas. Sands, silts, and clay-pebble conglomerate of glacial-fluvial origin (Equality Formation) overlie the till. The most recent formation, Lake Michigan, consists of as much as 17 m of lacustrine clay and silt. The lower members of the Lake Michigan Formation are very fine-grained and differ significantly from underlying sediments in mineralogy. The clays of the lower sediments in the Lake Michigan Formation contain significantly more expandable clay minerals than illite. The underlying Equality Formation is characterized by higher amounts of illite. The three younger members (the Winnetka, Lake Forest and Waukegan) of the Lake Michigan Formation also have a higher illite content than do the earlier members and are more similar to the tills surrounding the southern margin of Lake Michigan. The younger members also contain more calcite and dolomite, perhaps because of a warming of the lake before and during deposition of the younger members. Average values reported by Gross *et al.*[31] for the Waukegan Member are 49% illite, 29% chlorite, and 22% for other expandable clay minerals.

The Waukegan Member (Figure 12) of the Lake Michigan Formation is the surficial sediment over most of the southern basin. Glacial till is present at the water interface or under a thin cover of sand or gravel derived from erosion of the till in near-shore areas and on the shallow lake floor off Chicago. Lake floor outcrops of other stratigraphic units are of small extent. Except for beach sand the Waukegan is the coarsest member of the Lake Michigan Formation, averaging 8% sand, 41% silt, and 50% clay. It is thickest in a narrow westward sloping belt along the eastern shore, thinning both basinward and towards shore. The distribution of Waukegan sediments apparently results from a fractional settling of suspended particles according to the size and density of particles, the prevailing lake currents and travel distance from shore. Turbidity currents and downslope slumping of soft sediments may be significant in localized areas, but stratigraphic and seismic studies provide no evidence for these means of sediment transport. No major rivers flow into the western side of southern Lake Michigan. Consequently, the rate of sedimentation of Waukegan sediments is much lower on the western rim of the south basin than on the eastern side.

Lineback and Gross[26] differentiate two major facies in the Waukegan member, a gray silt facies that lies in the center and along the eastern side of the southern basin and a brown silt facies along the western side of the lake and the Mid-lake High (Figure 12). The sediments are similar in grain size and water content in the two facies but the gray silt facies contain a greater percentage of organic carbon. Electrode potentials in the gray silt facies are, almost without exception, much lower than those in the brown silt facies (Figure 13). In fact, the facies boundary constructed from electrode potential values is virtually identical to that obtained from geologic observations. The waters of the hypolimnion in southern Lake Michigan are saturated or nearly saturated with oxygen[32] and an oxidized layer (microzone) exists at every location sampled in the gray silt facies. Electrode potentials below the microzone are negative or only slightly positive, whereas the brown sediments are positive throughout the sediment column. The electrode potentials shown (Figure 13) were obtained at a sediment depth interval of 3 cm or below a watery floc that exists at the surface in many areas.

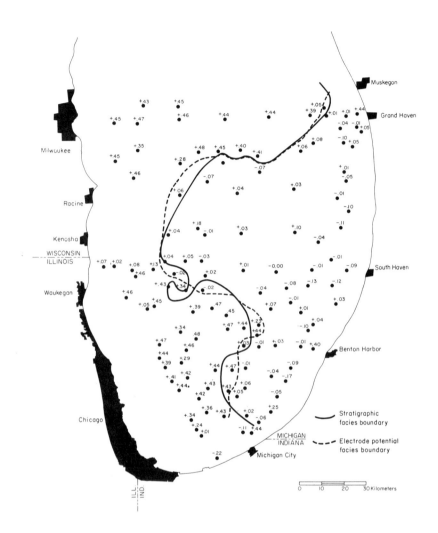

Figure 13. Boundary between the gray silt facies and the brown
            silt facies of the Waukegan Member as determined
            from stratigraphic observations and electrode-
            potential measurements. Electrode potentials
            (volts) were obtained at a depth of 3 cm below
            the sediment-water interface or below a watery
            floc which exists at the interface in many areas.
            Values were recorded after any drift in potential
            had either ceased or was very slow. At all
            stations the sediment was moist enough to provide
            good contact with the electrodes.

The thickness of the flocculated sediment varies from one to greater than five centimeters. The electrode potentials recorded were not equilibrium measurements and were sensitive to positioning of the electrode. Therefore, the potentials recorded are rough measurements and should not be used for calculations.

The existence of two distinct facies in the surficial sediments of southern Lake Michigan is of considerable geochemical significance. The brown color is attributable to the presence of finely divided hydrated iron oxides.[1,33] The color of the gray silt facies appears to be related to reduction of oxides. The reducing conditions probably are due to bacterial decomposition of organic matter in the sediments. Microbial populations and organic carbon content are greatest at the sediment-water interface and are reduced substantially in underlying sediments. At a station at Grand Haven, Michigan, seven miles west of the Grand River inflow, the organic carbon content and chemical oxygen demand of the sediments are reduced substantially at a sediment depth of 2.5 to 4 cm; these parameters are approximately three times higher in the uppermost 1.5 cm sediment interval than at a 10 to 12 cm depth of burial.[34] The organic carbon content increases from a mean of 1.4% in sediments below a 5 to 15 cm surface zone of organic carbon accumulation to a mean of 2.5% in the uppermost sediment interval (Table 24). Kemp *et al.*[35] attribute similar surface accumulations in Lakes Erie, Huron, and Ontario to increased organic loading in recent years and to decomposition of organic matter by microorganisms prior to burial in the sediments. The organic matter of the sediments consists primarily of stable compounds such as bitumens, humic acids, and fulvic acids.[36] Concentrations of organic carbon in the uppermost sediment intervals are directly related ($r = 0.65$) to the less-than-2 μ clay size fraction of the sediment, indicating the importance of topography and depositional patterns of sediments in the distribution of organic matter in the southern lake basin.[37]

There is a distinct geographic distribution of "accumulating" trace elements in the sediments of the southern basin of Lake Michigan[19] that corresponds to the distribution of water insoluble organochlorine insecticides.[37] The largest accumulations of these trace constituents in fine-grained surficial sediments occur in the central, south

central, and eastern parts of the basin which is
the region of the gray silt facies (distributions
of lead, bromine, and zinc are illustrated in Figures
14-16). Concentrations of "accumulating" trace
elements are, on the average, 2.2 times higher in
the gray silt facies than in other surficial
sediments.[26] Accumulations of organochlorine
insecticides are also higher in the gray silt
facies than in other regions.[37]

The rivers that drain southern Michigan appear
to be major contributors of trace elements and,
together with erosion of the eastern and southeastern shores, are the principal sources of sediment
to the basin. Trace elements entering the lake
through runoff or erosion of other areas or through
atmospheric inputs apparently are sorbed on suspended
particles, which are transported by water and
deposited eventually in greater abundance in fine-
grained sediments of the gray silt facies than in
other areas. The highest surficial sediment concentrations of trace elements and organochlorine
insecticides are in the central or deepest parts of
the southern basin, where the most finely divided
particles occur.

Efficient uptake of trace elements directly
from water cannot occur at the two-dimensional
sediment-water interface. Consequently, the observed
enrichment of certain trace elements in the surficial
sediments results either from deposition of particles
with associated trace elements, by diffusion from
deeper layers of the sediments, or by redistribution
through hydrodynamic processes. To ascertain if
deposition of particles satisfactorily explains the
surface accumulation of trace elements, determinations
of trace (and major and minor) constituents of the
suspended particles one meter from the lake floor were
conducted (Table 24). With the exception of bromine,
mean trace element concentrations in the suspended
matter equal or exceed the amounts in the surficial
sediments.

Zinc concentrations are substantially higher
in suspended particles than in surficial sediments.
Since a minimum core interval of 3 cm was used, the
difference could be attributed to the sampling procedure. An additional explanation is the pH-dependence
of zinc sorption on sediments and suspended matter.
Considerable differences in zinc-sorbing capacity
of suspended matter with small changes in pH near
neutral ranges are well documented.[38,39] Lake
Michigan water is greater than pH 8.2, compared to

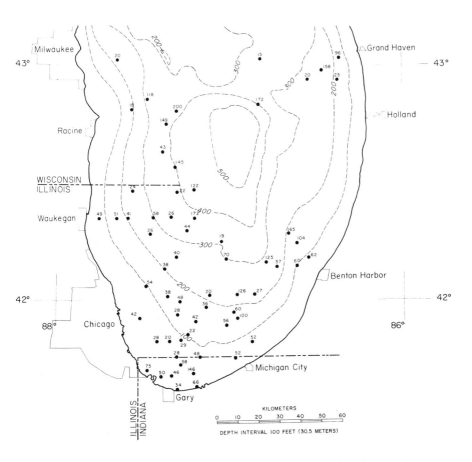

Figure 14.   Concentrations (μg per g) of lead in the most
recent sediments of southern Lake Michigan
(data from Reference 19 or previously
unpublished).

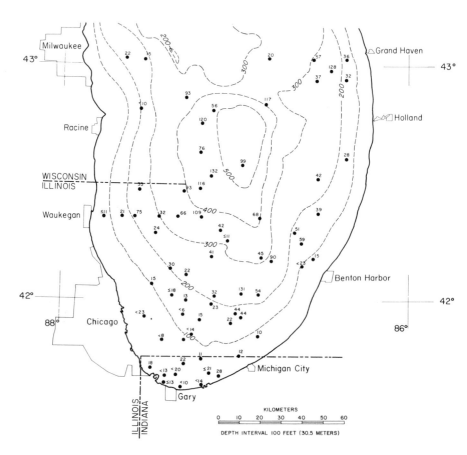

Figure 15. Concentrations (μg per g) of bromine in the
most recent sediments of southern Lake Michigan
(data from Reference 19 or previously
unpublished).

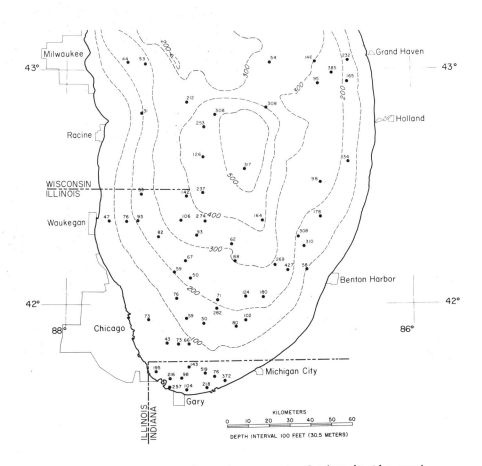

Figure 16. Concentrations (μg per g) of zinc in the most
recent sediments of southern Lake Michigan
(data from Reference 19 or previously
unpublished).

a pH of about 7.2 for surficial sediments. The pH
of lake water is within the expected theoretical
range for a calcareous lake in equilibrium with a
fixed partial pressure of CO_2 of the atmosphere;[40]
maximum sorption of zinc by river silts occurs in
this pH range.[39] Dilution after incorporation in
sediments could also account for the difference in
the concentrations of zinc in suspended matter and
sediments.

Lead contents of suspended particles in the
central portions of the basin exceed 80 µg per g,
which is equivalent to the mean concentration found
in surficial sediments of the basin. Suspended
matter in the hypolimnion generally contains less
lead in nearshore areas than in offshore regions,
which may be due to a larger average size of particles in the nearshore environment. The chemistry
and mechanisms of interaction of lead with various
sediment parameters are discussed in detail later
in this review.

Concentrations of bromine in suspended matter
are lower than in bottom sediments. The primary
natural source of bromine is believed to be rainfall,
with the ultimate derivation being sea spray, but
significant atmospheric additions may also result
from combustion of leaded gasoline containing
ethylene dibromide.[41,42] Bromine is readily lost
from exhaust particulates after emission.[43-45] In
surface waters bromine is dissolved but can associate
with hydrous oxide or organic or clay-mineral
fractions of the suspended matter. Hydrous oxides
of iron are able to sorb considerable amounts of
anions, particularly at neutral or lower pH values
where their surfaces are more positively charged.
The lower concentrations of bromine in suspended
matter, as compared to sediments, may be due to the
decreased ability of hydrous oxides to sorb bromine
at the higher pH ($\sim$8.2) of Lake Michigan waters.
This is true only if hydrous oxides are a dominant
factor in sorption of bromine in the lake water.
The ability of clays to sorb anions (like bromine)
is generally low, as is the complexing capability
of organic matter. The other possible mechanism
for bromine enrichment in surficial sediments of
southern Lake Michigan is specific incorporation
of bromine by aquatic organisms.

The chemical and/or biological mechanisms
whereby trace elements are cycled and incorporated
into lake sediments are poorly understood. The
concentrations of trace elements found to be

accumulating in the surficial fine-grained sediments of Lake Michigan are generally more closely related to the organic carbon content than to clay size material (as per cent less than 2 μ size), water depth, or amounts of manganese or aluminum. Correlation coefficients and number of samples (in parentheses) for trace element concentrations in the surficial sediments versus organic carbon are as follows: arsenic, 0.70 (17); bromine, 0.67 (38); chromium, 0.75 (42); copper, 0.73 (32); mercury, 0.71 (26); lead, 0.82 (42); and zinc, 0.65 (32). "Accumulating" trace elements are also related to iron content of surficial sediments: arsenic, 0.76 (17); bromine, 0.65 (38); chromium, 0.81 (42); copper, 0.54 (32); lead, 0.48 (42); and zinc, 0.62 (32). Concentrations of B, Be, Co, Ni, and V show no accumulation near the water interface in southern Lake Michigan sediments and are poorly correlated with contents of iron and organic carbon.

The concentrations of iron are slightly greater (an average of 18% of the amount present) in surficial sediments than in underlying older sediments (Table 24). Iron can exist in sediments in clay minerals, in other detrital silicates, in hydrous oxides, and in complexes with organic matter. The higher concentrations of iron in sediments near the water interface and in suspended matter of the hypolimnion may indicate an increased rate of supply of any of these components. Mineralogic studies of Lake Superior sediments, where similar surface enrichment of iron occurs, show that the higher iron concentrations do not result from changes in the clay minerals or in other detrital silicates.[1] Reduction, upward migration, and reprecipitation at the water interface of previously sedimented iron may also be important.[46-48] Manganese concentrations are closely related ($r ≈ 0.75$) to iron content of the sediments[19] and an average increase of 40% of the amount of manganese present was found in the surficial sediments. Accumulations of manganese in Lake Michigan sediments near the water interface have been reported previously by Nussmann[1] and Callender.[49] Mean concentrations of other major constituents, with the exceptions of SiO_2 and organic carbon, are either less than or not substantially different from the concentrations in the underlying older sediments.

CHEMISTRY OF LEAD AND MECHANISMS OF ITS INTERACTION WITH VARIOUS SEDIMENT COMPONENTS

Sorption and transport of trace elements in lakes depend on a combination of physicochemical and biological factors. Regression analysis indicates

that organic matter and iron oxide contents are
related to concentrations of As, Br, Cr, Cu, Hg,
Pb, and Zn in modern surficial sediments of southern
Lake Michigan. However, the relative influence of
these and other sediment parameters on distribution
of trace elements is presently unknown.

This section considers the chemistry and
mechanisms of interaction of lead with various
sediment components; discussion of all elements
analyzed would be beyond the scope of this review.
Solubility, interactions of lead with organic
matter, hydrous oxides and clays, and assimilation
by aquatic organisms are considered. Information
on other trace elements is included where appropriate.

Solubility Considerations

Hydrolysis of lead has been studied extensively.[50,5]
In a review of hydrolytic products of metal ions,
Olin[52] describes the predominance of ionic species
of lead as a function of pH. The fractions of
Pb(II) existing in different isopolycationic forms
has been reported[53] for concentrations ranging from
0.001 to 0.100 M Pb(II). Experiments discussed in
the above references are for controlled laboratory
conditions and high lead concentrations; consequently,
the findings reported are not directly applicable to
studies of natural waters. The major predicted ionic
species below pH 7 in natural waters containing 10^{-6}
M lead is Pb^{+2}. However, in the presence of dissolved
CO_2, carbonates of lead precipitate.[53] Bilinski and
Stumm[54] believe that Pb^{+2}, $PbCO_3$, $Pb(CO_3)_2^{-2}$, $PbOH^+$,
and $Pb(OH)_2$ can occur in natural waters and report
stability constants for these complexes. The main
species of lead in sea water are thought to be Pb^{+2},
$PbOH^+$, and $PbCl^+$.[55] Sillen[55] indicates that at the
pH of seawater (8.1), hydroxide-complexing is important for all ions of oxidation number greater than
2, and that the more abundant chloride ions must
compete with hydroxide to form complexes with metals.
From thermodynamic considerations, it appears that
PbO_2, but not $PbSO_4$, is stable in marine environments.[55]
The mineral plattnerite (PbO_2) indicates alkaline
oxidizing conditions[40] but is a rare mineral on the
earth's surface.[56] At equilibrium, PbO_2 may exist
in solid solution, e.g., with MnO_2, rather than in
the pure state.[55] This phenomenon also may arise
from adsorption onto surfaces or incorporation into
the ubiquitous ferromanganese substances present in
marine sediments.

In the presence of sulfide and reducing conditions, galena (PbS) is a stable mineral. In oceans,
local precipitation of sulfide is a possible control
mechanism for lead (and other elements) but is
probably not the chief control because the concentrations are unrelated to sulfide solubilities.[57]
Formation of carbonate, hydroxy, and mixed (hydroxy-
carbonate) compounds of lead is favored in lake
environments. In the presence of increased concentrations of dissolved carbonates, $PbCO_3$ (cerussite)
is a stable compound, the stability shifting to
$Pb_3(OH)_2(CO_3)_2$ (hydrocerussite) with increasing
pH. Formation of $PbCO_3$ *in vitro* from PbO_2 under
increased partial pressures of CO_2 caused by
anaerobic fermentation of plants has been reported,[58]
but the probability of $PbCO_3$ forming by this mechanism
in lakes is unknown.

Eh-pH diagrams illustrating stability fields of
compounds of lead and several other trace elements
are presented by Garrels and Christ,[40] Pourbaix,[59]
and, for natural waters, by Hem and Durum.[60]
Solubility calculations for lead from data collected
in a nationwide reconnaissance of minor element concentrations in surface waters of the United States
are discussed by Hem and Durum.[60] The authors
conclude that although most natural waters are below
saturation, many are near equilibrium with respect
to lead carbonate. Ocean waters are also reported
to be undersaturated with respect to $PbCO_3$ and
$Pb_3(PO_4)_2$.[57] Lack of mineralogic evidence for lead
compounds in bottom sediments of lakes, coupled
with a low concentration of lead in water, indicate
that the chemistry of lead in lakes and other surface
waters cannot be adequately described by solubility
considerations alone. Other mechanisms are emphasized in the recent literature, including association
with such sediment components as organic matter,
hydrous oxides, and clays. In natural waters, most
available lead(II) has been reported to be in association with suspended materials.[54] Of the Pb(II)
species, $Pb(OH)_2$(aq.) and $PbCO_3$(aq.) are strongly
adsorbed at interfaces. In solutions that do not
contain CO_3^{2-}, Pb(II) species are not adsorbed to
a significant extent.[54] Such adsorption phenomena
in natural waters alter the distribution of lead
and limit the utility of simple solubility
considerations.

Organic Matter

Organic matter appears to be an important factor in complexation of heavy metals in lake water and sediments. Barsdate and Matson,[61] employing anodic stripping voltammetry, report a high apparent-stability constant (K = 10^{24}) for the lead-organic complex in an organic-rich lake water. Concentrations of less than 0.1 µg per liter of free or weakly-complexed species of lead, copper, and zinc were present; however, upon oxidation of organic substances with persulfate, significant quantities of each metal were determined. In lake waters from the same region (Alaska) with low organic matter contents, organometal complexes, if present, were in low abundance compared to the concentration of free metals.[61] Release of manganese, copper, and zinc has also been reported in seawater upon destruction of organic matter with persulfuric acid.[62] In equilibrium studies, colloidally dispersed organic matter has been found to be more effective in removing heavy metals than hydrous oxides of iron.[63] Sufficient concentrations of organic ligands are present in oligotrophic Lake Michigan waters to complex lead brought in by atmospheric precipitation.[17] Concentrations of lead dissolved in distilled water were found to decrease from 20 µg per liter to approximately 0.4 µg per liter in less than one hour when mixed with Lake Michigan waters. The importance of organic matter in complexation of heavy metals in soils,[64,65] in lakes,[66,67] in rivers,[68] in oceans,[69] and in the transport of lead and other heavy metals[70,71] is well documented.

A direct evaluation of the relative influence of organic matter and other sediment components on retention of trace elements by lake sediments has not been attempted. The ability of stabilized organic matter in sediments to form complexes is generally attributed to the so-called humic and fulvic acids. Thus, discussion of the salient characteristics of these sediment components, their stability with heavy metals, and their properties in terrestrial, lake, and marine environments must be considered.

Concentrations of lead in surface soils are generally greater than in underlying soils,[72,73] a phenomenon which may be attributed to increased amounts of organic matter in surface horizons. Higher recoveries of EDTA-extractable lead and copper are obtained from surface horizons than from deeper soils.[74]   Organic matter may, in fact, dominate retention of heavy metals by soils. Wei[75] reports that the exchange capacity of soil organic matter is satisfied before significant sorption of heavy metals by clay minerals occurs.

Measurements of stability constants of humic and fulvic acids derived from soils with heavy metals have recently been attempted.[76-78]   Schnitzer[76] reported the order and log K values for several divalent cations with fulvic acid. Log K values were found to be higher at pH 5 than at pH 3.5, due primarily to increased ionization of the functional group COOH.   The apparent pK value of the COOH group in fulvic acid was 4.5.   Schnitzer and Hanson[79] found the log K values reported by Schnitzer[76] to be consistently high. Two methods were employed by the authors[79] and the results were in good agreement.

At pH 3
Cu > Ni > Co > Pb,   Ca > Zn > Mn > Mg
3.3   3.1   2.9   2.6   2.6   2.4   2.1   1.9

At pH 5
Ni = Co > Pb > Cu > Zn,   Mn > Ca > Mg
4.2   4.2   4.1   4.0   3.7   3.7   3.4   2.2

Log K values for iron (Fe^{3+} = 6.1) and aluminum (Al^{3+} = 3.7) were determined at pH 1 and pH 2.35 respectively to avoid precipitation and formation of polynuclear compounds. Judging from the decreases in pH of aqueous solutions of humic acid upon addition of inorganic salts, van Dijk[80] concludes that at pH 5, Pb, Cu, and Fe^{3+} (in that order) are firmly bound and that there are no large differences in bond strengths for other divalent ions, *viz*. Ba, Ca, Mg, Mn, Co, Ni, Fe, and Zn (slightly increasing in this order).

Stability constants reported for metal-soil organic matter complexes are generally lower than those for complexes of the same metal ions with commercial chelating agents such as EDTA and HEDTA.[65] Essentially two types of reactions are envisioned by Schnitzer[76] between metals and fulvic acids, the major one involving both the acidic COOH and phenolic groups and the minor one involving less acidic COOH groups. Alcoholic -OH groups play no part in the metal-fulvic acid interactions.

Fulvic acids have a higher number of functional groups than do humic acids (*i.e.*, more oxygen in COOH, OH and C = O groups) and they have lower molecular weights. Fulvic acids are more mobile than humic acids and consequently are of greater importance in the transport of heavy metals in waters. In electrophoresis, both humic and fulvic acids move towards the anode and, in fact, many fractions (15) of humic acids have been separated as a consequence of their relative mobilities.[81] Within the pH range of 2.0 to 8.6, the electrophoretic behavior of Al, Ag, Cu, Fe, Pb, and Tl compounds of humic acids indicates the existence of complexation products between the reactive materials.[82] Other lines of evidence for complexation products between soil humic substances and heavy metals have been discussed by Stevenson and Ardakani.[65]

The ability of an element to form complexes is determined by the structure of the electronic shells of the atoms, the size of its ionic radius, its valence, and its polarization. Ions of chalcophillic elements (like lead) having eight electron internal shells have a high complex-forming capacity.[70] An increase in valence and decrease in ionic radius favors complex formation. Using Mellor and Maley's data for the stability of complexes of several divalent metal ions, Irving and Williams[83] report that stability increases with atomic number to Cu(II) in the first transition series after which there is marked decrease for Zn(II). Similarly, in the second transition series Pd(II) > Cd(II), and in the third Pt(II) > Hg(II) and Pb(II). These relationships led the authors to conclude that perhaps complete filling of the "M-shell" is important. These findings are of particular significance to this discussion because several of the elements are environmental pollutants. The following order is referred to as the Irving-Williams series by soil organic chemists:

Pd > Cu > Ni > Co > Zn > Cd > Fe > Mn > Mg

Similarities in mechanism of interaction with trace elements and perhaps even in affinity for different metal cations are possible for humic and fulvic acids in soils and lake sediments. However, some evidence, based largely on structural differences in humic material, exists to the contrary. Humic substances in lakes are principally autochthonous, with aquatic organisms being the precursor material.[66,67] Humic acids in Great Lake sediments may be unlike those in soils because of the difference in source material and the microorganisms involved in decomposition of organic matter. The fact that properties of humic acids differ as a function of environment is illustrated by the comparison made by Rashid et al.[84]

between humic acids of terrestrial and marine origin.
As compared to their properties in soils, humic acids
in ocean sediments have a higher molecular weight,
lower cation exchange capacity, conspicuously low
phenolic hydroxyl or high carbonyl and quinone
content, lower metal-holding capacity, and similar
concentration of amino groups. Although a lower
metal-holding capacity is generally assumed to be
due to the high salinity factor of the marine
environment, it is evident that differences in
structure and cation exchange capacity do exist.

HYDROUS OXIDES

The ability of hydrous oxides of iron,
manganese, and aluminum to sorb cations and anions
is well established.[85-88] Parks[89] provides an
extensive list of isoelectric points of solid oxides
and hydroxides. Under oxidizing conditions, redox-
sensitive hydrous oxides of iron and manganese in
sediments are excellent scavengers of trace elements;
however, under reducing conditions they are
solubilized and may result in increases in concentrations of cations and anions in overlying waters.[46-48]
In ocean sediments, the ubiquitous ferromanganese
minerals act as hosts for Pb, Cu, Ni, Co, and other
rare earth elements.[90] The ferromanganese solids
contain lead in concentrations as high as several
tenths of one per cent, whereas the lead concentration
of seawater averages only 0.03 µg per liter. The
importance of hydrous oxides in distribution of trace
elements in soils is reviewed by Jenne.[91] In some
soils, most lead exists as inclusions in iron and
aluminum hydroxide minerals and $CaCO_3$.[92]
There is a widespread distribution of hydrous
oxides in fresh water environments. The anion
(phosphate) sorbing capacity of several Wisconsin
lakes has been attributed primarily to hydrous oxides
dominated by iron.[93,94] Most solid phases in natural
waters contain oxides and hydroxides.[88] The hydrous
oxides in lake sediments are believed to be short-
range-order, particularly for redox sensitive
elements like iron and manganese.[93,94] This is
important for surface-active substances, like
hydrous oxides, where reactivity is primarily due
to surface area and charge. Short-range-order
hydrous oxides are capable of sorbing much greater
amounts of cations and anions than are their mineral
counterparts. The capacity and point of zero charge
varies for hydrous oxides of different elements and

types of minerals of each element. There is also
evidence for a lower but definite sorption of
several cations and anions at or on either side of
the point of zero charge, suggesting specificity
and covalency in bonds. This is particularly true
for transitional elements. Marked specific adsorption of the transitional metal ions Ni^{2+}, Cu^{2+}, and
especially Co^{2+} has been observed in several prep-
arations of manganese dioxide.[95] Reversibility in
sorption depends on the ionic radius of the solvated
metal ions. In general, the smaller the crystalline
ionic radius, the greater the hydrated radius and
consequently the less the specificity in adsorption.
For δ-MnO_2 surfaces, Murray[96] finds alkaline earths
to be reversibly sorbed, having the following order:

$$Mg^{2+} < Ca^{2+} < Sr^{2+} < Ba^{2+}$$

whereas, transitional elements are irreversibly
sorbed, the order being

$$Co^{2+} > Mn^{2+},\quad Cu^{2+} > Zn^{2+} > Ni^{2+}$$

The extent of sorption of many trace elements,
including lead, on preformed ferric oxides is very
high compared to the extent of precipitation.[58]
Sorption is greater for freshly-formed hydrous
oxides of iron than for aged oxides, and sorption
of lead increases with pH over the range of pH 5
to 7.[97] In this pH range, hydrous oxides of iron
are positively charged; consequently, lead appears
to be specifically sorbed.

Chemical extractants used by soil scientists
also reveal an importance of hydrous oxides in
retention of trace metals. The most common extractants used are acid ammonium oxalate and acetic
acid. According to Mitchell[74] oxalate extracts
about 100 times more Co, Ni, Cu, and Pb from surface
horizons of soils than does acetic acid. In soils,
about 60% of the total iron and cobalt, 50% of the
copper and lead, 80% of the manganese, 40% of the
vanadium, and 20% of the chromium and nickel are
extracted by acid ammonium oxalate.[74] Le Riche and
Weir,[98] using ultraviolet light in conjunction with
acid ammonium oxalate, found oxide fraction to be
rich in trace elements (Pb, Co, Cu, Mn, and V).
This method dissolves goethite and other oxides of
iron but not gibbsite or boehmite. However, these
methods of extraction overestimate the amounts of
trace elements associated with hydrous oxides
because they also extract organically-bound trace
elements. Nevertheless, they do indicate the

important role played by hydrous oxides in binding of lead and other trace metals.

From the standpoint of mobility of trace metals, redox-sensitive hydrous oxides of iron and manganese are more important than are the oxides of aluminum. Several elements, including lead, have been found to be highly mobile upon the addition of fermenting plant material.[58]   The released metals were present as organic complexes because they could not be removed from the fermented extracts by cation-exchange resins. Analogous conditions may be expected in sediments of lakes having seasonal variations in dissolved oxygen concentrations and decomposition of freshly settled detritus. Such alternating conditions of reduction and oxidation result in the release and resorption of trace metals from hydrous oxide surfaces and may lead to a dynamic cycling of some trace elements in lake environments.

Clay Minerals

Concentrations of trace elements in Dahomey soil profiles vary with clay content, being highest in clayey horizons.[99]   In a study of several soil types the highest concentrations of lead and several other trace elements also occurred in heavy clay fractions.[100] Hirst[101] attributes the transport of lead and other trace elements to marine basins to clay minerals. Lead within the lattices of degraded clay minerals, primarily illite and montmorillonite, was considered to be the mode of association. Wedepohl[102] also suggests that most lead in marine sediments is associated with clay minerals.

The affinity and the relative importance of clay minerals in lake sediments for trace elements is poorly understood. The ion-exchange properties of clay are similar to the properties of cation-exchange resins, but the charge characteristics of clays originate from isomorphous replacements and the broken edges of crystal surfaces. The basic principles governing the selectivity characteristics of clays for different cations are valence, hydrated ionic radius, electronegativity, and the free energy of formation. Ionic potential (charge/radius) is a useful property for predicting the affinities of clays for different cations. The order of difficulty in displacement of cations is approximately:[74]

Cu > Pb > Ni > Co > Zn > Ba > Rb > Sr > Ca > Mg > Na > Li

but relative positions in such a series vary with concentration in solution and the nature of the exchange-active material. Preferences of different clay minerals for heavy metals has been studied under such variables as pH, selectivity, and affinity of different ground and unground clay minerals. Kaolinites retain lead at pH 5, but at pH 1.5 lead is completely released.[92]   Lead seems to be retained preferentially by Utah bentonite over calcium, for which the mass action equilibrium constant is 1.4.[103] For cobalt and zinc, with similar reactivities, Tiller and Hodgson[104] and Hodgson[64] give the following order of reactivity for unground clay minerals:

> hectorite > vermiculite > nontronite > montmorillonite > halloysite > kaolinite.

For ground minerals with the same elements the order is

> muscovite > phlogopite > talc > biotite, vermiculite > pyrophyllite.

Grinding has a serious effect on the reactivity of minerals for cobalt, zinc, and probably other cations due to an increase in the surface areas of clay minerals and formation of broken crystal edges.

Lead replaces potassium in its structural positions owing to similar ionic size.[99]   Isomorphic replacement of several other cations by lead is described by Goldschmidt.[56]   Lead on montmorillonite is rendered nonexchangeable by drying the clay,[105] a property shared by potassium due to fixation and collapse of lattices.

Complexing agents seem to alter the affinity of clay minerals for different cations. For kaolins, the natural affinity is Th > La > Ca > K but the order is completely reversed (K > Ca > La > Th) when citrate or fluoride ions are present.[74]   Several complexing agents are present in lake, river, and marine environments. Whether they affect the exchange properties of clay minerals or retention of lead and other heavy metals is not known.

Although clay minerals can be important in retention and transport of trace elements, their relative role must be evaluated with caution. Despite an emphasis on clay minerals as vehicles for transport of heavy metals to oceans, Hirst[101] indicates that limonitic concretions average 42 µg per g of lead which, allowing for lead in the clay impurities, amounts to 88 µg per g in iron oxides. More recent literature emphasizes the importance of organic matter and hydrous oxides in transport of heavy metals.

Calcite

Calcium carbonate is an integral part of the sediments of many lakes, including Lake Michigan. Its role in retention of lead by soils has been discussed previously.[92] Despite differences in ionic size (Pb^{2+}, 1.32 Å and Ca^{2+}, 1.06 Å) lead is found to isomorphically substitute for calcium in apatites and aragonites.[56,106] In human bones lead has been reported to replace calcium.[107] This property may also be important in calcareous lake sediments and soils. The zero point of charge for calcite is reported to lie between the pH values of 8.0 and 9.5,[108] with surfaces having a positive charge below pH 8 and negative charge above 9.5. Therefore, in calcareous lake sediments lead could precipitate, be sorbed, or substitute for calcium at calcite surfaces.

Aquatic Organisms

Aquatic flora and fauna are capable of concentrating many trace elements to levels much higher than those existing in water. This is most apparent for photosynthetic autotrophs, which obtain all their nutrients directly from water. The concentration factors depend upon the physicochemical interactions of each element with other environmental parameters and with organisms, as well as the nutritional requirements of individual species. In lakes having aquatic flora and fauna as the dominant source of organic matter or in which waters are highly productive and sedimentation is rapid, the influence of organisms on trace element distribution may be significant. Organic matter in sediments of the Great Lakes is primarily of autochthonous, mainly diatom, origin. Harlow[109] estimates an annual input of allochthonous organic carbon to Lake Erie of only 10%. Plankton (plant and animal) constitutes the major biomass of the Great Lakes and concentration factors and turnover rates for trace elements are higher for these forms than for other aquatic organisms. Consequently, plankton are more apt to influence transport of trace elements than are other forms of life.

Excretory products of zooplankton are a concentrated source of many trace elements[110] and may facilitate their transport to sediments. Settling follows death or molting of plankton and the efficiency of trace element transport from water to sediments depends in part on the rate of decay of the dead organism, test, molted exoskeleton, or excretory product. Decomposition of plankton and organic detritus proceeds during settling. Bacterial activity in the surficial sediments promotes further decay at the sediment-water interface and, after burial, may result in increases in trace element concentrations in the sediments. Preliminary results indicate that more than 90% of the organic matter in the Great Lakes is mineralized before burial.[111]

The major mechanism for concentration by marine plankton of alkali metals and perhaps the heavier alkaline earths is probably ion-exchange, although some elements appear to be sorbed mainly by chelation The order of concentration factors for divalent catic in the marine environment is not readily correlated with the order of stability of complexes, *viz*.

Cu > Ni > Pb > Co > Zn > Cd > Mn > Mg

as suggested by Goldberg.[113]   Bowen[114] found the order of affinites in marine organisms to be as follows:

Plankton: Zn > Pb > Cu > Mn > Co > Ni > Cd
Brown algae: Pb > Mn > Zn > Cu, Cd > Co > Ni

The heavier divalent metals appear to be accumulated to a greater extent by marine organisms than are lighter ones, which may be related to their greater polarizability. Of particular interest is the great affinity for lead, which possesses no known biological function. The order of concentration factors for Lake Michigan phytoplankton (based on data from Reference 115) is

Mn > Zn > Cu > Co > Mg

and corresponds, for the divalent cations analyzed, to the order of concentration factors for brown algae in seawater as reported by Bowen.[114]   It is doubtful, however, that any single factor is responsible for the order of concentration factors observed. Apparent affinities for trace elements among various molluscan species are considered by Pringle *et al.*[116] to depend upon concentrations available in the environment, physicochemical properties of each element, organic ligands available for chelation, stability of metal-organic ligands formed, and processes of transport and storage.

Considering the selective nature of trace element concentration by plankton and the fact that most organic matter in Lake Michigan sediments is of autochthonous origin, it is reasonable to assume that a relationship exists between trace element contents of plankton and affinities of different trace elements for sedimentary organic matter. It is interesting therefore to compare the order of trace element concentrations in Lake Michigan phytoplankton (from data in Reference 115) with the order of regression coefficients for organic carbon versus trace element contents of surficial sediments, *viz.*

Phytoplankton: Zn > [Br], Cu > As > Cr > Co
Sediments (uppermost interval): Zn > Pb > Cr,
 [Br] > Cu > As > Ni > Co.

With the exception of chromium, cations are in the same relative positions in each sequence.

CONCLUSION

Surface enrichment of arsenic, bromine, chromium, copper, mercury, lead and zinc, attributable in part to man's activities, exists in modern surficial sediments of southern Lake Michigan. These trace elements are apparently sorbed, either directly or through biological processes, to suspended particles, which are transported by water and deposited eventually in geographic areas of active sedimentation. The relative influences of specific sediment components on sorption and retention of the above elements are unknown, but regression analysis indicates a relationship between trace element content and amounts of organic matter and iron oxide in the sediments. Recent literature on heavy metal transport emphasizes the importance of metal interactions with organic matter and hydrous oxides of iron and manganese.

ACKNOWLEDGMENTS

This research was supported in part by project A-052-ILL, Office of Water Resources Research. The participation of S. S. Shukla was made possible by the support of NSF RANN Grant GI-31605. Statistical analyses were performed by J. A. Schleicher. The manuscript was reviewed by J. T. O'Connor, F. J. Stevenson, W. A. White, and D. L. Gross.

REFERENCES

1. Nussmann, D. G. "Trace Elements in the Sediments of Lake Superior," Ph.D. Thesis, Geology, Univ. of Michigan, Ann Arbor, Mich. (1965).

2. Beeton, A. M. "Eutrophication of the St. Lawrence Great Lakes," *Limnol. Oceanog. 10*:240 (1965).

3. Beeton, A. M. "Changes in the Environment and Biota of the Great Lakes," In *Eutrophication: Causes, Consequences, Correctives*. Proc. of a Symposium (Washington, D.C.: National Academy of Sciences, 1969), p. 150.

4. Powers, E. F. and J. C. Ayers. "Water Quality and Eutrophication Trends in Southern Lake Michigan," In *Studies on the Environment and Eutrophication of Lake Michigan*, Ayers, J. C. and D. C. Chandler, eds. Great Lakes Res. Div., Univ. Mich. Spec. Rep. 30., Ann Arbor (1967), p. 142.

5. Robertson, A., and W. Alley. "A Comparative Study of the Lake Michigan Macrobenthos," *Limnol. Oceanog. 11*: 576 (1966).

6. Kittrell, F. W. "Statement - Committee on Nuclear Power Plant Waste Disposal," Second Session of the Conf. in the Matter of Pollution of Lake Michigan and its Tributary Basin, Chicago, Ill. (1969), p. 564.

7. Stoermer, E. F. "An Historic Comparison of Offshore Phytoplankton Populations in Lake Michigan," In *Studies on the Environment and Eutrophication of Lake Michigan*, Ayers, J. C. and D. C. Chandler, eds. Great Lakes Res. Div., Univ. Mich. Spec. Rep. 30, Ann Arbor (1967), p. 47.

8. Stoermer, E. F., and J. J. Yang. "Plankton Diatom Assemblages in Lake Michigan," Great Lakes Res. Div., Univ. Mich. Spec. Rep. 47, Ann Arbor (1969).

9. Cratty, A. H. "Statement - United States Department of Agriculture," Third Session of the Conf. in the Matter of Pollution of Lake Michigan and its Tributary Basin, Milwaukee, Wis. (1970), p. 45.

10. Leland, H. V. and N. F. Shimp. "Trace Elements in Southern Lake Michigan," In *Chemistry and Biology of Trace Metals in the Environment*, Laitinen, H. A. and R. L. Metcalf, eds., Univ. of Illinois, Urbana (1971), p. 196.

11. Lake Michigan Enforcement Conference. "Report on Selected Trace Metals in the Lake Michigan Basin," Pesticide Technical Committee, U. S. Environmental Protection Agency, Chicago, Ill. (1972).

12. Shukla, S. S. and H. V. Leland. "Heavy Metals: A Review on Lead," (submitted for publication in the "Annual Literature Review," *J. Water Poll. Control Fed.*, 1973).

13. Winchester, J. W. and G. D. Nifong. "Water Pollution in Lake Michigan by Trace Elements from Pollution Aerosol Fallout," *Water, Air and Soil Pollution 1*:50 (1971).

14. Robbins, J. A., E. Landstrom, and M. Wahlgren. "Tributary Inputs of Soluble Trace Metals to Lake Michigan," *Abst. 15th Conf. Great Lakes Res.*, Madison, Wis., April 5-7 (1972), p. 120.

15. Cogley, A. C. Department of Energy Engineering, Univ. of Illinois, Chicago Circle (1973), personal communication.

16. Winchester, J. W. "A Chemical Model for Lake Michigan Pollution: Considerations on Atmospheric and Surface Water Trace Metal Inputs," In *Nutrients in Natural Waters,* Allen, H. E. and J. R. Kramer, eds. (New York: John Wiley & Sons, 1972), p. 317.
17. Matson, W. R. Environmental Science Associates, Boston, Mass. (1971), personal communication.
18. Shimp, N. F., H. V. Leland, and W. A. White, "Distribution of Major, Minor, and Trace Constituents in Unconsolidated Sediments from Southern Lake Michigan," Illinois Geol. Survey Environ. Geology Note 32 (1970).
19. Shimp, N. F., J. A. Schleicher, R. R. Ruch, D. B. Heck, and H. V. Leland. "Trace Element and Organic Carbon Accumulation in the Most Recent Sediments of Southern Lake Michigan," Illinois Geol. Survey Environ. Geology Note 41 (1971).
20. Kennedy, E. J., R. R. Ruch, and N. F. Shimp. "Distribution of Mercury in Unconsolidated Sediments from Southern Lake Michigan," Illinois Geol. Survey Environ. Geology Note 44 (1971).
21. Ruch, R. R., E. J. Kennedy, and N. F. Shimp. "Distribution of Arsenic in Unconsolidated Sediments from Southern Lake Michigan," Illinois Geol. Survey Environ. Geology Note 37 (1970).
22. Frye, J. C. and N. F. Shimp. "Major, Minor, and Trace Elements in Sediments of Lake Pleistocene Lake Saline Compared with Those in Lake Michigan Sediments," Illinois Geol. Survey Environ. Geology Note 60 (1973).
23. Schleicher, J. A. and J. K. Kuhn. "Phosphorus Content of Unconsolidated Sediments from Southern Lake Michigan," Illinois Geol. Survey Environ. Geology Note 39 (1970).
24. O'Connor, D. J. and J. A. Mueller. "A Water Quality Model of Chlorides in Great Lakes," *J. Sanitary Engr. Div., ASCE, 96*:955 (1970).
25. Leland, H. V., N. F. Shimp, and W. N. Bruce. "Distribution of Trace Elements and Chlorinated Pesticides in the Most Recent Sediments of Southern Lake Michigan," Abstracts, 34th Annual Meeting, Amer. Soc. Limnol. Oceanog., Winnipeg, Canada (1971).
26. Lineback, J. A. and D. L. Gross. "Depositional Patterns, Facies, and Trace Element Accumulation in the Waukegan Member of the Lake Pleistocene Lake Michigan Formation in Southern Lake Michigan," Illinois Geol. Survey Environ. Geology Note 58 (1972).
27. Lineback, J. A., N. J. Ayer, and D. L. Gross. "Stratigraphy of Unconsolidated Sediments in the Southern Part of Lake Michigan," Illinois Geol. Survey Environ. Geology Note 35 (1970).
28. Lineback, J. A., D. L. Gross, and R. P. Meyer. "Geologic Cross Sections Derived from Seismic Profiles and Sediment Cores from Southern Lake Michigan," Illinois Geol. Survey Environ. Geology Note 54 (1972).

29. Lineback, J. A., D. L. Gross, R. P. Meyer, and W. L. Unger. "High-Resolution Seismic Profiles and Gravity Cores of Sediments in Southern Lake Michigan," Illinois Geol. Survey Environ. Geology Note 47 (1971).

30. Gross, D. L., J. A. Lineback, W. A. White, N. J. Ayer, C. Collinson, and H. V. Leland. "Preliminary Stratigraphy of Unconsolidated Sediments from the Southwestern Part of Lake Michigan," Illinois Geol. Survey Environ. Geology Note 30 (1970).

31. Gross, D. L., J. A. Lineback, N. F. Shimp, and W. A. White. "Composition of Pleistocene Sediments in Southern Lake Michigan, U.S.A.," 24th Internat. Cong. Proc., Sec. 8, Montreal (1972), p. 215.

32. Federal Water Pollution Control Administration. "Lake Michigan Basin - Physical and Chemical Quality Conditions," Great Lakes Region, U.S. Dept. of the Interior, Chicago, Ill. (1968).

33. Murray, R. C. "The Petrology of the Cary and Valders Tills of Northeastern Wisconsin," *Am. Jour. Sci. 251:* 140 (1953).

34. Kupke, J. E. "Zinc Concentrations as Related to Organic Matter Stability in Lake Michigan Sediments," M.S. Thesis, Civil Engr., Univ. of Illinois, Urbana (1973).

35. Kemp, A. L. W., C. B. J. Gray, and A. Mudrochova. "Changes in C, N, P, and S in the Last 140 Years in Three Cores from Lakes Ontario, Erie, and Huron," In *Nutrients in Natural Waters,* Allen, H. E. and J. R. Kramer, eds. (New York: John Wiley & Sons, 1972), p. 251.

36. Kemp, A. L. W. "Organic Matter in the Sediments of Lakes Ontario and Erie," Proc. 12th Conf. Great Lakes Res., Ann Arbor, Mich (1969). p. 237.

37. Leland, H. V., W. N. Bruce, and N. F. Shimp. "Chlorinated Hydrocarbon Insecticides in Sediments of Southern Lake Michigan," *Environ. Sci. & Technol.* 1973), in press.

38. O'Conner, J. T. and C. E. Renn. "Soluble-Adsorbed Zinc Equilibrium in Natural Waters," *J. Amer. Water Works Assoc. 56:*1055 (1964).

39. O'Conner, J. T., C. E. Renn, and I. Wintner. "Zinc Concentrations in Rivers of the Chesapeake Bay Region," *J. Amer. Water Works Assoc. 56:*280 (1964).

40. Garrels, R. M. and C. L. Christ. *Solutions, Minerals, and Equilibria* (New York: Harper & Row, 1965).

41. Tiffany, M. A., and J. W. Winchester. "Surface Water Inputs of Iodine, Bromine, and Chlorine to Lake Huron," Proc. 12th Conf. Great Lakes Res., Intern. Assoc. Great Lakes Res., 789 (1969).

42. Tiffany, M. A., J. W. Winchester, and R. H. Loucks. "Natural and Pollution Sources of Iodine, Bromine and Chlorine in the Great Lakes," *J. Water Pollution Control Fed. 1319* (1969).

43. Tel Haar, G. L., and M. A. Bayard. "The Composition of Airborne Lead Particulates," *Nature 232:*553 (1971).

44. Jernigan, J. L., B. J. Ray, and R. A. Duce. "Lead and Bromine in Atmospheric Particulate Matter on Oahu, Hawaii," *Atmos. Environ. 5:*881 (1971).

45. Moyers, J. L., W. H. Zoller, R. A. Duce, and G. I. Hoffman. "Gaseous Bromine and Particulate Lead, Vanadium and Bromine in a Polluted Atmosphere," *Environ. Sci. & Technol. 6:*68 (1972).

46. Hutchinson, G. E. *A Treatise on Limnology,* Vol. 1 (New York: John Wiley and Sons, Inc., 1957).

47. Mortimer, C. H. "The Exchange of Dissolved Substances Between Mud and Water in Lakes. Part I and II," *J. Ecol. 29:*280 (1941).

48. Mortimer, C. H. "The Exchange of Dissolved Substances Between Mud and Water in Lakes. Part III and IV, Summary and References," *J. Ecol. 30:*147 (1942).

49. Callender, E. "Geochemical Characteristics of Lakes Michigan and Superior Sediments," Proc. 12th Conf. Great Lakes Res., Intern. Assoc. Great Lakes Res. (1969), p. 124.

50. Olin, A. "Studies on the Hydrolysis of Metal Ions. 25. The Hydrolysis of Lead(II) in Perchlorate Medium," *Acta Chem. Scand. 14:*126 (1960).

51. Olin, A. "Studies on the Hydrolysis of Metal Ions. 28. Application of the Self-Medium Method to the Hydrolysis of Lead(II) Perchlorate Solution," *Acta Chem. Scand. 14:* 814 (1960).

52. Olin, A. "Studies on the Hydrolysis of Metal Ions," *Svensk Kem. Tidskr. 73:*482 (1961).

53. Tyree, S. Y., Jr. "The Nature of Inorganic Solute Species in Water," In *Equilibrium Concepts in Natural Water Systems,* Adv. Chem. Ser. No. 67 (Washington, D.C.: American Chemical Society, 1967), p. 183.

54. Bilinski, H. and W. Stumm. "Pb(II) - Species in Natural Waters," EAWAG No. 1, Swiss Federal Institutes of Technology, Federal Institute for Water Resources and Water Pollution Control (January, 1973).

55. Sillen, L. G. "The Physical Chemistry of Sea Water," In *Oceanography,* Sears, M., ed. Publication No. 67 (Washington, D.C.: Amer. Assoc. Adv. Science, 1961), p. 549.

56. Goldschmidt, V. M. *Geochemistry* (London: Claredon Press, 1954).

57. Krauskopf, K. B. "Factors Controlling the Concentrations of Thirteen Rare Metals in Sea-Water," *Geochim. Cosmochim. Acta 9:*1 (1956).

58. Kee, N. S. and C. Bloomfield. "The Solution of Some Minor Element Oxides by Decomposing Plant Materials," *Geochim. Cosmochim. Acta 24:*206 (1961).

59. Pourbaix, M. *Atlas of Electrochemical Equilibria in Aqueous Solutions,* Tr. from French by J. A. Franklin (New York: Pergamon Press, 1966), p. 485.

60. Hem, J. D. and W. H. Durum. "Occurrence and Solubility of Lead in Surface and Ground Water," *J. Water Works Control. Fed.* (1973) in press.

61. Barsdate, R. J. and W. R. Matson. "Trace Metals in Arctic and Sub-Arctic Lakes with Reference to the Organic Complexes of Metals," In *Radioecological Concentration Processes,* Aberg, B. and F. P. Hungate, ed. Proc. Intern. Symp., Stockholm, 1966 (London: Pergamon Press, 1967), p. 711.

62. Slowey, J. F., L. M. Jeffery, and D. W. Wood. "Evidence for Organic Complexed Copper in Sea Water," *Nature 214:* 377 (1967).

63. Barsdate, R. J. "Transition Metal Binding by Large Molecules in High Latitude Waters," In *Proc. Symp. Organic Matter in Natural Waters, 1968.* Hood, D. W., ed. Inst. Marine Sci., Univ. of Alaska (1970), p. 485.

64. Hodgson, J. F. "Chemistry of the Micronutrient Elements in Soils," *Advance Agron. 15:*119 (1963).

65. Stevenson, F. J. and M. S. Ardakani. "Organic Matter Reactions Involving Micronutrients in Soils," In *Micronutrients in Agriculture,* Mortvedt, J. J., P. M. Giordano, and W. L. Lindsay, eds. (Madison, Wis.: Soil Science Society of America, Inc., 1972), p. 79.

66. Ishiwatari, R. "An Estimation of the Aromaticity of a Lake Sediment Humic Acid by Air Oxidation and Evaluation of It," *Soil Sci. 107:*53 (1969).

67. Otsuki, A. and T. Hanya. "Some Precursors of Humic Acid in Recent Lake Sediment Suggested by Infra-red Spectra," *Geochem. Cosmochim. Acta 31:*1505 (1967).

68. Lamar, W. L. "Evaluation of Organic Color and Iron in Natural Surface Waters," *Geol. Survey Prof. Paper 600-D:* D24 (1968).

69. Rashid, M. A. and L. H. King. "Molecular Weight Distribution Measurements on Humic and Fulvic Acid Fractions from Marine Clays on the Scotian Shelf," *Geochim. Cosmochim. Acta 33:*147 (1969).

70. Manskaya, S. M. and T. V. Drozdova. *Geochemistry of Organic Substances,* trans. from Russian, Shapiro, L. and I. A. Breger, eds. (New York: Pergamon Press, Inc., 1968).

71. Vinogradov, A. P. *The Geochemistry of Rare and Dispersed Chemical Elements in Soils,* trans. from Russian by Consultant Bureau, Inc., New York, N.Y., Second Edition (1959).

72. Swaine, D. J. and R. L. Mitchell. "Trace-Element Distribution in Soil Profiles," *J. Soil Sci. 11:*347 (1960).

73. Wright, J. R., R. Levick, and H. J. Atkinson. "Trace Element Distribution in Virgin Profiles Representing Four Great Soil Groups," *Soil Sci. Soc. Amer. Proc. 19:* 340 (1955).

74. Mitchell, R. L. "Trace Elements in Soils," In *Chemistry of the Soil,* Bear, F. E., ed. (New York: Reinhold Publishing Corp., 1964), p. 320.

75. Wei, L. S. "The Chemistry of Soil Copper," Ph.D. Thesis, Agronomy, Univ. of Illinois, Urbana (1959).

76. Schnitzer, M. "Relations Between Fulvic Acid, a Soil Humic Compound and Inorganic Soil Constituents," *Soil Sci. Soc. Amer. Proc. 33:*75 (1969).

77. Schnitzer, M. and S. I. M. Skinner. "Organo-Metallic Interactions in Soils: 5. Stability Constants of Cu^{2+}, Fe^{2+}, and Zn^{2+} Fulvic Acid Complexes," *Soil Sci. 102:* 361 (1966).

78. Schnitzer, M. and S. I. M. Skinner. "Organo-Metallic Interactions in Soils: 7. Stability Constants of Pb^{2+}, Ni^{2+}, Mn^{2+}, Co^{2+}, Ca^{2+}, and Mg^{2+} Fulvic Acid Complexes," *Soil Sci. 103:*247 (1967).

79. Schnitzer, M. and E. H. Hanson. "Organo-Metallic Interactions in Soils: 8. An Evaluation of Methods for the Determination of Stability Constants of Metal-Fulvic Acid Complexes," *Soil Sci. 109:*333 (1970).

80. van Dijk, H. "Cation Binding of Humic Acids," *Geoderma 5:*53 (1971).

81. Kaurichev, I. S., E. A. Fedorov, and I. A. Shnabel. "Use of Continuous Paper Electrophoresis in the Separation of Humic Acids," *Soviet Soil Sci. 10:*1050 (1960).

82. Mücke, D. and H. Kleist. "Paper-Electrophoretic Investigations on Metallic Compounds of Humic Acids," *Albrecht-Thaer-Arch. 9:*327 (1965); Abs. No. 83, *Soils & Fertilizers 29:*31 (1966).

83. Irving, H. and R. J. P. Williams. "Order of Stability of Metal Complexes," *Nature 162:*746 (1948).

84. Rashid, M. A., D. E. Buckley, and K. R. Robertson. "Interactions of a Marine Humic Acid with Clay Minerals and a Natural Sediment," *Geoderma 8:*11 (1972).

85. Hsu, P. H. "Interaction Between Aluminum and Phosphate in Aqueous Solution," In *Trace Inorganics in Water,* Adv. Chem. Ser. No. 73 (Washington, D.C.: American Chemical Society, 1968), p. 115.

86. Morgan, J. J. and W. Stumm. "The Role of Multivalent Metal Oxides in Limnological Transformations, as Exemplified by Iron and Manganese," Proc. 2nd Inter. Water Poll. Res. Conf., Tokyo, Japan (1964), p. 103.

87. Parks, G. A. "Aqueous Surface Chemistry of Oxides and Complex Oxide Minerals," In *Equilibrium Concepts in Natural Water Systems,* Adv. Chem. Ser. No. 67 (Washington, D.C.: American Chemical Society, 1967), p. 121.

88. Stumm, W. and J. J. Morgan. *Aquatic Chemistry* (New York: Wiley-Interscience, 1970).
89. Parks, G. A. "The Isoelectric Points of Solid Oxides, Solid Hydroxides, and Aqueous Hydroxo Complex Systems," *Chem. Rev. 65*:177 (1965).
90. Goldberg, E. D. "Chemistry - The Oceans as a Chemical System," In *The Sea*, Vol. 2 "The Composition of Sea-Water Comparative and Destriptive Oceanography," (New York: Interscience Publishers, 1963), p. 3.
91. Jenne, E. A. "Controls on Mn, Fe, Co, Ni, Cu, and Zn Concentrations in Soils and Water: The Significant Role of Hydrous Mn and Fe Oxides," In *Trace Inorganics in Water*, Adv. Chem. Ser. No. 73 (Washington, D.C.: American Chemical Society, 1968).
92. Iordanov, N. and M. Povlova. "Geochemistry of Lead in Soils," *Izv. Inst. Obshch. Neorgan. Khim. bulg. Akad. Nauk 1*:5 (1963); *Chem. Abstr. 61*:1655 (1964).
93. Shukla, S. S., J. K. Syers, J. D. H. Williams, D. E. Armstrong, and R. F. Harris. "Sorption of Inorganic Phosphate by Lake Sediments," *Soil Sci. Soc. Amer. Proc. 35*:244 (1971).
94. Williams, J. D. H., J. K. Syers, S. S. Shukla, R. F. Harris, and D. E. Armstrong. "Levels of Inorganic and Total Phosphorus in Lake Sediments as Related to Other Sediment Parameters," *Environ. Sci. & Technol. 5*:1113 (1971).
95. Murray, D. J., T. W. Healy, and D. W. Fuerstenau. "The Adsorption of Aqueous Metal on Colloidal Hydrous Manganese Oxide," In *Adsorption from Aqueous Solution*, Adv. Chem. Ser. No. 79 (Washington, D.C.: American Chemical Society, 1968).
96. Murray, J. W. "Trace Metal Interactions at the Manganese Dioxide Solution Interface," presented before Division of Water, Air, and Waste Chemistry, Amer. Chem. Soc., New York (August, 1972), p. 172.
97. Gadde, R. R. and H. A. Laitinen. "Study of the Sorption Characteristics of Synthetic Hydrous Ferric Oxide." (submitted for publication in *Soil Science*, 1973).
98. Le Riche, H. H., and A. H. Weir. "A Method of Studying Trace Elements in Soil Fractions," *J. Soil Sci. 14*:225 (1963).
99. Pinta, M. and C. Ollat. "Physico-Chemical Investigations of Trace Elements in Tropical Soils. I. A Study of Some Soils in Dahomey," *Geochim. Cosmochim. Acta 25*:14 (1961); Abs. No. 2945, *Soils & Fertilizers 24*:420 (1961).
100. Erviö, R. and K. Virri. "Trace Element Contents of the Soils of Central Uusimar," *Annal. Agric. Fenn. 4*:178 (1965); Abs. No. 1575, *Soils & Fertilizers 29*:241 (1966).
101. Hirst, D. M. "The Geochemistry of Modern Sediments from the Gulf of Paria - II. The Location and Distribution of Trace Elements," *Geochim. Cosmochim. Acta 26*:1147 (1962).

102. Wedepohl, K. H. "Untersuchungen zur Geochemie des Bleis," *Geochim. Cosmochim. Acta 10*:69 (1956).

103. Krishnamoorthy, C. and R. Overstreet. "An Experimental Evaluation of Ion-Exchange Relationships," *Soil Sci. 69*: 41 (1950).

104. Tiller, K. G. and J. F. Hodgson. "The Specific Sorption of Cobalt and Zinc by Layer Silicates," *Clays and Clay Minerals 9*:393 (1962), Proc. 9th National Conf. on Clays and Clay Minerals, Purdue Univ. Lafayette, Ind. (1960).

105. MacKenzie, R. C. "Retention of Exchangeable Ions by Montmorillonite," *Inter. Clay Conf. 1*:183, Proc. Conf. held at Stockholm, Sweden, August 12-16 (1963).

106. Rankama, K. and T. G. Sahama. *Geochemistry* (Chicago: Univ. of Chicago Press, 1950).

107. Bryce-Smith, D. "Lead Pollution - A Growing Hazard to Public Health," *Chem. Brit. 7*:54 (1971).

108. Somasundaran, P. and G. E. Agar. "The Zero Point of Charge of Calcite," *J. Colloid Int. Sci. 24*:433 (1967).

109. Harlow, G. L. "Task Which Lies Ahead in the Lake Erie Basin," Proc. 23rd Ind. Waste Conf., Purdue Univ. (1968), p. 856.

110. Boothe, P. N. and G. A. Knauer. "The Possible Importance of Fecal Material in the Biological Amplification of Trace and Heavy Metals," *Limnol. Oceanog. 17*:270 (1972).

111. Kemp, A. L. W. "Organic Carbon and Nitrogen in the Surface Sediments of Lakes Ontario, Erie and Huron," *J. Sediment. Petrol. 41*:537 (1971).

112. Lowman, F. G., T. R. Rice, and F. A. Richards. "Accumulation and Redistribution of Radionuclides by Marine Organisms," In *Radioactivity in the Marine Environment* (Washington, D.C.: National Academy of Sciences, 1971), p. 161.

113. Goldberg, E. D. "Biogeochemistry of Trace Metals," *Geol. Soc. Amer. Mem. 67*:345 (1957).

114. Bowen, H. J. M. *Trace Elements in Biochemistry* (New York: Academic Press, 1966).

115. Copeland, R. A. and J. C. Ayers. *Trace Element Distributions in Water, Sediment, Phytoplankton, Zooplankton, and Benthos of Lake Michigan: A Baseline Study with Calculations of Concentration Factors and Buildup of Radioisotopes in the Food Web* (Ann Arbor, Mich.: Environmental Research Group, Inc., 1972).

116. Pringle, B. H., D. E. Hissong, E. L. Katz, and S. T. Mulawka. "Trace Metal Accumulation by Estuarine Mollusks," *J. Sanitary Div., ASCE 94*:455 (1968).

5. PATHWAYS OF MERCURY IN A POLLUTED NORTHWESTERN ONTARIO LAKE

F. A. J. Armstrong and A. L. Hamilton. Fisheries Research Board of Canada, Freshwater Institute, Winnipeg, Manitoba

INTRODUCTION

The Wabigoon River-Clay Lake-English River-Winnipeg River system (Figure 17) in western Ontario and eastern Manitoba appears to be one of the waterways in Canada most heavily contaminated with mercury. In Clay Lake, about 50 miles downstream of Dryden, fish have been found to contain up to 16 µg/g of mercury.[1]

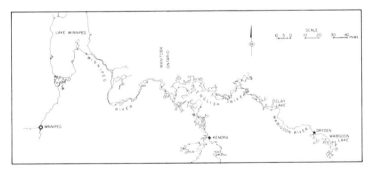

Figure 17.   Wabigoon-English-Winnipeg River system, showing position of Clay Lake.

The source of this mercury is believed to be a chlorine-alkali plant at Dryden, in operation since 1962, with a capacity of 11,000-12,000 tons of chlorine per year. Since early 1970 the effluent of this plant has been regulated, and discharges

131

have been reduced to 0.007 lb (3 g) mercury per ton
of chlorine produced. The discharge before this
was stated to be 0.21 lb (95 g) mercury per ton of
chlorine. Calculation shows that during the eight
years of unregulated operation, 1962-1969, the total
discharge of mercury in the waste waters from this
plant may have amounted to 20,000 lb (9,000-11,000
kg).

When Wobeser *et al.*[2] first reported unacceptably
high levels of mercury in freshwater fish from
Canadian waters there was almost no information
available on mercury levels in other freshwater
organisms. In 1967, Johnels *et al.*[3] had included
some invertebrate organisms in their search for
indicators of mercury pollution in Sweden, and
Hannaerz, another Swedish worker, had conducted a
number of experiments on the accumulation of mercury
by freshwater organisms.[4] In 1969 Matida and Kumada[5]
had reported on the distribution of mercury in water,
bottom mud, and aquatic organisms from marine, estuarine and river habitats in Japan. However, there
were apparently no published North American surveys
of mercury levels in invertebrate organisms. The
surveys reported here are part of a continuing study
of the lake and its biota and have been performed in
order first to measure the quantity of mercury in
the sediments and its distribution; second, to
determine levels of mercury in invertebrates and
small vertebrates in the contaminated ecosystem and
to attempt to relate differences in mercury concentrations to the life histories of the organisms;
and third, to identify organisms which might be
suitable as indicators of mercury pollution. Most
specimens were collected in Clay Lake itself but
some attention was given to animals from other parts
of the system.

DESCRIPTION OF CLAY LAKE

Clay Lake is situated on the Wabigoon River in
northwestern Ontario (Figure 17). It has an area
of 7400 acres (3000 HA) and a bathymetric map prepared by the Ontario Department of Lands and Forests
indicates it has a maximum depth of 75 feet (24 m)
and mean depth of 25 feet (8 m) with a volume of
185,000 acre feet (2.4×10^8 m^3). The lake at 50°03'N
93°30'W probably takes its name from the deposits of
clay in the area. Deposits of gently undulating
sediments with rock outcrops are found throughout
the Wabigoon River drainage area[6] and give the river

and Clay Lake high turbidity. The 12 year (to 1965) mean inflow from the Wabigoon River is 1580 cfs (56 m^{3}sec^{-1})[7] so that the theoretical renewal time of the water in the lake is about 50 days.

SAMPLING AND MATERIALS

Preliminary to the main sampling program for sediments, samples were taken with a Mackereth piston corer at the locations indicated by the letters A, B, and K through N in Figure 18. Sections of these

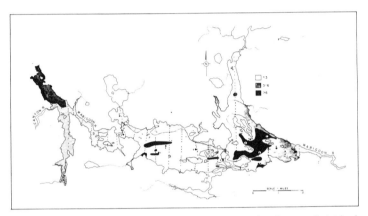

Figure 18.   *Concentration of mercury (µg/g dry weight) in sediments of Clay Lake, Ontario.*

cores were analyzed for mercury in order to determine the variation of mercury content with depth. Another series of cores at positions C through J, which were taken later with an Ekman dredge as described below, were subjected to similar analysis. The analyses are shown in Table 26. High concentrations were found in the upper 4 to 6 cm whereas concentrations below 6 cm were usually less than 1 µg/g. From this it was estimated that mercury in the upper 6 cm accounted for 92-95% of the total above a background level of about 0.1 µg Hg/g. For survey purposes it seemed sufficient to have analyses of the sediments to this depth only.

An Ekman dredge was used to collect samples for the main survey, bringing up portions of the top 10 cm of sediment. From these portions an undisturbed cylindrical core representing the upper 6 cm was removed with a plastic tube of 4 cm diameter. This portion was placed in a polyethylene envelope,

Table 26

Variation with Depth of Mercury Content of Sediment Cores from Clay Lake
(µg Hg/g, dry weight basis)

Depth (cm)	A	B	C	D	E	F	G	H	J	K	L	M	N
0-2	8.42	8.30	0.95	6.53	6.34	6.65	5.17	2.38	7.50				
2-3	8.04	6.38	0.91	7.98	6.31	7.25	4.27	2.59	7.32				
3-4	7.14	4.48	0.62	8.55	6.42	7.81	3.55	2.07	6.96	5.92	6.40	4.58	6.04
4-5	5.52	1.58	0.43	6.57	6.12	7.81	3.43	1.61	4.10				
5-6	5.52	0.20		4.05	3.61	6.34	1.06	0.02	0.32				
6-7	3.93	0.20		2.32	1.15	2.42	0.24	0.88	<0.01	1.79	2.29	0.17	0.16
7-8	0.66	0.16				0.14	0.27	0.13	<0.01	0.57	0.65	0.06	0.08
8-9	--	0.18								0.42	0.22	0.06	<0.01
9-10		--										<0.01	<0.01
10-11	0.04	0.10											
20-22		0.06											
30-32		0.02											
42-44		0.05											

labelled, frozen as soon as possible and stored at
-20°C until analyzed. During July and August 1971
27 transects of the lake were made, and a total of
283 samples collected.

Water samples were collected with a Van Dorn
2 liter sampler and transferred to polyethylene
bottles. Analysis for the quantity and mercury content of suspended matter was performed within 24
hours on samples to which no preservative had been
added. Samples to be analyzed for total mercury
content were preserved by addition of 1 ml sulfuric
acid (36N) per liter.

For direct determination of sedimentation rate,
clusters of four traps were placed at three locations
in the central basin of the lake for approximately
77 days from the end of May to mid-August 1972. Each
trap consisted of a $10\frac{1}{2}$-inch (26.5 cm) diameter
polyethylene funnel with its stem inserted in the
neck of a 1 liter polyethylene bottle. Four traps
were fixed in a cross-shaped wooden frame made
buoyant with a sealed empty 5-gallon plastic bottle,
and attached to an anchor with the rims of the
funnels about 1 m from the lake bottom. The collected material was separated by centrifugation,
dried and weighed. For calculation of estimated
depth of sedimentation, compacted sediments were
assumed to contain 1.2 g dry material per cm^3. This
value was obtained by measurement on sediment cores
from the lake.

CHEMICAL ANALYSIS

Sediment samples were thawed, homogenized by
mixing (using a blender to reincorporate water
exuded during freezing), and air-dried for 2-3 days,
usually in a current of air in a fume hood. They
were pulverized to about 100 mesh (125 μm) in a
mechanical mortar, passed through a screen of 1 mm
mesh, transferred to stoppered bottles and stored.
Three subsamples, usually of 0.2-0.5 g, were taken
for analysis and digested with 10 cc aqua regia for
10 minutes at about 100°C. The solutions and residues were transferred to measuring cylinders, made
to known volume and mixed. This extraction procedure
is that used in the Dow Chemical Company analysis
for muds and sediments.[8] After the suspended solids
had settled, portions of the liquid were analyzed
for mercury by the flameless AAS method of Armstrong
and Uthe.[9]

Because of the possibility of losses of mercury during sample preparation and extraction, considerable attention was paid to monitoring the analytical method, as outlined below.

(1) *Inhomogeneity of sample:* Samples, particularly those from the margins of the lake and around islands, often contained coarse sand and rock particles as well as pieces of organic material, making it difficult to take small representative portions for analysis. After the samples had been dried and passed through the 1 mm mesh screen the coarse material that did not pass through this mesh (about 1% of the total) was collected and was then subdivided by screening it through a 2 mm mesh. The coarse and fine fractions were found to contain 0.08 and 0.27 µg Hg/g respectively. Because these concentrations are low and because there is so little of this material it was concluded that removal of this fraction was justified.

(2) *Effect of drying:* Samples were analyzed before and after drying to determine whether or not mercury was lost during the drying process. The analysis of the wet sample was converted to dry weight basis by using the loss of weight on drying. Table 27 shows that no significant differences were found.

Table 27

Mercury Content of Sediment Samples
Before and After Drying
(µg Hg/g dry weight)*

Sample no.	Wet sample	Dry sample
1	6.67	6.74
2	4.60	4.61
3	4.58	4.36
4	0.92	1.23

*Means of three analyses

(3) *Recovery of added mercury:* Samples of wet sediment from an uncontaminated lake were spiked with 0.5-2.0 µg/g dry weight of mercury as mercuric chloride or as methyl mercuric chloride, and were then air-dried and analyzed. Table 28 shows that recoveries of mercury were satisfactory.

Table 28

Recovery of Mercury Added to Wet Sediment*

Compound	Mercury added μg Hg/g dry wt.	Mercury found μg Hg/g dry wt.	Recovery %
HgCl$_2$	0	0.04	–
	0.484	0.51	97
	0.971	1.00	99
	1.884	1.88	98
CH$_3$HgCl	0	0.08	–
	0.563	0.68	107
	1.083	1.19	102
	2.237	2.11	91

*Means of four analyses

(4) *Completeness of extraction:* Extraction leaves a great deal of insoluble material and some tests were made to see if any appreciable quantity of mercury remained in it. Some residues after aqua regia extraction were opened up by treatment with hydrofluoric and sulfuric acids and the solutions obtained were analyzed for mercury. In seven such analyses the amount of mercury found amounted to 0-0.98% (mean 0.67%) of the total present. This fraction was considered negligible, and the extraction by aqua regia was accepted as being virtually complete.

A number of sediment samples were also analyzed for carbon and nitrogen content using a Carlo Erba Elemental Analyzer Model 1102, and for iron by an absorptiometric method using 1:10 phenanthroline.[10] Some samples were subjected to particle size analysis by Jackson's pipette method.[11] Suspended matter in water samples was separated from about 300 cc on Millipore matched-weight monitor filters of 0.8 μm porosity. The filters were air-dried and weighed to obtain the weight of suspended matter, and then the filter and adhering material were digested with nitric and sulfuric acids and permanganate and the mercury content determined. For this determination the water samples were not acidified and were analyzed within 24 hours of collection in order to minimize change on storage. In the first set of such samples the filtrate was then analyzed by the method of Chau and Saitoh[12] for dissolved mercury. In a later experiment samples were acidified when collected, and the suspended matter was filtered off but not analyzed. The filtrates were analyzed for mercury after first being irradiated for 12 hours with UV radiation to decompose any organic mercury compounds which may have been present.[13]

BIOLOGICAL SURVEYS

An initial survey of the biota in Clay Lake was conducted in June, 1970. Benthic invertebrates living in the deeper parts of the lake were collected with an Ekman dredge and the mud samples were then sieved through a 400 µm mesh. Individual organisms were picked out with forceps and sorted into taxonomic groups. Littoral invertebrates were collected with dip nets and seines and then sorted into taxonomic groups. Planktonic organisms including small minnows, were collected with a towed plankton net (apertures 73 µm) and then put through a series of graded sieves to sort the organisms into size categories. Crayfish were collected in August and September of 1970, throughout the summer of 1971 and again in the spring of 1972. Most crayfish were taken with minnow traps baited with raw meat. Those collected in 1970 were dissected for analysis of various organs, while in 1971 and 1972 the specimens were measured and then the abdominal muscle was removed and stored. All samples were frozen for storage and were later analyzed by the method of Armstrong and Uthe.[9] Results were reported as µg Hg/g wet weight (ppm).

RESULTS

The horizontal distribution of mercury in the upper 6 cm of the sediments of the lake is shown in Figure 18. Concentrations vary from about 0.1 µg Hg/g at the mouth of the Wabigoon River at the eastern end and at the margins and around islands (that is to say, where river scour or wave action have removed fine material) to 7.8 µg/g in the deeper part of the eastern basin. The mean value for the samples analyzed is 3.1 µg/g.

The total amount of mercury in the top 6 cm in the lake sediments has been estimated from the areas and mean values within the contours to be about 4,400 lb (2000 Kg). As discussed above, this value

may be underestimated by 5-10%, since the analyses
are of the top 6 cm of sediment only.

It is noticeable that although the contours
enclosing the highest concentrations of mercury are
elongated in the direction of water movements through
the lake, there is nevertheless considerable deposition of mercury in the northern arm. It is probable
that sediment has been carried into this arm by
currents induced by wind action and fluctuations of
water level. It is also worth noting that although
the depth of the eastern basin (maximum 6 m) is a
good deal less than that of the central basin
(maximum 24 m), mercury concentrations are highest
in the eastern basin.

Particle Size Analysis

Results of particle size analysis are shown in
Tables 29 and 30. Table 29 includes mercury content
of the inorganic fractions and shows that the clay
fraction is highest in mercury content. To inves-
tigate further the relationship between the amount
of fine material and mercury concentration, additional analyses were performed (Table 30). There
is a weak positive correlation ($r = 0.23$) between
mercury concentration and silt content, but this is
not significant ($P > 0.1$). There is a slightly
stronger one ($r = 0.49$) between mercury concentration
and clay content, which is highly significant ($P >
0.02$). It should be noted that the correlations
would appear very much stronger if only the eastern
and central basins (with high mercury concentrations)
were considered.

Twenty-five samples with a range of mercury
concentrations from 0.14 to 7.83 µg/g were analyzed
for carbon, nitrogen and iron content. The results
are shown in Table 31. Since these sediments are
free of carbonate, the carbon and nitrogen content
is a measure of the amount of organic matter present. There are highly significant positive
correlations between total mercury and carbon
($r = 0.69$ $P < 0.001$), nitrogen ($r = 0.56$ $P < 0.01$) and
iron ($r = 0.671$ $P < 0.001$). The first two of these
correlations may be the result only of the known
affinity of mercury compounds for organic matter.
The correlation of mercury and iron may result from
a coprecipitation phenomenon, or more probably from
the fact that both metals are associated with
organic matter.

Table 29

Particle Size Analysis of Sediment Samples from Locations P, R, and S in Clay Lake

Location	Inorganic Material %	Sand 2mm – 50μm		Silt 50 – 2μm		Clay <2μm	
		% of inorganic	μg Hg/g	% of inorganic	μg Hg/g	% of inorganic	μg Hg/g
P	97	73	<0.01	20	0.51	7	5.93
R	90	0	–	39	1.07	61	6.10
S	89	0	–	10	1.20	90	4.78

Table 30

Particle Size Analysis of Sediments from Clay Lake

Location	Sample No.	Inorganic material %	Sand 2mm-50µm % of in-organic	Silt 50-2µm % of in-organic	Clay <2µm % of in-organic	Mercury content µg Hg/g
East	3 SM 4	95	80	14	6	1.10
basin	5 SN 4	92	<1	49	50	5.92
	7 SN 8	92	<1	36	64	4.92
	16	95	<1	39	61	4.5
	18	92	<1	46	53	3.3
	24	94	<1	34	66	4.2
	8 SN 7	93	<1	34	66	6.18
	8	97	0	38	62	6.98
	11	94	<1	45	55	4.00
	14	96	0	33	67	4.98
	23	95	<1	34	66	3.46
Central	11 NS 3	95	<1	28	72	4.66
basin	14 SN 3	95	<1	24	76	3.1
	13	95	<1	16	84	6.3
	19	92	<1	23	77	4.95
	16 SN 5	94	0	17	83	4.8
	12	95	<1	16	84	6.3
	20 SN 4	94	0	17	83	5.23
	9	95	1	38	61	1.29
West	22 NS 6	96	1	24	75	2.5
basin	25 SN 4	95	51	14	35	0.95
	29 SN 7	90	3	28	69	2.9
	30 WE 5	89	1	23	76	2.5

Table 31

Analysis of Sediments from Clay Lake (dry weight basis)

Location	Sample No.		Carbon %	Nitrogen %	Iron %	Mercury µg/g
East	2 SN	2	0.8	0.07	0.91	0.21
basin	3 SN	4	1.3	0.09	0.67	1.10
	4 SN	4	2.7	0.15	2.56	5.57
	5 SN	6	4.2	0.27	3.34	4.93
	7 SN	7	4.3	0.31	4.17	7.34
	8 SN	6	4.6	0.35	4.20	7.83
		12	2.7	0.19	4.42	4.23
		25	3.1	0.28	4.50	3.36
	9 SN	4	0.6	0.10	1.19	0.52
		9	3.7	0.30	4.64	2.68
Central	11 WS	13	3.9	0.37	4.79	2.37
basin	12 NS	2	1.8	0.16	1.56	0.88
		6	3.3	0.24	4.64	5.41
	15 NS	6	3.9	0.26	4.52	4.43
		16	4.2	0.27	4.95	6.21
	16 SN	11	4.7	0.36	5.32	5.30
	18 SN	6	4.6	0.34	5.06	6.00
		10	3.7	0.29	5.53	3.99
	20 SN	10	3.9	0.32	4.35	1.51
West	24 SN	1	0.9	0.12	0.71	0.19
basin		6	4.1	0.31	4.74	2.55
	26 SN	3	0.6	0.07	2.41	0.14
	28 SN	5	4.8	0.37	4.68	1.98
		8	4.9	0.38	5.13	3.06
River	R4		1.2	0.09	1.18	1.41

Two samples of high total mercury concentrations were analyzed for methyl mercury by the gas chromatographic method of Uthe, Solomon and Grift[14] and gave values of 0.01 and 0.03 μg Hg/g. These represent about 1 and 6% of the mercury present.

The quantity of suspended matter and the mercury content in the water was measured in a series of samples from positions eastward from a point about 2 miles up the Wabigoon River to the narrows near point G in Figure 18. The results are given in Table 32. The quantity of suspended material decreases along the lake as suspended material settles

Table 32

*Suspended Material in Water of Wabigoon River and Clay Lake**

Location		Quantity of suspended matter (mg/l)	Mercury content μg Hg/g
Wabigoon River	1	12.5	4.4
	2	14.7	2.5
	3	21.0	2.5
	4	18.0	2.0
	5	19.5	4.1
East basin	6	20.5	3.4
	7	16.2	2.3
Central basin	8	7.7	2.9
	9	8.7	2.1
	10	6.4	5.8
West basin	11	6.9	3.2
	12	7.2	3.6

*October 19, 1971.

out. By noting the difference in the amounts of suspended material at the inlet and outlet ends of the lake it is possible to make a rough estimate of the rate of sedimentation at the time these samples were taken. If these are 20 and 7 mg/l respectively and if the volume of $2.4 \times 10^8 m^3$ passes through the lake in 50 days, it may be calculated that the mean sedimentation rate for the lake is 0.6-0.7 mm/year. Confirmation of these indirect results was obtained by direct measurement between May and August 1972.

Measured rates of sedimentation at three positions
in the central basin of the lake were 1.85, 2.90 and
3.22 g/m^2 per day, which, if they continued at this
rate throughout the year, would correspond to annual
deposits of 0.6, 0.9 and 1.0 mm consolidated sediment.
     Some analyses of water and of suspended matter
are shown in Table 33. The February 1972 samples
were taken near the centers of the eastern and central basins, the deeper samples being about 1 m from
the bottom in each case. At the time, water movement was probably at a minimum and the water column
should have been thermally stable. It is possible
that the higher concentrations of suspended matter
in the central basin indicate that material was
settling out. Mercury in the suspended matter accounted for 11-75% of the mercury in the water.

Analysis of Food Chain Organisms

     The initial survey of food chain organisms
indicated that some taxa accumulated much more
mercury than others. Organisms living or feeding
in the water column had only about one-tenth the
concentration of mercury as organisms that lived in
or attached to the bottom substrate (Figure 19).
It also appeared that organisms feeding on phytoplankton, zooplankton or attached algae had only
about one-tenth as much mercury as detritus feeders,
omnivores or taxa that fed primarily on both benthic
invertebrates (Figure 20).
     The crayfish *Orconectes virilis* (Hagen) con-
tained substantially higher concentrations of mercury
than any other invertebrate collected. The 1970
adult specimens had concentrations of approximately
10 µg/g in the abdominal muscle with concentrations
in other organs ranging from 0.9 to 6.0 µg/g
(Figure 21). The claw muscle of one of these
specimens was analyzed for methyl mercury and the
result was within 10 per cent of the total mercury
value.
     Analysis of a large sample of crayfish collected
in June, July and August 1971 from a bay on the
southern shore of the eastern basin indicated a
definite relationship between body weight and mercury
concentration in abdominal muscle (Figure 22).
     A survey of mercury concentrations in the
abdominal muscle of crayfish collected in August 1972
from the Wabigoon, Eagle, English and Winnipeg Rivers
clearly indicated that there was a major source of
mercury pollution originating in the vicinity of

Table 33

Analysis of Water and Suspended Matter from Clay Lake*

Location	Date	Depth	Suspended matter mg/l	μg Hg/g	Hg in solution μg/l	Hg in suspension μg/l	μg/l
East basin	9 Feb 1972	0m	10.8	3.10	0.11	0.03	0.14[1]
		5m	9.9	2.34	0.05	0.02	0.07[1]
	18 Feb 1972	0m	10.1	7.0	0.09[2]	0.07	0.16
		5m	9.7	4.3	0.13[2]	0.04	0.17
Central basin	9 Feb 1972	0m	6.6	1.56	0.05	0.01	0.06[1]
		17m	16.1	--	0.05	--	--
	18 Feb 1972	0m	7.0	1.7	0.08[2]	0.01	0.09
		17m	15.5	5.7	0.03[2]	0.09	0.12
Inlet	20 Aug 1972	0m	1.4	2.8	0.02	0.04	0.06[1]
Outlet	20 Aug 1972	0m	2.3	3.2	0.02	0.02	0.04[1]

*[1]By addition; [2]By difference.

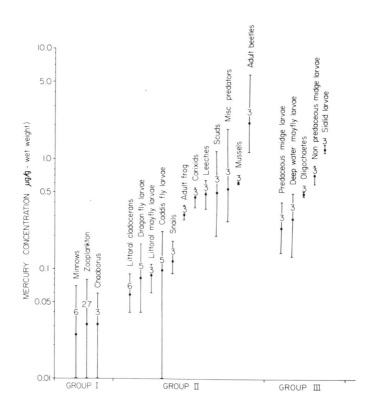

Figure 19.   The relationship between mercury concentration
and habitat for organisms in Clay Lake, Ontario,
June, 1970.   Group I: organisms living and/or
feeding primarily in the water column; Group II:
organisms from the shallow littoral part of the
lake;   Group III: organisms from the profundal
sediments.   Numbers indicate sample size while
the lines and points refer to the range and
geometric mean respectively.

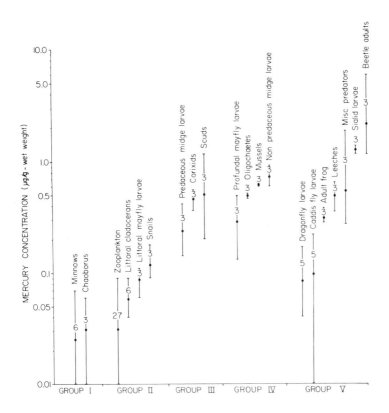

Figure 20. The relationship between mercury concentrations
and food selection for organisms in Clay Lake,
Ontario, June 1970. Group I: organisms feeding
primarily on zooplankton; Group II: organisms
feeding primarily on phytoplankton or periphyton;
Group III: omnivorous organisms; Group IV:
detritus feeders; Group V: organisms predaceous
on other benthic invertebrates. Numbers indicate
sample size while the lines and points refer to
the range and geometric mean respectively.

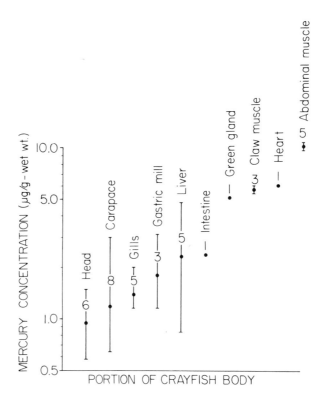

Figure 21.   *Mercury concentrations in different body parts of mature crayfish (Orconectes virilus) from Clay Lake, Ontario, 1970. Numbers indicate sample size while the lines and points refer to the range and geometric means respectively.*

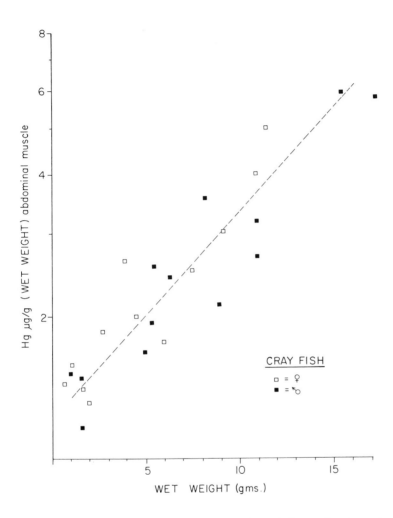

Figure 22.   Relationship between body weight and concentration
of mercury in abdominal muscle for crayfish
(Orconectes virilis) collected during June, July
and August 1971 in Clay Lake, Ontario.   Each
point represents a pooled sample consisting of
four specimens.

Dryden, Ontario (Figure 23). Four samples (each
consisting of four individuals) collected in the
Wabigoon River above Dryden, Ontario, averaged between 0.18 and 0.27 µg/g on a wet weight basis.
Five similar samples from the Eagle River which
flows into the Wabigoon about midway between Dryden
and Clay Lake yielded similar values (0.09 to 0.27
µg/g). In contrast 13 samples from Clay Lake and
the Wabigoon River immediately above and below the
lake yielded values between 1.43 and 7.36 µg/g. A
single sample from the English River downstream
from its junction with the Wabigoon River had a
value of 1.34 µg/g. Values for eight samples taken
in the Winnipeg River varied from 0.12 to 0.56 µg/g.

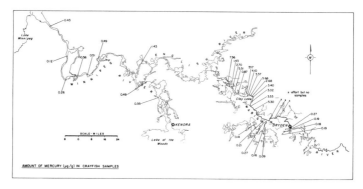

Figure 23.   Concentrations of mercury in the abdominal muscle
            of mature crayfish (Orconectes virilis) collected
            in the Wabigoon, English, and Winnipeg River
            systems during August 1971. Each value represents
            the arithmetic mean of the values found for four
            individual specimens.

DISCUSSION

     The total quantity of mercury in sediments in
the lake is somewhat more than 2,000 kg, or about
20% of our estimate of the total amount discharged
during the first eight years of operation of the
chlorine-alkali plant in Dryden. Deposition of
mercury has been greatest at the eastern end of the
lake, although the water is shallower and has more
rapid circulation than in the central basin. Deposition of mercury has also occurred in the arms
of the lake.

Elevated mercury concentrations are found in the top 5-6 cm of the sediments, a much greater depth than that calculated from the estimated rate of deposition during eight years, which is probably 5-6 mm only. As Jernelov[15] has shown, fish and clams can cause surface sediments to be mixed down to a depth of several centimeters, and no doubt fish and invertebrates living in Clay Lake have helped to mix the surface sediments. The evidence collected so far indicates that the 5 or 6 mm of material which we estimate may have been deposited during these eight years has in fact been mixed to about 10 times that depth. Since the average concentration in the upper 6 cm is about 3 µg Hg/g, we calculate that the concentration of mercury in the newly deposited material was about 30 µg/g. Unfortunately it is no longer possible to check this nor is it possible to predict whether the mercury-rich layer will deepen with time as the mixing process continues.

The suspended matter in the lake water contains about 3 µg Hg/g, which is about one-tenth of the concentration (if our suppositions are valid) in material deposited when the effluent from the chlorine-alkali plant was uncontrolled. Nevertheless it is close to the mean concentration in the top part of the sediment, and on settling out cannot greatly change this concentration.

It is probable that previously deposited material is still being washed out from the river. However, it is nevertheless instructive to calculate the concentrations in water and in suspended matter which currently result from the controlled discharge at the Dryden plant. This controlled discharge, estimated at 0.007 lb (3 g) of mercury per ton of chlorine manufactured, represents a total daily input of about 100 g mercury per day, with a river flow (at Clay Lake) of 56 m^3 per second. This in turn represents a concentration in the water of 0.02 µg/l or in the suspended matter (at 20 mg/l) of about 1 µg if all mercury was observed. These concentrations are not greatly different from what we have measured.

At the earlier unregulated discharge of 0.21 lb mercury per ton of chlorine the corresponding concentrations would be 0.6 µg/l and about 30 µg Hg/g, in quite remarkable agreement with our estimate of the concentration of mercury in sediment deposited before control. It seems that even with the present controls, there is little hope of any rapid improvement in the degree of pollution of the sediments.

For any improvement it is of course necessary that existing deposits should be covered by material of low mercury content. Several centimeters of depth may be required[15] because of disturbance by animals, and such an accumulation might take some centuries if our estimates of the sedimentation rate in the lake are correct. Moreover, improvement is dependent on the newly deposited material having a low mercury content so that it is itself harmless, and this can hardly be said to be the case when the material being deposited in Clay Lake at present has a mercury content of more than 2 μg/g.

Mercury Concentration in Organisms

Results of this investigation have demonstrated that organisms in a mercury-contaminated freshwater ecosystem accumulate very different concentrations of mercury. Undoubtedly many factors are involved, but these differences seem to be generally related to both food habits and habitat selection. In this survey it was apparent that organisms living in bottom sediments or attached to littoral vegetation or the water surface had much higher mercury concentrations than organisms living and/or feeding in the water column. This does not correspond with the findings of Matida and Kumada[5] who found that in a marine habitat the mercury content of planktonic microorganisms was extremely high (up to 279 μg/g on a dry weight basis). In contrast, zooplankton organisms collected in Clay Lake never had concentrations greater than 0.1 μg/g on a wet weight basis.

Organism Food Selection

Mercury concentration also appeared to be related to food selection and in most cases omnivorous organisms and organisms feeding on detritus or bottom dwelling invertebrates had much higher mercury concentrations than either herbivorous organisms or organisms feeding on zooplankton (Figure 20). The most conspicuous exceptions were the samples of dragon fly larvae and caddis fly larvae where the mercury concentrations were much lower than expected. Both samples were collected within 200 meters of a large, unpolluted stream flowing into the Wabigoon system and the specimens may have been recent arrivals to Clay Lake. Hannaerz[4] concluded that there was no general connection between mercury concentration in invertebrates and their trophic level although he

found that predaceous insect larvae, such as dragon
flies and *Sialis* did accumulate more mercury than
organisms feeding on detritus or decaying plant
material. He concluded that several factors including metabolic rate and feeding habits were
responsible for differences in rates of mercury
accumulation.

A series of experiments designed to provide
some indication of the factors affecting the uptake
and elimination of mercury by selected freshwater
organisms was conducted in 1971.[16] These experiments, designed to simulate conditions in Clay Lake,
indicated that mercury uptake via the food chain
was probably much more important than direct uptake
from either water or sediment.

Crayfish Study

The more detailed survey of mercury concentration in the crayfish *Orconectes virilis* showed that
mercury accumulates in some parts of the body more
than others with the highest levels recorded being
found in abdominal muscle. Similarly Johnels *et al.*[3]
found that the mercury concentration in the abdominal
muscle of a Swedish crayfish (*Astacus fluviatilis*)
was approximately five times as high as that in the
carapace and above twice that in the liver. Most of
the mercury concentrated by the crayfish in Clay
Lake is in the methylated form; there apparently
is a definite age-mercury concentration relationship
in this species, with the highest concentrations
being found in the larger and hence supposedly older
specimens. A complicating factor is that mercury
levels in the crayfish from this part of Clay Lake
appear to be decreasing. As indicated (Figure 21)
levels in abdominal muscle of adult crayfish averaged
10 µg/g wet weight in specimens collected in 1970.
The mean value for a large collection of similar-
sized specimens collected in May of 1971 was 6.6
µg/g. As indicated in Figure 22 none of the values
(pooled samples of abdominal muscles from four
specimens) collected in June, July and August exceeded 6 µg/g. The corresponding average value for
specimens collected in June 1972 was 3.98 µg/g.
Consequently the relatively low concentrations of
mercury in the abdominal muscle of smaller crayfish
may be partly a reflection of improving conditions.
The smaller (and younger) specimens were 0+ and 1+
individuals who would have spent their entire lives
in an environment that was apparently receiving

substantially less mercury than in previous years.
In contrast the older specimens (2+ and 3+) would
have spent the earlier parts of their lives in an
environment that apparently was receiving a greater
load of mercury.

Implications of Lower Hg Levels

The evidence indicating that mercury levels are
dropping in these organisms in Clay Lake is certainly
not conclusive, but it has interesting implications,
and additional surveys will be conducted in 1972 and
1973 to establish whether or not this trend continues.
Experiments conducted in 1971[16] indicate that contaminated Clay Lake sediments in an otherwise clean
environment had only a very minor impact on the rate
of mercury uptake by crayfish. In contrast Clay
Lake water had a major impact on the rate of mercury
uptake, and crayfish fed ground fish from Clay Lake
exhibited an even more dramatic elevation in mercury
concentrations. These experiments provide some explanation as to why mercury levels in some freshwater
organisms could show some improvement even though
the amount of mercury in the sediments of Clay Lake
remains essentially unchanged. Nevertheless this
immense reservoir of mercury in Clay Lake probably
ensures that it will be decades, at least, before
mercury concentrations in crayfish approach the
levels found in those from tributaries unaffected
by the Dryden chlorine-alkali plant. The 1971
survey (Figure 23) indicates that mercury concentrations in crayfish from Clay Lake and vicinity exceed
levels found in crayfish collected above Dryden and
in the Eagle River by a factor of about 20 times.
The pattern certainly suggests that the major source
is in the immediate vicinity of Dryden where the
chlorine-alkali plant has been in operation.

Crayfish as Mercury Pollution Indicator

Present indications are that the crayfish
Orconectes virilis is a very useful indicator of
mercury pollution. This species clearly concentrates
mercury to very high levels and it is large enough
that a single specimen provides a tissue sample
suitable for analysis. In addition this abundant
and widely distributed species is easy to collect
and individuals are much less migratory than the
fish species usually used as indicators. Comparisons with the material reported on by Bligh[1] indicate

that mercury levels in crayfish abdominal muscle are
very similar to levels in the muscle of northern
pike (*Esox lucius*), one of the most widely used
biological indicators of mercury pollution.

In conclusion there is an aspect of mercury
pollution which should be considered when improvement measures are discussed. This is the near
impossibility of eliminating mercury from organisms
in a closed environment once the metal has entered
the food chain. It is recycled by predation and
scavenging, and can only be removed if organisms
are bodily removed or if it can be excreted and
perhaps sequestered by some nonliving component of
the environment. It is known that the excretion
of mercury by fish and crayfish may be immeasurably
slow.[14,16]

REFERENCES

1. Bligh, E. G. Fisheries Res. Board Canada. Manuscript
   Report Series No. 1088 (1970).
2. Wobeser, G., N. O. Nielson, R. H. Dunlop, and F. M.
   Atton. J. Fisheries Res. Board Canada *27*, 830-834 (1970).
3. Johnels, A. G., T. Westermark, W. Berg, P. I. Persson,
   and B. Sjöstrand. Oikos *18*, 323-333 (1967).
4. Hannaerz, L. Natl Nature Conservancy Board, Drottningholm,
   Sweden, Report No. 48 (1968).
5. Matida, Y., and H. Kumada. Bull. Freshwater Res. Lab.
   19(2), 72-93 (1969).
6. Zoltai, S. C. Forest sites of Regions 5S and 4S, Northwestern Ontario, vol. 1, Ontario Department of Lands and
   Forests Technical Series (1965).
7. German, M. J. "A Water Pollution Survey of the Wabigoon
   River," Ontario Water Resources Commission. M.S. Report
   (1969).
8. *Determination of Mercury by Atomic Absorption Spectro-
   photometric Method* (Midland, Michigan: Dow Chemical Co.,
   1970).
9. Armstrong, F. A. J., and J. F. Uthe. Atomic Absorption
   Newsl. *10*, 101-103 (1971).
10. Armstrong, F. A. J. J. Marine Biol. Assoc. U.K. *36*, 509-
    517 (1957).
11. Jackson, M. L. *Soil Chemical Analysis, Advanced Course*
    (University of Wisconsin: Department of Soil Science,
    1956).
12. Chau, Y. K., and S. Saitoh. Environ. Sci. Technol. *4*,
    839-841 (1970).
13. Goulden, P. D., and B. K. Afghan. Technical Bulletin
    No. 27 Inland Waters Branch, Department of Energy Mines
    and Resources Ottawa (1970).

14. Uthe, J. F., J. Solomon and B. Grift. J. Assoc. Offic. Anal. Chemists *55* (3), 583-589 (1972).
15. Jernelov, A. Limnol. Oceanog. *15*, 958-960 (1970).
16. Hamilton, A. L. Fisheries Res. Board Canada, Manuscript Report Series No. 1167 (1972).

6. INTERACTIONS AND CHEMOSTASIS IN AQUATIC CHEMICAL SYSTEMS: ROLE OF pH, pE, SOLUBILITY, AND COMPLEXATION

Francois Morel, Russell E. McDuff, and James J. Morgan.
W. M. Keck Laboratory of Environmental Engineering
Science, California Institute of Technology, Pasadena,
California

The speciation of metal ions and ligands in
natural waters is a subject of considerable interest
in aquatic ecology, chemical oceanography, water
quality management, and other related fields. The
biological availability of a metal such as copper,
iron or manganese is expected to depend upon the
fraction of the metal that is "free," as opposed to
that which is bound to a ligand in a solution complex, adsorbed on a solid surface, or existing as a
distinct precipitate. A number of experiments have
been reported in which the addition of organic
chelators to sea waters has improved phytoplankton
growth.[1,2] Investigations employing the rather
sensitive anodic stripping voltammetry technique[3-5]
provide evidence for the existence of a number of
metal ions (Zn, Cu, Pb, Cd) in forms other than as
the free ion in natural waters. An interesting
feature of the phytoplankton growth experiments
described by Barber[6] is the strong response of the
system to rather small additions of chelators or
metals.
    One would like to get some chemical insight
into the basis for the rather dramatic biological
response shown by the marine phytoplankton experiments. How stable are the metal-ligand interactions?
In order to understand the chemostatic properties of
aquatic systems it seems essential to characterize
different systems in terms of their dominant chemical
variables: major ions, oxidation-reduction status,

157

acid-base components, minor ions, and strong complexing components. A systematic approach--from the
abundant ions and their "simple" interactions with
the rest of the system to the less abundant ions
and their "complex" interactions with the rest of
the system--should prove valuable in defining the
relative importance of different variables in determining the stability of aquatic chemical systems.
     A systematic treatment of the status of metal
ions in aqueous solutions was presented by Stumm.[7]
His work emphasized the role of coordination chemistry
and developed the quantitative features of competition
between metal ion hydrolysis and metal ion coordination with such inorganic and organic ligands as
fluoride, phosphate, sulfate, oxalate, citrate, and
nocardamine. He discussed the idea of simple metal
ion buffer systems, consisting of a metal-ligand
solution at an appropriate pH. Basically, Stumm's
approach involved calculations for isolated metal-
ligand sub-sets within some larger system. It represents in part an extension of the methodology
developed earlier by Schwarzenbach[8] and Ringbom[9]
for treating portions of complicated solution systems
encountered in analytical chemistry. Some amplifi-
cations of Stumm's original treatment have been
presented by Stumm and Morgan.[10] More recently,
Morel and Morgan[11] have described a numerical method
for computing equilibria in aqueous systems and have
applied it to compute the equilibrium distribution
(solids and solution species) for a hypothetical
model system of 20 metals and 31 ligands. The system
involved over 700 complexes and over 80 possible
solids. The numerical method used has now been extended to treat redox reactions.[12] A further extension permits the computation of the interaction
intensities for the various components in a complex
system, as described later in this paper.[13]
     From the computation of chemical equilibrium
models it is possible to decide, for example, in
what complex form a given metal ion is likely to be
found in a body of water containing a certain set
of reacting ligands and competing metals. If we
now suppose that one of the conditions of the
equilibrium problem--say the analytical concentration
of another metal--is varied, the free concentration
and, perhaps, the speciation of the metal ion under
study will also vary. The problem is then to define
the interactions between the metal that we are
studying and the constituent that is varied. A
quantification of this interaction is a measure of
the instability of the system.

In general, all the constituents of a natural chemical system--abundant and trace metal ions, inorganic and organic ligands--are related to each other through complex formation and solid precipitation reactions. The variation of any constituent will give rise to variations in, at least, some of the other constituents. Families of interdependent constituents can then be defined for a given domain in the range of possible concentrations. A measure of the chemostatic properties of the system will be given by the magnitude of all the possible interactions in the system.

This paper will be organized in a didactic way from the simple to the complex. We shall first study what the typical possible interactions can be in simple systems. Results of systematic equilibrium computations representing exclusively the inorganic fraction of natural waters will then be presented. Finally, typical organics will be introduced in the model computations in order to demonstrate how they modify the preceding results.

GENERAL CONSIDERATIONS

In equilibrium systems containing several metals and several ligands, it is expected that the free concentration as well as the distribution of any metal or ligand shall depend on the total concentration of all the other constituents of the system. Metals and ligands combine to form complexes and solids each one of which is a node in the network of interactions. This interdependence of the different chemical constituents of a system can be quite strong as exemplified in the relatively simple example of Figure 24. Changing any one of the four analytical concentrations drastically affects all the free concentrations. This particular system has been chosen as an example because calcium and magnesium both form strong complexes with carbonate and sulfate and by choosing sufficiently high concentrations it is easy to make the system very interdependent. For other choices of concentrations, the interactions between the same constituents can become much weaker.

In order to study chemical interaction in equilibrium systems, we shall distinguish between different "orders of interaction": the zeroth order is the effect of changing the total concentration of a constituent on the free concentration of this same constituent; the first order interaction is the

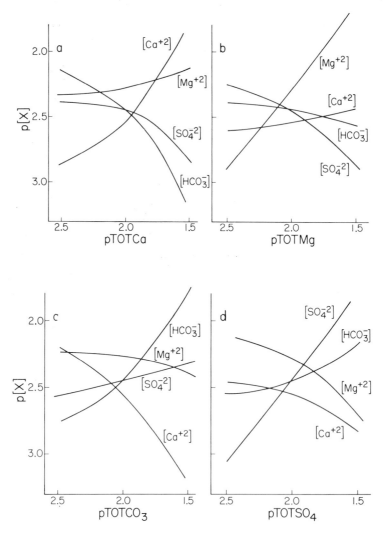

Figure 24.   *Interdependence of free concentrations in a*
Ca-Mg-CO₃-SO₄ system.   In each figure three of
the total concentrations are maintained at 10^{-2}M
and the fourth one is varied from $10^{-2.5}$ to $10^{-1.5}$M.
pH = 7, pε = 12.   (a) Ca, (b) Mg, (c) CO₃, and
(d) SO₄.   This figure illustrates that interactions
of even *order are positive (effects of metals on*
metals and ligands on ligands) while those of odd
order are negative (effect of metals on ligands
and vice versa).   The figure also illustrates the
attenuation of the effect as the order of inter-
action increases.

effect that changing the free concentration of a
metal (*e.g.*, by a 0th order effect) has on the free
concentration of a ligand or vice versa. Higher
orders of interactions are obtained by reverberation
of a series of first order effects through the system
of metals and ligands. To keep the terminology
simple, we shall consider the order of interaction
to be an additive property; the sum of a 0th and 1st
order interaction will also be called a 1st order
interaction (*e.g.*, TOTM* on [L]†); the sum of three
1st order interactions will be called a 3rd order
interaction; etc.

In Figure 25a, part 1 shows the free metal ion
concentration increasing proportionally to the total
concentration when the main form of the metal is a
complex with a superabundant ligand. Part 2 shows
the sudden increase in the free concentration of the
metal resulting from the exhaustion of the complexing
ligand. This is the typical titration effect and it
is the only case where there is amplification from
the relative change in total concentration to the
relative change in free concentration. It can be
easily demonstrated that the amplification effect
takes place only if the complex formed is strong,
i.e., if there is no need for a large excess of metal
to titrate the ligand. In this case the maximum
slope (d p[M] / d pTOTM)‡ is proportional to the
square root of the formation constant. If the complex is weak, so that titration takes place when the
metal is in large excess of the ligand, the maximum
slope of the curve is 1 (assuming a 1:1 complex).
Part 3 is similar to part 1 except that the effect
is directly on the free concentration, which is now
the main form, instead of a complex. The perfect
buffering capacity afforded by the precipitation of
a solid with a superabundant ligand is shown in part
4. Upon exhaustion of the ligand, another titration
effect (type 2) is obtained. If the pH is well buffered an effect of type 4 is also obtained upon
precipitation of an oxide or hydroxide.

Increasing the free concentration of a ligand
can directly affect a metal which forms an important
complex or solid with this ligand as shown in Figure
25b which describes first order interactions. At
very low free concentration of the ligand, part 1,

---

*TOTX is the total (analytical) concentration of the component X.

†[X] is the free concentration of the component X.

‡p [X] = $-\log_{10}$ [X].

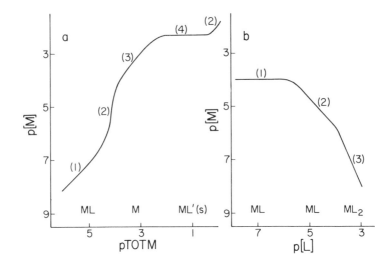

Figure 25.   *Typical zero- and first-order interactions in*
metal-ligand systems. Figure 25a illustrates
schematically the typical possible patterns of
variation in free concentration of a metal when
its total concentration is increased. Part 1:
Proportional increase of [M] (pM decrease) with
TOTM (pTOTM decrease) when the metal is mainly
present in a complex formed with a superabundant
ligand. Part 2: Strong increase of [M] when the
metal "titrates" the ligand (it is supposed that
the complex ML is "strong"). Part 3: p[M] vs
pTOTM when the metal is essentially in the free
form. Part 4: Perfect buffering effect when M
is present as a solid formed with a superabundant
ligand. Figure 25b illustrates variation of [M]
with the free concentration of a reacting ligand,
[L].   At low free ligand concentration (Part 1,
high pL) the free metal is independent of the
ligand, and pM is constant. Part 2 shows linear
decrease of [M] with [L] when ML is the main
form of the metal. Part 3 illustrates the
linear dependence of [M] on [L]^2 when the main
form of the metal is an ML$_2$ complex.

the metal is controlled in some other way and its
free concentration is constant. At some critical
concentration it starts being titrated. Parts 2 and
3 show how the free concentration of this metal decreases as the free concentration of the ligand
increases. The change in slope is due to a hypothetical change in predominance from a 1:1 complex
to a 1:2 complex. It is worthwhile to note that
for first order effects, the slope (d p[M] / d p[L]
is always between 0 and -1 (considering only 1:1
complexes). This can be demonstrated easily in
simple cases.

Higher order effects can be obtained if a series
of first order effects reverberates through the system of metals and ligands. This is illustrated in
Figure 24b for example: TOTMg has 0th order effect
essentially of type 3 on [Mg] although there is a
partial titration of SO_4 which in turn decreases by
1st order effects of types 1 and 2. A 2nd order
effect through the formation of $MgSO_4$ and dissociation of $CaSO_4$ is the main reason for the increase
of [Ca]. In turn the 3rd order effect on CO_3 ([Ca]·
[CO_3] = K_S) decreases [CO_3]. ([HCO_3] is plotted
instead of [CO_3] because it is the major species.)
At high concentration of Mg, the 1st order effect of
Mg on CO_3 through the formation of $MgHCO_3$ becomes
comparable with the 3rd order one.

Figure 24 illustrates two important features of
reverberation phenomena: (1) interactions of even
order are positive while those of odd order are
negative, for example, both free metals increase
and both free ligands decrease when any total metal
increases; (2) the absolute magnitude of the interaction decreases as the order of the interaction
increases, for example, in Figure 24c the effects
of total carbonate on [HCO_3^-], [Ca^{+2}], [SO_4^{-2}] and
[Mg^{+2}] decrease in that order as they are 0th, 1st,
2nd and 3rd order effects, respectively. The first
result follows directly from the constant sign of
the slopes in Figure 25a and 25b. The second result
is obviously true when first order effects have a
slope greater or equal to -1 which is the case when
the stoichiometry is favorable. It is of course
conceivable to obtain an amplification effect: one
mole of metal binds two moles of ligand which release
four moles of another metal, etc. However, even in
such an extreme case, the interactions will quickly
decrease as soon as the complex which causes the
interaction is a small part of the next constituent
to be affected.

Natural water systems which contain many metals and ligands are likely to demonstrate similar chemical interdependence characteristics as those described above. Some subsystems such as the $Ca-Mg-CO_3-SO_4$ system itself must be relatively interdependent, while others must be relatively loosely connected: surely the calcium concentration depends negligibly on the concentration of lead over the range encountered in nature. One of the purposes of this investigation is to define the interdependent subsystems and their domain of interdependence: over what pH and concentrations ranges are the constituents strongly connected? The emphasis will be put on the study of trace metal ions as their control is of prime importance to the management of natural waters. The problem is then to define the set of variables that will permit an accurate description of the speciation of a given metal ion. The isolation of such a subsystem should simplify the study of interactions and thus the study of homeostatic properties of natural waters from a chemical equilibrium point of view.

INORGANIC SYSTEMS

The organic component of fresh natural waters is poorly known analytically and its complexing properties are in the domain of hypothesis and investigation. In order to make our study as general as possible, we shall first study model systems devoid of organics and show later how the presence of organics is likely to modify the results.

Table 34 lists the main chemical constituents of fresh natural waters with their range of concentration and a typical value. In order to study all the possible interactions that could affect the trace metal ions, systematic computations covering the whole range of concentrations for calcium, magnesium, carbonate, sulfur and chloride as well as pH from 5 to 9.5 were performed. In those computations the concentrations of the other metals and ligands were chosen at their average value for natural waters. More computations varying the concentration of one or several of those intermediary or trace constituents were performed when the results of the first series of computations suggested a set of conditions that could make such studies interesting. For simplicity not all metal ions and inorganic ligands were included in the model. For example, nitrate which essentially forms no important

Table 34

Approximate Ranges and Typical Values
for Some Metals and Ligands in Fresh Waters*

| Constituent | -log molar concentration | | |
	Lower Range	Upper Range	Typical
Ca	4.5	2	3
Mg	5	2.5	3.5
Na	4.5	2	3.5
Fe	6	4	5
Mn	9	4.5	5.5
Cu	7.5	5	6
Ba	8	6	7
Cd	9	5.5	6
Zn	9	6	7
Ni	8	5.5	6.5
Hg	11	7.5	9
Pb	9	6	7
Co	9	6	7.5
Ag	10	8.5	9
Al	9	4	5
CO_3	4	2	3
S	5	2.5	4.5
Cl	5.5	2.5	3.5
F	6	4	5.5
NH_3	6.5	5	5.5
PO_4	7	4.5	5
SiO_3	5	3	4

*Based on perusal of information in references 15, 16, and 17.

complex was omitted. Potassium whose coordination
chemistry is also rather uninteresting was not
included either.

Two different oxidation-reduction levels were
chosen for the computations. A typical oxidizing
condition representing well-mixed lakes during the
wintertime or the surface waters during the summer-
time was imposed by fixing the $p\varepsilon$* at 12. To model
the bottom of a lake during the summer stratification
a $p\varepsilon$ of -4 was chosen. Since the total concentrations used in the model correspond to samples of
surface water, keeping the same values for the
reducing conditions approximates the situation of
the bottom waters in the spring, at the end of the
winter mixing and at the beginning of the summer
stratification. As observed by Lee[14] an enrichment
of the bottom layers in those metals which have
formed insoluble oxides or hydroxides during the
winter is expected to take place during the summer.
We shall discuss the effect of an increase in iron
and manganese concentrations in the model of the
reduced system. A limited number of redox reactions
are considered in order to model the behavior of
real systems. For example the oxidation of the
ferrous or the sulfide ions was allowed but not that
of the ammonium or any organic ions. All the computations were made at fixed pH according to previously
described methods.[11-13]

Tables 35 and 36 show the results of typical
computations for a well-oxygenated model and a
reducing model at average concentrations. Table 37
lists the main forms of the trace metal ions found
for all the computed cases for pH's from 5 to 9.5
and various main constituent concentrations. The
species are listed in the order in which they are
found with increasing pH. The general pattern is
well known: the free ions tend to predominate at
low pH, the carbonate and then the oxide, hydroxide
or even silicate solids precipitate at higher pH's.
In the reducing environment insoluble sulfides
dominate the scene. The general result is that in
most cases the trace metal ions are either free or
controlled by solids with superabundant ligands.
Only four important soluble complexes were found
in the computations: copper carbonate, mercuric
chloride, mercuric sulfides and silver chloride.

---

*$p\varepsilon$ = $-\log_{10}[e]$ where $[e]$ is the effective activity of electrons
in the system.[10]

From the general considerations on interactions (Figure 25) we then expect the zeroth order inter-actions for the metals to demonstrate perfect buffer-ing in the case of solids or simple proportionality in the case of the free or complex forms. The first order effect of the ligands on the metals should be important for both solids and complexes.

The only cases which might give rise to titra-tion phenomena are the possible complexation of carbonate and sulfate by calcium and magnesium, and the precipitation of sulfide by iron. Aluminum could conceivably titrate silicate but the formation of aluminum silicates is certainly kinetically con-trolled in natural systems and we shall not inves-tigate this possible interaction.

In order to study interactions with trace metal ions we are mainly interested here in knowing how the free carbonate and sulfate concentrations depend on total calcium, total magnesium, total carbonate and total sulfate. Figures [26] and [27] show the variations of $[CO_3^{2-}]$ and $[SO_4^{2-}]$ respectively as a function of their total concentrations under all the possible extreme concentrations for the other main constituents. The results are what one would expect from general principles of interactions and are in accord with the results of Figure 24. The maximum free concentration is obtained for each ligand when the total metals are minimum (curves 7 and 8 in both Figures 26 and 27). For such conditions the total concentrations of the other ligand have no measurable effect since the second order effects mediated by complexation with nonabundant metals have to be negligible. Each free ligand is minimum when the metals are maximum and the other ligand minimum (curve 2 on both Figures 26 and 27). Here the second order effects become important and the in-fluence of carbonate on sulfate (Figure 27) or vice versa (Figure 26) is quite noticeable as seen by the gap between curves 1 and 2. The precipitation of calcium carbonate creates an interesting situation as seen in the middle part of Figure 26. In that region the free carbonate is exclusively a function of the free calcium concentration. As a result the effect of magnesium on carbonate becomes strictly a third order effect through $MgSO_4$, $CaSO_4$ and $CaCO_3(s)$: at low total sulfate concentration no third order effect can be mediated and free carbonate is inde-pendent of magnesium; at high total sulfate concen-tration the third order effect of magnesium on carbonate becomes quite large as seen by the gap between curves 1 and 3 in Figure 26.

Table 35

Results of an Equilibrium Computation for Oxidizing Conditions: pE = 12, pH = 7

	Total Conc	Free Conc	CO_3	S	Cl	F	NH_3	PO_4	SiO_3	OH
			3.00	4.50	3.50	5.50	5.50	5.00	4.00	--
Ca	3.00	3.02	6.41	4.60	3.50	5.52	7.60	11.84	12.95	7.00
Mg	3.50	3.51	4.81	5.32	--	7.43	10.82	7.56_s	--	8.52
Na	3.50	3.50	5.40	5.71	--	7.12	10.91	7.55	--	7.81
Fe*	5.00	16.60	8.61	7.30	--	--	--	--	--	--
		15.40	--	17.20	17.90	16.18	21.60	15.44	13.45	7.40_s
Mn	5.50	10.00	11.41	12.30	12.30	--	16.80	12.94	--	13.20_s
Cu	6.00	6.70	6.40_s	9.00	9.50	10.81	8.40	9.24	--	7.69
Ba	7.00	7.00	--	--	--	--	--	--	--	13.00
Cd	6.00	7.19	10.50_s	9.50	8.09	11.61	12.20	15.63	--	9.29
Zn	7.00	7.01	--	9.31	9.21	11.12	12.31	10.45	--	9.61
Ni	6.50	6.51	--	8.81	9.61	10.83	11.31	--	--	8.31
Hg	9.00	16.20	--	18.41	9.00	20.12	13.88	--	--	25.70

Pb	7.00	7.69	--s	9.70	9.39	--	--	--	--	*7.89*
Co†	7.50	7.52 *25.12*	--	9.62	10.62	25.52	13.12	--	--	*9.02*
Ag	9.00	9.22	--	12.72	9.41	--	13.42	--	--	*13.92*
Al	5.00	13.45	--	15.26	--	11.63	--	--	--s	*9.25*
H	--	7.00	3.01	9.30	--	9.32	5.50	6.09	4.05	--

Numbers are negative logarithms of molar concentrations in solution. The presence of a solid is indicated by s. A blank signifies the absence of a computable species. Italic numbers indicate species or solids that amount to more than 1% of the total metal.

*The first number corresponds to Fe^{+3}, the second to Fe^{+2}.
†The first number corresponds to Co^{+2}, the second to Co^{+3}.

Table 36

Results of an Equilibrium Computation for Reducing Conditions: $p\epsilon = -4$, $pH = 7$

	Total Conc → / Free Conc ‡	CO_3	S	Cl	F	NH_3	PO_4	SiO_3	OH
		3.00	4.50	3.50	5.50	5.50	5.00	4.00	--
		6.41	7.23 / 11.23	3.50	5.52	7.60	11.84	12.95	7.00
Ca	3.00 / 3.01	4.81	7.95	--	7.43	10.82	7.56_s	--	8.51
Mg	3.50 / 3.51	5.40	8.34	--	7.12	10.91	7.55	--	7.81
Na	3.50 / 3.50	8.61	9.93	--	--	--	--	--	--
Fe*	5.00 / 25.37 / 8.17	--	13.10_s	10.67	24.94	14.37	24.21	22.21	9.77
Mn	5.50 / 5.52	6.93	10.45	7.82	--	12.32	8.46	--	8.72
Cu	6.00 / 27.07	26.77	32.00_s	29.87	31.18	28.77	29.61	--	28.07
Ba	7.00 / 7.00	--	--	--	--	--	--	--	13.00
Cd	6.00 / 14.87	18.17	19.80_s	15.76	19.28	19.87	23.31	--	16.97
Zn	7.00 / 15.07	--	20.00_s	17.27	19.18	20.37	18.51	--	17.67
Ni	6.50 / 11.67	--	16.60_s	14.77	15.98	16.47	--	--	13.47
Hg	9.00 / 42.47	--	10.62_s	35.27	46.38	40.14	--	--	--

Pb	7.00	19.47	--	24.10$_s$	21.17	--	--	--	--	19.67
Co$^+$	7.50	13.07 / 46.67	--	17.80$_s$	16.17	47.07	18.67	--	--	14.57
Ag	9.00	20.13	--	13.97$_s$	20.32	--	24.34	--	--	24.83
Al	5.00	13.45	--	17.89	--	11.63	--	--	--$_s$	9.25
H	--	7.00	3.01	4.72	--	9.32	5.50	6.09	4.05	--

Numbers are negative logarithms of molar concentrations in solution. The presence of a solid is indicated by s. A blank signifies the absence of a computable species. Italic numbers indicate species or solids that amount to more than 1% of the total metal.

*The first number corresponds to Fe^{+3}, the second to Fe^{+2}.
$^+$The first number corresponds to Co^{+2}, the second to Co^{+3}.

Table 37

Predominant Trace Metal Species Under All Conditions
of Computation in the Models

	Species Accounting for more than 90%	Species Accounting for a few per cent
Fe	$Fe(OH)_2^+$, $FePO_4(s)$, $Fe(OH)_3(s)$ $FeCO_3(s)$, $FeS(s)$, $FeSiO_3(s)$	
Mn	Mn^{++}, $MnCO_3(s)$, $MnO_2(s)$ $MnS(s)$	$MnHCO_3^+$, $MnSO_4$, $MnCl^+$
Cu	Cu^{++}, $Cu_2CO_3(OH)_2$ $CuCO_3$, $Cu(OH)_2(s)$ $CuS(s)$	$CuSO_4$
Ba	Ba^{++}, $BaSO_4(s)$	
Cd	Cd^{++}, $CdCO_3(s)$, $Cd(OH)_2(s)$ $CdS(s)$	$CdSO_4$, $CdCl^+$
Zn	Zn^{++}, $ZnCO_3(s)$, $ZnSiO_3(s)$ $ZnS(s)$	$ZnSO_4$, $ZnCl^+$
Ni	Ni^{++}, $Ni(OH)_2(s)$ $NiS(s)$	$NiSO_4$
Hg	$HgCl_2$, $Hg(OH)_2(s)$ $HgS(s)$, $Hg(liq)$, HgS_2^{2-}, $Hg(SH)_2$	
Pb	Pb^{++}, $PbCO_3(s)$, $PbO_2(s)$ $PbS(s)$	$PbSO_4$, $PbCl^+$
Co	Co^{++}, $CoCO_3(s)$, $Co(OH)_3(s)$ $CoS(s)$	$CoSO_4$, $CoCl^+$
Ag	Ag^+, $AgCl$ $Ag_2S(s)$	
Al	$Al_2Si_2O_7(s)$, $Al(OH)_3(s)$	AlF^{+2}, AlF_2^+

The species of each metal are listed in the order they are
found with increasing pH.

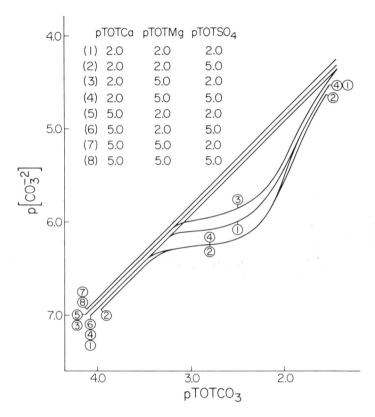

Figure 26. *Variation of free carbonate concentration, $[CO_3^{2-}]$, with total carbonate concentration, TOT CO_3, at various total concentrations of Ca, Mg, and SO_4. pH = 7.5, pε = 12. Eight cases, representing extremes of concentrations in natural systems, are shown on the figures. The flattening of the curves for cases 1, 2, 3, and 4 corresponds to the precipitation of $CaCO_3$ (compare part 4, Figure 25a). The linear variations of $p[CO_3]$ with pTOT CO_3 (cases 5, 6, 7, 8 over the entire range, cases 1, 2, 3, 4 in the low concentration range) illustrate typical zero-order effects (compare part 1, Figure 25a). The gaps between curves 2 and 8 indicate the __maximum__ indirect effects (1st, 2nd, 3rd order) of Ca, Mg, and SO_4 on $[CO_3^=]$.*

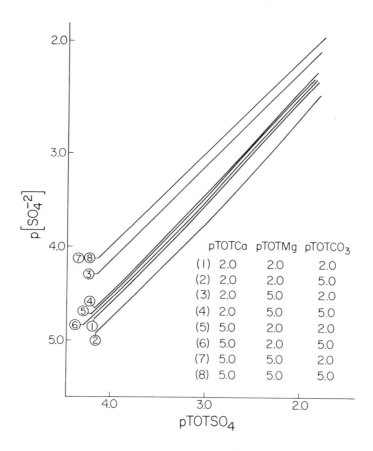

Figure 27.   *Variation of free sulfate concentration, [SO₄⁼],*
with total sulfate concentrations, TOTSO₄, at
various total concentrations of Ca, Mg, and CO₃.
pH = 7.5, pε = 12. Eight cases, representing
similar extremes of concentration as in Figure 26,
are shown. The lower six curves exhibit a <u>weak</u>
titration effect of Mg²⁺ by SO₄⁼ at concentrations
of 10⁻³M and greater. The gaps between the curves
are indicative of higher (1st, 2nd, 3rd) order
effects of Ca, Mg, and CO₃ on [SO₄]. The gap
between curves 2 and 7 is considerably greater
than a half-order of magnitude.

The combined *indirect* effects are more than half an order of magnitude for sulfate and one order of magnitude for carbonate in the region where calcium carbonate precipitates. Such effects are fairly large when compared with the range of two to three orders of magnitude obtained as a result of zero order effects. pH has essentially no influence on the curves of Figure 27. Carbonate is, of course, a very strong and well known function of the pH; the indirect effects are decreased at lower pH and very much increased at higher pH. For example, at pH 8.5 and $TOTCO_3 = 10^{-3}M$, the gap between curves 2 and 7 is three orders of magnitude, most of it being due to the first order effect of calcium. In summary, the free carbonate and sulfate concentrations depend strongly on the total concentrations of the four main constituents of fresh natural waters: calcium, magnesium, carbonate and sulfate. As a consequence calcium and magnesium will demonstrate fairly important high order effects on trace metal ions when the latter are controlled by carbonate or sulfate. For example, at medium pH copper, cadmium and lead are less likely to form carbonate solids in hard waters than in soft waters.

Because of the extreme insolubility of most of the sulfide solids, the possible titration of sulfide by iron at low pε is likely to be the most dramatic of the second order effects among metals. Figure 28 shows the results of computations similar to those of Table 36 where the total iron concentration is varied from $10^{-6}M$ to $10^{-3.5}M$. The ordinate shows the logarithm of the relative increase of the free concentration of the sulfide-forming metals. When the total iron is increased from $10^{-4.8}M$ to $10^{-4.2}M$, the free concentration of iron and of the related metals increase by almost four orders of magnitude (only two for silver, which forms Ag_2S). This is quite an amplification indeed and such a release of trace metal ions from sediments might actually take place when the bottom layers of lakes become enriched in iron during the summer stagnation.

For pH's higher than 7.2 at pε = -4 most of the free sulfur is in the sulfate form. It is thus possible to obtain second order effects on sulfide-forming metals by titrating the sulfate. The most likely candidate to do so is magnesium. An example of such a titration is shown in Figure 36. Note that the effect is much smaller than the preceding one, only half an order of magnitude or so. This second order interaction supposes that the dissolution

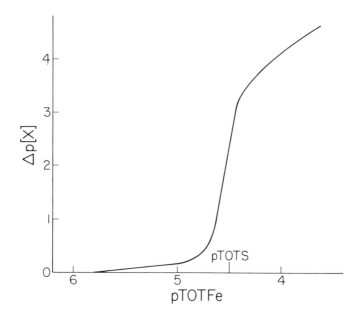

Figure 28.   *Titration of sulfide by iron in a 16-metal, 8-ligand inorganic system similar to that described in Table 36 under reducing conditions.* $p\varepsilon = -4$, *pH = 7. The ordinate represents the change in the logarithm of the free concentration of Fe^{2+}. Since sulfides precipitate, it is also the change in the log of the free concentration of sulfide (with a minus sign), Cu^{2+}, Cd^{2+}, Zn^{2+}, Ni^{2+}, Hg^{2+}, Pb^{2+}, Co^{2+}, and Ag^+ (with a factor of two, since $Ag_2S(s)$ is formed). The effect of the titration is quite dramatic where TOTFe $\simeq$ TOTS (i.e. $10^{-4.5}M$). At that point the slope $\partial p[X]/\partial pTOT$ Fe is approximately 20.*

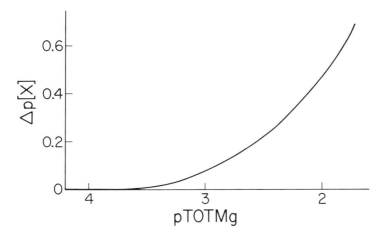

Figure 29.   *Titration of sulfate by magnesium in the inorganic system of Table 36, except that pH = 8. The ordinate represents the decrease in the free concentration of sulfate, $[SO_4^2]$. Since redox equilibrium is assumed, it also represents a decrease in $p[S^=]$, and thus the corresponding increase in the concentration of the sulfide-forming metals (Cu^{2+}, Cd^{2+}, Zn^{2+}, Ni^{2+}, Hg^{2+}, Pb^{2+}, Co^{2+}, Ag^+). Note that the effect is much weaker than that shown in Figure 28 (about half-order of magnitude).*

of sulfides will not be too slow and also postulates that the oxidation of sulfide to sulfate will be fast enough. If these conditions are met the phenomenon can be important in a given lake upon enrichment of the bottom layers in magnesium. Both of the preceding cases show that the metallic sulfides might be somewhat more soluble in lakes that are richer in iron and magnesium.

In summary the inorganic constituents of fresh natural waters are relatively independent of each other except for the obvious first order metal-ligand relationship. Only two examples of relatively important high order effects have been demonstrated: the titration of carbonate and sulfate by calcium and magnesium which can indirectly affect the carbonate- and sulfate-forming trace metal ions and the titration of sulfide by iron which can dramatically increase the free concentration of sulfide-forming metal ions at low pε.

   A way to measure the magnitude of the interactions between constituents of the system is given
by the study of "interaction intensities":

$$\delta_{X,Y} = \partial \ p[X] \ / \ \partial \ p\text{TOTY} \tag{1}$$

These numbers can be computed from the previously
referenced program with the addition of a simple
output routine.[13] It should be noted that contrary
to the commonly used pH buffering intensities these
interaction intensities are nondimensional numbers
and that the order of derivation is reversed. Large
interaction intensities, as defined here, correspond
to very strong dependence of [X] upon TOTY while a
large pH buffering intensity corresponds to a weak
dependence of the pH upon changes in acid or base
concentrations. Tables 38 and 39 show the interaction intensities of a selected group of metal ions
with respect to a selected group of metals and
ligands for both of the computations presented in
Tables 35 and 36 respectively. The results are in
perfect accord with the preceding discussion. The
free concentration of metals that are in the free
form, such as calcium and magnesium, depend almost
exclusively on their own total concentrations
($\delta_{M,M}$ = 1). The free concentration of metals present in a complex form with a superabundant ligand,
such as mercury ($HgCl_2$), depend both on their own
total concentration and on the ligand, and the
interaction intensities correspond to the stochiometric coefficients. Silver, at a high pε, gives
an example of a metal present partially in the free
form and partially in a complex form (AgCl). In
such a case it can be expected intuitively and
demonstrated algebraically on simple examples that
the interaction intensity $\delta_{M,L}$ is proportional to
the percentage of the total metal in the corresponding ML complex. Indeed $\delta_{Ag,Cl}$ = -0.4 corresponds to the proportion of the silver in the silver
chloride form. The free concentration of metals
entirely present in a solid form with a superabundant
ligand depend uniquely on the total concentration
of the ligand and the interaction intensity depends
upon the stochiometric coefficients. This is exemplified by copper, cadmium and lead which form
carbonates at high pε. Aluminum which is entirely
in the solid silicate form gives somewhat the same
result, but at the chosen concentrations the total
aluminum is one-tenth of the total silicate and the
latter is partially titrated so that the interaction
intensity $\delta_{Al,Si}$ = -1.1 is greater than 1 (in

Table 38

Table of Interaction Intensities ($\delta_{x,y} = \partial[X]/\partial TOTY$) for a Selected Set of Metal Ions
with Respect to a Chosen Set of Metals and Ligands

	TOTCa	TOTMg	TOTFe	TOTCu	TOTCd	TOTHg	TOTPb
Ca	*1.0*	4.1×10^{-4}	0	7.8×10^{-6}	1.6×10^{-5}	8.7×10^{-13}	1.6×10^{-6}
Mg	1.3×10^{-3}	*1.0*	0	6.3×10^{-6}	1.3×10^{-5}	7.0×10^{-13}	1.3×10^{-6}
Fe	0	0	0	0	0	0	0
Cu	7.8×10^{-3}	2.0×10^{-3}	0	2.5×10^{-4}	5.0×10^{-4}	2.8×10^{-11}	5.0×10^{-5}
Cd	1.6×10^{-2}	4.0×10^{-3}	0	5.0×10^{-4}	1.0×10^{-3}	5.5×10^{-11}	1.0×10^{-4}
Hg	8.4×10^{-7}	2.2×10^{-7}	0	2.7×10^{-8}	5.4×10^{-8}	*1.0*	5.4×10^{-9}
Pb	1.6×10^{-2}	4.0×10^{-3}	0	5.0×10^{-4}	1.0×10^{-3}	5.5×10^{-11}	1.0×10^{-4}
Ag	2.9×10^{-5}	1.2×10^{-5}	0	5.8×10^{-9}	1.2×10^{-8}	2.5×10^{-6}	1.2×10^{-9}
Al	5.4×10^{-10}	1.1×10^{-9}	0	1.1×10^{-14}	2.2×10^{-14}	6.7×10^{-21}	2.2×10^{-15}

	TOTAg	TOTAl	TOTCO$_3$	TOTS	TOTCl	TOTSiO$_3$
Ca	2.9×10^{-11}	5.4×10^{-12}	-1.6×10^{-2}	-4.9×10^{-3}	-1.4×10^{-7}	-5.4×10^{-11}
Mg	3.7×10^{-11}	3.4×10^{-11}	-1.3×10^{-2}	-6.1×10^{-3}	-1.1×10^{-7}	-3.4×10^{-10}
Fe	0	0		0	0	0
Cu	5.8×10^{-12}	1.1×10^{-13}	*-.50*	-5.0×10^{-5}	-4.4×10^{-6}	-1.1×10^{-12}
Cd	1.2×10^{-11}	2.2×10^{-13}	*-1.0*	-1.0×10^{-4}	-8.7×10^{-6}	-2.2×10^{-12}
Hg	2.5×10^{-5}	6.8×10^{-17}	-5.4×10^{-5}	-3.4×10^{-8}	*-2.0*	-6.8×10^{-16}
Pb	1.2×10^{-11}	2.2×10^{-13}	*-1.0*	-1.0×10^{-4}	-8.7×10^{-6}	-2.2×10^{-12}
Ag	*1.0×10^{-20}*	9.4×10^{-15}	-1.2×10^{-5}	-1.9×10^{-4}	*-.40*	-9.4×10^{-15}
Al	9.4×10^{-20}	*.11*	-2.2×10^{-11}	-1.5×10^{-11}	-1.0×10^{-15}	*-1.1*

The conditions are those of Table 35, $p\varepsilon = 12$, pH = 7. Italic numbers are those greater
than 0.1 in absolute value.

Table 39

Table of Interaction Intensities ($\delta_{x,y} = \partial[X]/\partial TOTY$) for a Selected Set of Metal Ions with Respect to a Chosen Set of Metals and Ligands

	TOTCa	TOTMg	TOTFe	TOTCu	TOTCd	TOTHg	TOTPb
Ca	*1.0*	1.1×10^{-4}	6.0×10^{-6}	6.0×10^{-7}	6.0×10^{-7}	6.0×10^{-10}	6.0×10^{-8}
Mg	3.5×10^{-4}	*1.0*	7.6×10^{-6}	7.6×10^{-7}	7.6×10^{-7}	7.6×10^{-10}	7.6×10^{-8}
Fe	6.0×10^{-4}	2.4×10^{-4}	*.52*	5.2×10^{-2}	5.2×10^{-2}	5.2×10^{-5}	5.2×10^{-3}
Cu	6.0×10^{-4}	2.4×10^{-4}	*.52*	5.2×10^{-2}	5.2×10^{-2}	5.2×10^{-5}	5.2×10^{-3}
Cd	6.0×10^{-4}	2.4×10^{-4}	*.52*	5.2×10^{-2}	5.2×10^{-2}	5.2×10^{-5}	5.2×10^{-3}
Hg	6.0×10^{-4}	2.4×10^{-4}	*.52*	5.2×10^{-2}	5.2×10^{-2}	5.2×10^{-5}	5.2×10^{-3}
Pb	6.0×10^{-4}	2.4×10^{-4}	*.52*	5.2×10^{-2}	5.2×10^{-2}	5.2×10^{-5}	5.2×10^{-3}
Ag	3.0×10^{-4}	1.2×10^{-4}	*.26*	2.6×10^{-2}	2.6×10^{-2}	2.6×10^{-5}	2.6×10^{-3}
Al	5.4×10^{-10}	1.1×10^{-9}	1.9×10^{-14}	1.9×10^{-15}	1.9×10^{-15}	1.9×10^{-18}	1.9×10^{-16}

	TOTAg	TOTAl	TOTCO$_3$	TOTS	TOTCl	TOTSiO$_3$
Ca	3.0×10^{-10}	5.4×10^{-12}	-1.6×10^{-2}	-1.9×10^{-5}	-3.7×10^{-8}	-5.4×10^{-11}
Mg	3.8×10^{-10}	3.4×10^{-11}	-1.3×10^{-2}	-2.4×10^{-5}	-7.3×10^{-9}	-3.4×10^{-10}
Fe	2.6×10^{-5}	1.9×10^{-14}	-1.2×10^{-5}	-1.7	-1.1×10^{-6}	-1.9×10^{-13}
Cu	2.6×10^{-5}	1.9×10^{-14}	-1.2×10^{-5}	-1.7	-1.1×10^{-6}	-1.9×10^{-13}
Cd	2.6×10^{-5}	1.9×10^{-14}	-1.2×10^{-5}	-1.7	-1.1×10^{-6}	-1.9×10^{-13}
Hg	2.6×10^{-5}	1.9×10^{-14}	-1.2×10^{-5}	-1.7	-1.1×10^{-6}	-1.9×10^{-13}
Pb	2.6×10^{-5}	1.9×10^{-14}	-1.2×10^{-5}	-1.7	-1.1×10^{-6}	-1.9×10^{-13}
Ag	1.3×10^{-5}	9.5×10^{-15}	-1.2×10^{-6}	$-.83$	-5.7×10^{-7}	-9.5×10^{-14}
Al	9.5×10^{-19}	*.11*	-6.2×10^{-11}	-6.0×10^{-14}	-2.8×10^{-17}	*-1.1*
			-2.2×10^{-11}			

The conditions are those of Table 36, $p\varepsilon = -4$, pH = 7. Italic numbers are those greater than 0.1 in absolute value.

absolute value) and $\delta_{Al,Al} = 0.11$ is appreciable. An even clearer example of a titration phenomenon is given by the metal ions forming insoluble sulfides at low $p\varepsilon$. The sum of the reacting metals is about forty percent of the total sulfur, most if it being iron. As a result the titration of sulfide is well under way so that $\delta_{S,S}$ is markedly greater than 1; $\delta_{S,S} = 1.7$ as it turns out. Since, for a 1:1 solid, $\partial p[M]/\partial p[S] = -1$, each of the $\delta_{M,S}$ is equal to -1.7 (except for $\delta_{Ag,S}$ which is -0.83 since Ag forms Ag_2S). All the other δ's in a given column are also equal since $\delta_{M,X} = \delta_{S,X} \cdot \partial p[M]/\partial p[S] = \delta_{S,X}$ ($= -\frac{1}{2}\delta_{S,X}$ for Ag).

The main feature of these tables of interaction intensities is that very few numbers are close to 1 (in absolute value), which demonstrates that the corresponding equilibrium systems are not strongly interdependent: most constituents have negligible dependence on most analytical concentrations.

Figure 30 a and b presents a case study of cadmium for the variations of speciation and inter-action intensities in function of pH in the model system at high $p\varepsilon$. The relations between the speciation and the interaction intensities are obvious at low pH's where $CdCO_3$ does not precipitate. At higher pH's the dependence of cadmium on carbonate becomes, of course, primordial: $\delta_{Cd,CO_3} = -1$. It can be shown that in the case of a solid ML, $\delta_{M,M} =$ TOTM/TOTL $\cdot \delta_{L,L}$. This is the reason why $\delta_{Cd,Cd}$ becomes constant and bears no relation to the importance of Cd^{++} as a species. In fact, at high pH, the only interaction intensity which is directly related to the speciation of cadmium is $\delta_{Cd,Cl}$ which decreases as the importance of cadmium chloride decreases. The influence of calcium is obviously mediated by the influence of carbonate as a second order effect. The effect of sulfate is mainly a third order effect [through $CaSO_4$, $CaHCO_3$ and $CdCO_3(S)$] which explains the parallelism between the $\delta_{Cd,S}$ and $\delta_{Cd,Ca}$ curves. Upon precipitation of calcium carbonate at high pH, the effect of titra-tion of calcium by carbonate becomes quite notice-able and both δ_{Cd,CO_3} and $\delta_{Cd,Ca}$ become greater than 1 (absolute value). The slight decrease in $|\delta_{Cd,CO_3}|$ at the very onset of precipitation of calcium carbonate is due to the weak buffering effect of phosphate on calcium because of the existence of apatite. This buffering effect is transmitted to carbonate as a second order interaction and to cadmium as a third order interaction.

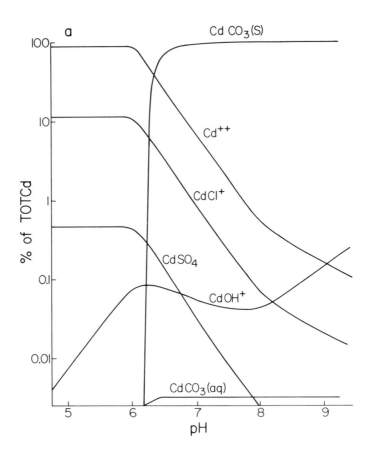

Figure 30.   *Case study for cadmium on the variation of*
            speciation and interaction intensities as a
            function of pH.   The equilibrium system is the
            same as that of Table 35 (16-metal, 8-ligand
            inorganic system, $p\epsilon = 12$).   At pH lower than
            6.2, where $CdCO_3(s)$ first precipitates, there
            is a direct correlation between the interaction
            intensities and the speciation of Cadmium.
            At higher pH (62 to 84), the interaction

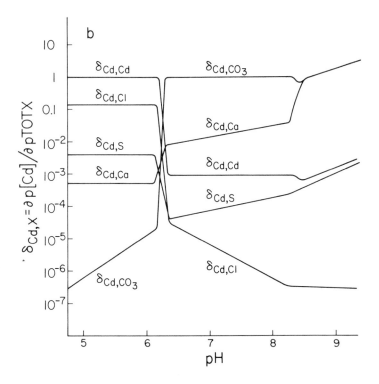

intensities are mainly correlated with higher
order effects, via CO_3 and Ca. At still higher
pH, $CaCO_3(s)$ precipitates and the titration of
Ca^{2+} by $CO_3^=$ manifests itself in interaction
intensities greater than unity in absolute value
for $\delta_{Cd,Ca}$ and δ_{Cd,CO_3}.

INFLUENCE OF ORGANIC LIGANDS

The previous studies of interactions using interaction intensities is, of course, applicable to any chemical equilibrium system regardless of the nature of its constituents, whether they are organic or inorganic. The limitation of the method comes from the use of partial derivatives: only infinitesimal changes can be predicted. One of the most interesting cases where large concentration changes should be considered is the addition of an organic ligand to an inorganic system, thus complexing and perhaps titrating a number of metal ions. The goal, then, is to define parameters that will allow us to predict, without resolving a new large chemical equilibrium problem, which metal ions will be titrated and when.

The study of the inorganic system has shown that interactions between the different metals are usually negligible. Thus we can study in general the complexation of a trace metal ion by an organic ligand by looking at first order effects only and without worrying about higher order indirect effects. We shall study later the exceptions to this rule, such as the high order interactions of the organic through calcium, magnesium or iron.

Let us compute the free concentration of a metal at the point of half titration as a function of the free concentration of the metal before the addition of any organic. If the metal is present mainly as a solid with a ligand in excess (or as a hydroxide or oxide solid), the free ligand will not change since it is controlled in some other way and the free metal concentration will stay constant up to the point of dissolution of the solid:

$$[M] = [M]_o \tag{2}$$

where $[M]_o$ is the free concentration of the metal in the absence of organics and $[M]$ its free concentration at the point of half titration. If the metal is mainly present as the free ion, the point of half titration is obviously given by:

$$[M] = \tfrac{1}{2}[M]_o \tag{3}$$

If the metal is present mainly as a complex with a ligand in excess, it is shown that, as in the last case, the point of half titration is given by:

$$[M] = \tfrac{1}{2}[M]_o \tag{4}$$

This result supposes that the inorganic complex
which contained most of the metal ion has a stoichiometric coefficient of 1 for the metal. This is
the case for all the important complexes that we
have encountered in the study of the inorganic
models. Otherwise the numerical factor would be
1/4 or 1/8, etc. instead of 1/2.

It can then be seen that upon titration by a
new ligand the free metal ion concentration changes
by a factor of at the most 1/2 between the beginning
and the mid-point of the titration. Such a change
is small compared with orders of magnitude effects
which result from changes in main parameters such
as pH. It is then possible, from the knowledge of
the system before the addition of organic, to make
qualitative and, at least, semiquantitative predictions on the effect of titration.

The complexation of a metal M by an organic Y
is described by:

$$X = M + Y \quad \text{(for a 1:1 complex)}$$

$$\therefore [X] = k \cdot [M] \cdot [Y] \tag{5}$$

At the point of half titration:

$$1/2 \text{ TOTM} = k \cdot \lambda \cdot [M]_o \cdot [Y] \tag{6}$$

where λ is 1 or 1/2 according to the preceding
discussion.

$$\therefore [Y]^{-1} = 2\lambda \cdot k \cdot \alpha \tag{7}$$

(2λ) is a factor between 1 and 2 and can usually
be ignored; k is the formation constant of X;
$\alpha = [M]_o/\text{TOTM}$ is the unbound fraction of the metal
M in the organic system. The right side of Equation 7 is independent of
the free concentration of the ligand [Y]. Thus as
TOTY increases and $[Y]^{-1}$ decreases, the metals will
be titrated in order of decreasing product $(2\lambda \cdot k \cdot \alpha)$.
This allows the prediction of the sequence in which
metal ions will be titrated upon addition of a new
ligand to the system just from the knowledge of k's
and α's. For example, observe the addition of
citrate to the well-oxygenated model of Table 35
The first six columns of Table 40 give the total
concentration of the metal ions, their free concentration in the inorganic model, the log of the
formation constants of the citrate complexes at
pH 7, the value 0.3 (= log 2) if the metal is controlled by a solid, the log of the product $(2\lambda k\alpha)$
and finally the expected order of titration of the

Table 40

Predicted Order of Titration of Metals by Citrate and Main Forms of Citrate in a Multi-metal, Multi-ligand System

	$pTOTM$	$p[M]_O$	$-pK$*	$log(2\lambda)$**	$log(2\lambda K\alpha)$†	order of titration	$-pK-p[M]_O$	order of controlling metals
Ca	3.00	3.02	14.8	0	14.78	7	11.78	3
Mg	3.50	3.51	14.1	0	14.09	8	10.59	-
Na	3.50	3.50	--	0	--	-	--	-
Fe III	5.00	16.60	27.6	0.3	16.30	4	11.00	-
Fe II	5.00	15.40	17.3	0.3	7.20	12	1.90	-
Mn	5.50	10.00	15.3	0.3	11.10	10	5.30	-
Cu	6.00	6.70	19.8	0.3	19.40	1	13.10	1
Ba	7.00	7.00	-2.3	0	-2.30	14	-9.30	-
Cd	6.00	7.19	15.8	0.3	14.91	6	7.42	-
Zn	7.00	7.01	13.2	0	13.19	9	6.19	-
Ni	6.50	6.51	16.4	0	16.39	3	9.69	-
Hg	9.00	16.20	--	0	--	-	--	-
Pb	7.00	7.69	16.5	0.3	16.11	5	8.81	-
Co II	7.50	7.52	0.1	0	0.08	13	-7.42	-
Co III	7.50	25.12	--	0	--	-	--	-
Ag	9.00	9.22	17.6	0	8.16‡	11	-0.84‡	-
Al	5.00	13.45	26.1	0.3	17.95	2	12.65	2
H	(∞)	7.00	17.0	0			10.00	4

Conditions of Table 35; $pTOTM$ and $p[M]_O$ are taken from Table 35.

*pK of formations represents a conditional (or effective) constant applicable at pH = 7.

**The value is 0.3 when there is a solid shown in Table 35.

†$log(2\lambda K\alpha) = log2\lambda - pK - p[M]_O + pTOTM$

‡Ag^+ forms a Ag_2 CIT complex, so that 2 $p[M]_O$ is used in the computation.

metal ions. Computation of the total equilibrium system at increasing concentrations of citrate confirms
the predicted result as shown in Figure 31a. It can
be noted that in order to know which metal ions will
actually be titrated by a strongly complexing organic
ligand, it is sufficient to count the metal ions in the
order previously obtained until TOTY$\leq$TOTM. This means
that trace metal ions listed after the major ions (Ca, Mg
maybe Fe) will not usually be titrated in natural systems.

The unbound fraction is thus a convenient parameter
to study the titration of metal ions by a given ligand
in a multimetal multiligand system. In effect, α is a
measure of the availability of the metal ion in that
system. Figure 32a and b show the unbound fractions
of the trace metal ions for the two typical systems
already presented for pH 7 in Tables 35 and 36. At
high pε there is a characteristic increase in pα's
(decrease in α's) as the pH increases. At low pε the
predominance of the insoluble sulfides produces a
reversal of slope at pH 7.2 where sulfate replaces
hydrogen sulfide as the main sulfur species. In
general the unbound fractions are larger at high pε
than at low pε.

If, rather than being interested in the sequence
of titration of trace metal ions by an organic, we
want to know what metal controls* the organic at a
given concentration, the relative influences of the
free and the total metal ion concentrations have to
be weighted in a different way than that provided
by the ratio α: a metal ion present in large amounts
can control the organic even if the corresponding
complex is only a small fraction of the total metal.

From Equations 2, 3, and 4 it is obvious that
up to the point of half titration (neglecting a
possible factor of 1/2)

$$[X] \simeq k [M]_o [Y] \tag{8}$$

and after titration:

$$[X] \simeq TOTM \tag{9}$$

Therefore, in order to find the metal ion that controls the organic at a given TOTY, it is sufficient
to find the metal ion for which $k[M]_o$ is maximum
among the set of metals present in excess of the
ligand (*i.e.*, TOTM $\geq$ TOTY). This method will of
course be correct only if the proton is included as
a metal, and since we suppose the pH to be fixed,
TOTH should be taken as infinite. Evidently, if
the product $k[M]_o$ is less than 1, the organic is

---

*A metal in excess controls the free concentration of a ligand
when the main form of the ligand is a metal-ligand complex.

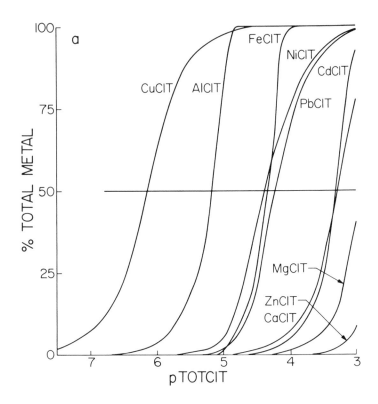

Figure 31.   *Titration of metal ions by* <u>added</u> *citrate in the 16-metal, 8-ligand inorganic system (pH = 7, pε = 12, same conditions as for Table 35. Figure 31 a shows percentage of the metal-citrate complexes with respect to the total metal. The sequence in which the metals are half-titrated (MCIT = 50% TOTM) is the same as that predicted in column 6 of Table 40.   Figure 31a*

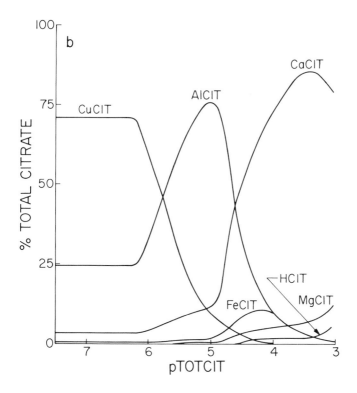

shows percentage of each citrate complex with respect to total citrate. Up to 10^{-6}M, citrate is controlled by Cu; from 10^{-6} to ca. 10^{-5}M, it is controlled by Al; from 10^{-4} to $\overline{10^{-3}}$M, it is controlled by Ca. These successive controls correspond to the predictions of column 8 in Table 40. Note the noticeable Al CIT complex below 10^{-6}M and the appreciable FeCIT complex at ca. 10^{-4}M.

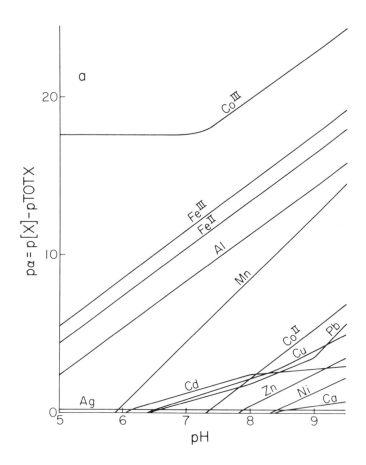

Figure 32.   *Unbound fraction (pα = log TOTM-log[M]) of the*
metal ions as a function of pH in the 16-metal,
8-ligand inorganic system, under oxidizing
conditions, pε = 12 (Figure 32a) and reducing
conditions, pε = -4 (Figure 32b). At high pε,
Mg and Na are essentially unbound at all pH
(pα≈0). At low pε, Mg, Na, Ca, Ba, and Mn

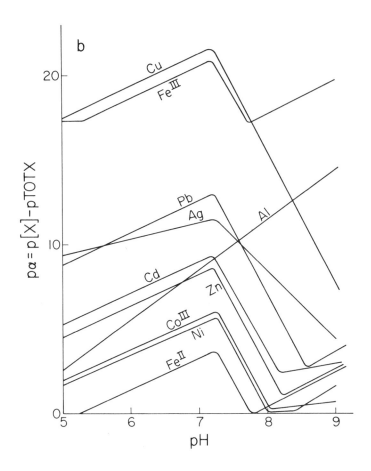

are essentially unbound at all pH (pα≈0), and
Hg is so completely bound (pα≈30) that it is
not shown on the graph. α is essentially a
measure of the "availability" of the metal for
further binding of the metal now in the multi-
metal, multi-ligand system.

essentially present in the free form. The results
of this treatment for the example already studied
are given in the last two columns of Table 40 and
are in accord with the computations of the total
system as seen by comparison with Figure 31b which
gives the speciation of citrate as a function of
total citrate concentration. It is interesting to
note that the product $k[M]_0$ for aluminum is barely
smaller than the one for copper and, as a result,
aluminum citrate is a noticeable part of the total
citrate at low concentration when copper is controlling citrate. A similar effect is seen for iron
which forms a noticeable citrate complex when calcium
starts controlling the citrate.

So far all the procedures outlined to predict the
influence of the introduction of ligands in natural
water systems are valid only if second order effects
between metals (third order effects for the new
ligands) are negligible. We have seen three cases
in which these effects are not negligible: the influence of calcium on other trace metal ions when
carbonates form; the influence of iron on other
solid sulfide forming metals at low $p\varepsilon$; and, under
both conditions, the influence of magnesium on the
same metals. It could be expected that, under the
right conditions, an added ligand could modify the
free concentration of a trace metal ion as a third
order effect through complexation with one of these
three major metals. Since third order interactions
are always negative, the effect would be to retard
the complexation of the trace metal ion, maybe
enough so that titration, which would have been
expected according to the preceding analysis, does
not actually take place.

More detailed explorations of this possible
phenomenon shows that it is in fact relatively im-
probable under the range of conditions encountered .
in natural waters. Calcium and magnesium have
indirect influence on other metal ions mainly when
they are present in large amounts, about 10^{-2}M.
Thus under such conditions in real systems, organic
compounds are unlikely to be present in sufficient
amounts to measurably alter the free concentrations
of these two metals. On the contrary, iron is present in natural waters over a concentration range
of 10^{-6}M to 10^{-4}M which is comparable to that expected for some abundant organics. We can modify
the system of Table 36 in order to demonstrate this
phenomenon. By choosing a total iron concentration
of 10^{-4}M which, according to Figure 28, is a point

of large second order interactions between the sulfide-forming metals, we should obtain a large third order effect upon titration of iron by an organic ligand, say citrate. In applying the previously described scheme to results of an equilibrium computation similar to that of Table 36 but with TOTFe = 10^{-4}M, it seems that nickel should be titrated readily after aluminum and iron. In fact, computations with increasing amounts of citrate show that nickel is never titrated up to TOTCit = $10^{-2.5}$M. The third order effect of the titration of iron by citrate decreases the free nickel by some four orders of magnitude and its unbound fraction becomes much too low to permit titration.

We have considered, so far, the addition of a new organic ligand* to a completely inorganic system. What happens to our analysis when the system already contains some organics? If the conditions of the previous analysis (no large high order interactions) are met, the very same schemes as those exemplified in Table 40 can be used to predict the effect of introducing a new ligand in the system. This will be the case only if all the ligands already present in the system, including the organics, are in great excess of the metals with which they form important complexes. In natural systems, organic ligands are usually present in concentrations comparable to those of the metals they bind. In consequence we can expect not only titration phenomena to take place but also high order effects to be important as was shown in the study of inorganic systems. "Retardation" of titration--similar to what happens to nickel upon addition of citrate under conditions of large iron concentration and low pε--is thus likely to be frequently observed in systems containing organics. Thus if the simplified treatment of Table 40 is applied to the addition of a new organic to a system already containing organics, it will relatively often be in error. One way to alleviate this difficulty would be to consider simultaneous addition of all the organics to the inorganic system. Using the results of the computations for the inorganic system, the controlling metal for each organic ligand is determined. This allows a good approximation of each free organic concentration. It is then easy to predict the result of

---

*Obviously the analysis does not depend upon the organic character of the ligand and is applicable to any complexing agent.

titration including competition among the organic
ligands.

Such a study can of course be quite time-
consuming, and solving the large new equilibrium
problem using the computer might be just as simple.
Table 40 presents the result of such a computation
where four organics (citrate, glycine, NTA, and
cysteine) have been added at 10^{-5}M total concentrations to the system of Table 35. Comparison between
Table 40 and Table 35 illustrates the importance
of organic ligands in keeping metals in solution:
none of the three carbonates found in the inorganic
system is allowed to precipitate in the new system.
As shown by the increase in italic numbers in the
table, the general result of the addition of organics
is to partition the total metals in a larger number
of relatively important complexes. (Italic numbers
represent species that are greater than one percent
of the total metals.)

In terms of interactions in the system, we expect the presence of organics to mediate more high
order effects. Table 42, which presents some of the
interaction intensities for the computation of Table
41 strikingly demonstrates this increase in the interdependence of the system. Comparison with Table 38
shows a great increase in the number of relatively
large interaction intensities (those in italic are
greater than 0.1). Even for the relatively smaller
numbers it can be seen that many of them are now in
the one percent range while there were very few in
Table 38. For example, let us take cadmium which
is mainly present in an NTA complex form. First of
all the free cadmium concentration depends upon the
total NTA concentration; but, because NTA is largely
titrated by the metal ions present in the system,
the corresponding interaction intensity is greater
than one in absolute value: $\delta_{Cd, NTA} = -1.3$.
Furthermore the interaction intensities of cadmium
corresponding to those metals that bind a large part
of the total NTA, such as calcium or copper, are
fairly large ($\delta_{Cd,Ca} = 0.68$; $\delta_{Cd,Cu} = 0.13$), demonstrating strong second order interactions. Other
noticeable (in the one percent range) second order
effects mediated by NTA on cadmium are demonstrated
for metals such as magnesium, zinc or lead. Finally
a relatively large third order effect--through calcium
and NTA--is obtained for carbonate. In other words,
the size of the subsystem that allows a complete
analysis of the speciation and interactions of cadmium is much larger in the organic than in the

inorganic system: Cd, CO_3 and, to a lesser extent, Ca for the inorganic system of Table 35; Cd, NTA, Ca, Cu and, to a lesser extent, Mg, Zn, Pb, and CO_3, for the organic system of Table 41. Furthermore the dependence upon the main complexing ligand has been increased because of a titration phenomenon (δ_{Cd,CO_3} = -1.0 in Table 38, while $\delta_{Cd,NTA}$ = -1.3 in Table 42) and the buffering effect afforded by the precipitation of a solid, typical of inorganic systems, is now nonexistent ($\delta_{Cd,Cd}$ = 10^{-3} in Table 38 while $\delta_{Cd,Cd}$ = -1.1 in Table 42). Behavior similar to that of cadmium is demonstrated in Table 42 for copper and lead.

In short, many pathways of interdependence among the constituents of the chemical equilibrium system are mediated by the presence of organics and the magnitude of interactions are often larger than in the inorganic system.

CONCLUSION

By systematic studies of chemical equilibrium models for natural aqueous systems, we have found that the speciation of the trace metal ions is usually simple: free ions, solid carbonates, sulfides or hydroxides, chloride or sulfate complexes, etc. Because the complexing or precipitating ligands are usually in large excess of the metals, they mediate few important interactions (competition) among those metals. A most notable exception to this rule is observed under reducing conditions when a sufficiently large iron concentration precipitates most of the sulfide and keeps other trace metal ions from precipitating.

For infinitesimal changes in analytical concentrations, interactions among constituents of a chemical equilibrium system can be quantified by the use of "interaction intensities": $\delta_{X,Y}$ = $\partial p[X]/\partial pTOTY$. In the case of large concentration changes such as the addition of a new ligand, we have shown simple ways of predicting, in general, the new equilibrium state. For such predictions the important parameters appear to be the free concentrations of the constituents and their unbound ratios: α_M = [M]/TOTM. This last parameter is, in effect, a measure of the availability of the metal M for complexation by an added ligand.

It has been noted by many authors[10] that organic ligands can play a major role in natural waters by complexing trace metal ions and keeping them in

Table 41

Results of an Equilibrium Computation for Oxidizing Conditions (pε = 12, pH = 7) for a System Containing Four Organic Ligands

	Total Conc →	Free Conc →	CO_3	S	Cl	F	NH_3	PO_4	SiO_2	CIT	GLY	NTA	$CYST$	OH
Ca	3.00	3.02	3.00	4.50	3.50	5.50	5.50	5.00	4.00	5.00	5.00	5.00	5.00	--
Mg	3.50	3.51	6.41	4.60	3.50	5.52	7.60	11.84	12.91	17.68	8.10	10.06	11.07	7.00
Na	3.50	3.50	4.81	5.32	--	7.43	10.82	7.55_s	--	5.90	9.92	5.28	--	8.52
Fe*	5.00	16.60 / 15.40	--	16.90	17.80	16.18	21.60	15.44	13.41	6.68	14.00	5.73	5.55	7.40_s
Mn	5.50	10.00	11.41	12.30	12.30	--	16.80	12.94	--	12.38	15.50	11.26	16.37	13.20_s
Cu	6.00	10.05	9.75	12.35	12.85	14.16	11.75	12.58	--	7.93	9.27	6.01	--	11.05
Ba	7.00	7.00	--	--	--	--	--	--	--	26.98	14.10	10.96	--	13.00
Cd	6.00	7.46	10.76	9.76	8.36	11.87	12.46	15.89	--	9.34	10.56	6.02	--	9.56
Zn	7.00	8.85	--	11.16	11.05	12.97	14.16	12.29	--	13.34	11.01	7.01	8.92	11.45
Ni	6.50	9.14	--	11.45	12.24	13.46	13.94	--	--	10.42	10.84	6.50	11.39	10.94
Hg	9.00	31.95	--	34.16	24.75	35.87	29.63	--	--	--	27.63	26.50	9.00	57.21

Pb	7.00	10.20	--	12.21	11.90	--	--	--	--	11.31	12.60	7.06	*7.88*	10.40
Co+	7.50	9.35 / 26.95	--	11.45	12.45	27.36	14.95	--	--	26.93	12.15	7.51	10.22	10.85
Ag	9.00	9.22	--	12.72	9.41	--	13.42	--	--	18.52	13.62	--	--	13.92
Al	5.00	13.49	--	15.30	--	11.67	--	--	--$_s$	*5.08*	--	7.94	--	9.29
H	--	7.00	3.01	9.30	--	9.32	5.50	6.08	4.01	7.69	5.00	6.66	5.37	--

CIT = citrate, GLY = glycine, NTA = nitrilotriacetate, CYST = cysteine. Numbers are negative logarithms of molar concentrations in solution. The presence of a solid is indicated by s. A blank signifies the absence of a computable species. Italic numbers indicate species or solids that amount to more than 1% of the total metal.

*The first number corresponds to Fe^{+3}, the second to Fe^{+2}.
†The first number corresponds to Co^{+2}, the second to Co^{+3}.

Table 42

Table of Interaction Intensities for a Selected Set of Metal Ions
with Respect to a Chosen Set of Metals and Ligands

	TOTCa	TOTMg	TOTCu	TOTCd	TOTZn	TOTHg	TOTPb
Ca	*1.0*	5.4×10^{-3}	7.0×10^{-4}	6.8×10^{-4}	6.9×10^{-5}	1.2×10^{-8}	6.1×10^{-5}
Mg	1.7×10^{-3}	*1.0*	7.2×10^{-5}	6.9×10^{-5}	7.0×10^{-6}	1.2×10^{-10}	6.3×10^{-6}
Fe	0	0	0	0	0	0	0
Cu	*.7*	2.3×10^{-2}	*1.1*	*.13*	1.3×10^{-2}	2.1×10^{-7}	1.1×10^{-2}
Cd	*$.68\times10^{-3}$*	2.2×10^{-2}	*.13*	*1.1*	1.2×10^{-2}	2.1×10^{-7}	1.1×10^{-2}
Hg	1.1×10^{-3}	3.7×10^{-5}	2.1×10^{-3}	2.1×10^{-4}	1.7×10^{-4}	*1.0*	1.7×10^{-3}
Pb	*$.61\times10^{-5}$*	2.0×10^{-2}	*$.11\times10^{-7}$*	*$.11\times10^{-6}$*	1.1×10^{-2}	*1.7×10^{-5}*	*1.0*
Ag	3.3×10^{-5}	1.2×10^{-5}	7.2×10^{-7}	6.3×10^{-6}	8.2×10^{-8}	1.2×10^{-12}	6.5×10^{-8}
Al	1.1×10^{-2}	7.0×10^{-4}	1.4×10^{-4}	3.9×10^{-5}	3.5×10^{-6}	6.0×10^{-11}	3.2×10^{-6}

	TOTAl	TOTCO$_3$	TOTS	TOTCl	TOTNTA	TOTCYST
Ca	1.1×10^{-4}	-1.6×10^{-2}	-4.8×10^{-3}	-3.0×10^{-6}	-7.1×10^{-3}	-5.8×10^{-6}
Mg	2.2×10^{-5}	-1.3×10^{-2}	-6.1×10^{-3}	-3.1×10^{-7}	-7.2×10^{-4}	-5.9×10^{-7}
Fe	0	0	0	0	0	0
Cu	1.4×10^{-3}	-1.1×10^{-2}	-3.5×10^{-3}	-5.6×10^{-4}	*-1.3*	-1.1×10^{-3}
Cd	3.9×10^{-4}	-1.1×10^{-2}	-3.6×10^{-6}	-5.0×10^{-3}	*-1.3*	-1.0×10^{-3}
Hg	6.0×10^{-7}	-1.8×10^{-5}	-5.8×10^{-6}	-9.5×10^{-7}	-2.2×10^{-3}	*-1.3*
Pb	3.2×10^{-4}	-9.7×10^{-3}	-3.1×10^{-4}	-5.0×10^{-4}	*-1.1*	-8.6×10^{-2}
Ag	5.7×10^{-9}	-6.6×10^{-7}	-1.9×10^{-4}	*-.4*	-7.3×10^{-6}	-6.2×10^{-9}
Al	*.1*	-1.8×10^{-4}	-5.7×10^{-5}	-1.7×10^{-7}	-3.6×10^{-4}	-3.0×10^{-7}

The conditions are those of Table 41 . Italic numbers are those greater than 0.1 in
absolute value.

solution. We have demonstrated that, in addition to this important effect on the speciation of metal ions, organic ligands are likely to mediate large interactions among those metal ions. For example, the presence of NTA couples the free concentrations of copper and cadmium (Tables 41 and 42). In order to predict the global effects of a given change in a natural chemical system it is thus essential to know the organic content of the system.

In ecological studies, general statements are often made upon the relations between complexity and stability, diversity and homeostasis, etc. In the far simpler chemical equilibrium systems we have dealt with, it is clear that no such general statement can be made. To be meaningful, the quantification of the concept of chemostasis, for a given chemical system, has to be the complete matrix of interaction intensities or equivalent quantities. To say that a system is more or less chemostable than another system is, thus, somewhat abusive. It is only in the case where we are interested in the effect of a given change on a given variable that the stability of two systems can be simply compared.

ACKNOWLEDGMENTS

This work was supported in part by the Environmental Quality Laboratory of Caltech through a grant from the National Science Foundation and in part by a grant from the Gulf Oil Corporation.

REFERENCES

1. Barber, R. T., and J. H. Ryther. J. Exp. Mar. Biol. Ecol. *3*, 191-199 (1969).
2. Siegel, A. "Metal-Organic Interactions in the Marine Environment" in *Organic Compounds in Aquatic Environments,* S. D. Faust and J. V. Hunter, Eds., (New York: Marcel Dekker, 1971).
3. Matson, W. R. "Trace Metals, Equilibrium and Kinetics of Trace Metal Complexes in Natural Media," Thesis, MIT (1968).
4. Fitzgerald, W. F. "A Study of Certain Trace Metals in Sea Water using Anodic Stripping Voltammetry," Thesis, MIT (January 1970).
5. Zirino, A., and M. L. Healy. Environ. Sci. Technol. *6*, 243-249 (1972).
6. Barber, R. T. "Organic Ligands and Phytoplankton Growth in Nutrient-Rich Seawater." Presented at American Chemical Society Meeting, Boston (April 1972) (to be published)

7. Stumm, W. "Metal Ions in Aqueous Solutions" in *Principles and Applications of Water Chemistry*, S. D. Faust and J. V. Hunter, Eds. (New York: Wiley, 1967).
8. Schwarzenbach, G. *Complexometric Titrations* (New York: Interscience, 1957).
9. Ringbom, A. *Complexation in Analytical Chemistry* (New York: Interscience, 1963).
10. Stumm, W., and J. J. Morgan. *Aquatic Chemistry* (New York: Wiley Interscience, 1970).
11. Morel, F., and J. J. Morgan. Environ. Sci. Technol. *6*, 58-67 (1972).
12. McDuff, R. E., and F. M. M. Morel. "REDEQL, a General Program for the Computation of Chemical Equilibrium in Aqueous Systems," Tech. Report EQ-72-01, W. M. Keck Laboratories, California Institute of Technology (1972).
13. Morel, F. M. M., and R. E. McDuff. "Computation of Interaction Intensities using the Chemical Equilibrium Program REDEQL." Tech. Report EQ-72-04, W. M. Keck Laboratories, California Institute of Technology (1972).
14. Delfino, J. J., and G. F. Lee. Water Research *5*, 1207-1217 (1971).
15. Livingstone, D. A. "Chemical Composition of Rivers and Lakes" in *Data of Geochemistry*, M. Fleischer, Ed., Geological Survey Professional Paper 440-G (1963).
16. Bowen, H. J. M. *Trace Elements in Biochemistry* (New York: Academic Press, 1966).
17. Turekian, K. K. "Rivers, Tributaries, and Estuaries" in *Impingement of Man on the Oceans*, D. L. Hood, Ed. (New York: Wiley-Interscience, 1971).

7. METAL-ORGANIC COMPLEXES IN NATURAL WATERS: CONTROL OF DISTRIBUTION BY THERMODYNAMIC, KINETIC AND PHYSICAL FACTORS

A. Lerman. Department of Geological Sciences, Northwestern University, Evanston, Illinois

C. W. Childs. Soil Bureau, Department of Scientific and Industrial Research, Lower Hutt, New Zealand.

INTRODUCTION

Life on earth is responsible for the occurrence of organic substances in natural waters. Strong association of some metal ions with organic substances in living systems makes it plausible that similar associations exist in natural waters in the form of metal-organic complexes in solution. Chlorophyll ($C_{55}H_{72-7}$ MgN_4O_{5-6}), an integral component of photosynthetic plants, is the most commonly identified metal-organic compound in natural waters. Among the numerous organic substances identified in fresh and ocean waters,[1-3] many form strong complexes with metal ions in solution.[4] An example is a-amino acids,[5] and the occurrence of amino acids in natural waters suggests that there are at least precursors of one type of metal-organic complexes. However, the lack of evidence on the occurrence of metal-organic complexes in natural waters has recently been reemphasized and the difficulties in analytical techniques aimed at direct identification of complexes in solution[3] (p. 283) have been attributed to this fact.

Concentrations of dissolved organic matter in natural waters, expressed as the amount of carbon per unit volume of water, are commonly in the range from 10^{-1} to 10 mg/l,[3] and typical ocean water values are 1.0 ± 0.5 mg/l or 1×10^{-3} moles C/liter.[1] Although on the molar basis the total concentrations of organic compounds and, hence, the concentrations

of potential metal-complexing agents, may be significantly higher than the concentrations of such
metals as Pb, Cu, Cd, Zn, Ni, and Co [1 x 10^{-10} to
5 x 10^{-8} moles/liter in ocean water[6]], the behavior
of metals in natural waters may be controlled to a
greater extent by any of the more abundant inorganic
ligands, such as SO_4^{2-}, HCO_3^-, CO_3^{2-}, OH^-. The difficulties in identification of metal-ligand complexes
in natural waters are well exemplified by the mode
of occurrence of ferric iron, Fe^{III}. In many rivers
and lakes, the concentrations of Fe^{III} (determined
on finely filtered samples of water) are appreciably
higher than the values at saturation with respect
to the ferric oxide and hydroxide mineral phases.
Possible explanations of this are the occurrence
of some iron-organic complexes in solution or the
presence of fine colloidal particles of ferric
hydroxide.

Natural water systems untouched by man are
usually regarded as existing in a stationary or
nearly-stationary state for periods ranging from
years to millenia, depending on the size and physical
characteristics of the system. The chemical composition of natural waters is the product of reactions
taking place between water and the inorganic and
organic matter near the surface of the earth. Introduction of man-made substances into natural
waters is a new geochemical process responsible for
the appearance of reactive chemical species in the
stable matrix of the natural waters that have been
in existence for a long period of time.

Concentrations of organic substances capable
of complexing metal ions in solution can be evaluated
and predicted when the chemical and physical characteristics of the water system are known. The
following three are the main classes of the chemical
and physical characteristics needed for even a very
preliminary assessment of metal-organic complexes
in a natural water system.

(1) thermodynamic properties of the metal-
organic complexes that are likely to exist in the
water of a given ionic composition; chemical exchange that may take place between complexed metals
in solution and metals held on the surfaces of
clay and biogenic particles; possible uptake of
metal-organic complexes by the biota and inorganic
phases from solution.

(2) mechanisms and rates of the reactions
responsible for the breakdown of organic ligands
in free and complexed forms in solution.

(3) physical characteristics of bodies of water into which organic substances are introduced, such as the residence time of water in the system, the presence or absence of density stratification of the water column, and mechanisms of dispersion within the body of water.

In this paper we examine the possible behavior of metal-organic complexes in some natural waters as affected by 1-3 given above. The information available on the nature and stability (the thermodynamics) of metal-organic complexes in aqueous solutions is considerably more complete than anything known on their kinetics and behavior in heterogeneous systems. Therefore we treat the behavior of metal-organic complexes in natural waters as bracketed by what we believe are reasonable lower and upper limits of the physical and chemical kinetic characteristics.

THERMODYNAMIC RELATIONSHIPS

Concentrations of chemical species in a solution at equilibrium are determined by the thermodynamics of the reactions involved. For the complex-formation reactions between metal ions (M_i) and organic ligand (L),

$$M_i + L \rightleftarrows M_iL, \tag{1}$$

the equilibrium constants are

$$K_i = \frac{(M_iL)}{(M_i)(L)}, \tag{2}$$

where parentheses denote concentrations, and charges of the species have been omitted. In Equation 2, the activity coefficients of the species are included in K_i. In natural waters, when the total concentrations of the metal-ions (M_i) and the organic ligand (L) are small, the equilibrium concentrations of the metal-organic complexes will usually depend to a large extent on equilibria such as (1) that involve only one metal ion and one ligand molecule. In general, concentrations of the species in solution are likely to be pH dependent, owing to the dissociation equilibria of the type

$$HL^{\nu-} \rightleftarrows H^+ + L^{(1+\nu)-}. \tag{3}$$

Many natural water systems are fairly well buffered
with respect to the pH by the carbonate and/or
silicate buffers, and reactions of the type given
by (1) may apply to systems of constant pH and
fixed ionic composition. The manner in which the
concentrations of metal-organic complexes (M_iL) in
waters containing different dissolved constituents
depend on the total ligand concentration (L_T) and
the stability constants of the individual complexes
(K_i) is discussed below, in order from an over-
simplified to a more realistic case.

Trace Concentration of
Organic Ligand

When an organic complexing agent is introduced
into a water system at concentration much lower than
the concentrations of any of the metal-ions it complexes, the organic ligand has only a very small
effect on the ionic composition of the solution.
The total concentration of a metal-ion ($M_{T,i}$) can
be expressed as the sum of concentrations of the
free (M_i) and complexed (M_iL) forms,

$$(M_{T,i}) = (M_i) + (M_iL). \tag{4}$$

When the organic ligand concentration (L) is very
low, (M_iL) is insignificant compared with (M_i) or
($M_{T,i}$). Implied in Equation 4 is that concentrations
of complexes with other ligands (*e.g.*, M_iCO_3, M_iSO_4)
are insignificant by comparison with the concentration
of M_iL. In this case, the total concentration of the
organic ligand (L_T), using Equation 2, is

$$(L_T) = (L) [1 + K_1(M_1) + \ldots + K_n(M_n)], \tag{5}$$

where (L) denotes the concentration of the free
ligand and the products $K_i(M_i)$ depend on concentrations of the complexed species, as given by Equation
2.
When free metal-ion concentrations in solution
have not been affected by introduction of a very low
concentration of an organic ligand (L_T), then the
concentrations of complexed species (M_iL) are simple
fractions of (L_T). From Equation 5,

$$(L_T) = (L) \left[1 + \sum_{i=1}^{n} K_i(M_i) \right], \quad \text{and} \quad (L) = f(L_T), \tag{6}$$

where f is a fraction having a constant value for the solution of a given ionic composition. From (6) and (2), concentrations of the complexed species are

$$(M_iL) = K_i(M_i)f(L_T). \tag{7}$$

Equation 7 may be used to compute the concentrations of M_iL species in the presence of trace concentrations of an organic complexing agent L, when equilibria of the type shown in (1) are the only significant ones which determine (L) and (M_i).

Higher Concentrations of Organic Ligand

When the total concentration of a complexing agent (L_T) is comparable to the concentration of any of the metal-ions it complexes, then the simple relationships (6) and (7) do not hold. In this case, using the mass balance of the metal-ions as previously,

$$(M_{T,i}) = (M_i) + (M_iL), \tag{8}$$

the concentrations of the free ligand (L) and n metal-L complexes (M_iL) are given by polynomials of degree $n+1$, the coefficients in which depend on $(M_{T,i})$, (L_T) and K_i, as in the following relationship:

$$(L_T) = (L)\left[1 + \sum_{i=1}^{n} \frac{K_i(M_{T,i})}{1 + K_i(L)}\right]. \tag{9}$$

The free ligand concentration, (L), can be computed from (9) when (L_T) and all the $(M_{T,i})$ and K_i are known. Similar relationships may be used to compute the concentrations of the metal-L complexes, (M_iL).

Computer Calculations of Concentrations of Metal-Organic Complexes in Solution

Since in most natural waters significant fractions of divalent and higher-valence metal-ions are complexed with carbonate, sulfate, phosphate, and hydroxide ligands, equations like 8 and 9 need to be expanded to include all the free and complexed species in solution. Such computations of the distribution of ionic species in a solution containing many metals and ligands can be done with the aid of digital computers using available programs.[7-9]

One of the convenient programs, program COMICS written in the FORTRAN language,[7] has been slightly modified to allow computation for solutions containing up to 20 metal-ions and 5 ligands.[10]   The essence of the COMICS method is the following.

A general form of a complex that can form in solution and contain all the metal-ions [M_i] and ligands [L_i] is

$$[M_1]_a [M_2]_b \cdots [M_n]_k [L_1]_r [L_2]_s \cdots [L_m]_t [OH]_v, \tag{10}$$

where subscripts a,...,v denote stoichiometric coefficients. Coefficients a,...,t may be either positive or zero, whereas v may be a positive integer (for hydrolyzed species), zero, or a negative integer (for protonated species).

The concentration of a generalized complex, C_j, characterized by an equilibrium constant K_j, may be written as

$$C_j = K_j (M_1)^a \cdots (M_n)^k (L_1)^r \cdots (L_m)^t (OH)^v. \tag{11}$$

The total concentration of a metal ion ($M_{T,i}$) is then

$$(M_{T,i}) = (M_i) + \sum_{j=1}^{n} P_{ij} C_j, \tag{12}$$

where P_{ij} is the number of M_i ions in the species j, and n is the total number of complexed species containing M_i in the system.

The computer reads from cards sets of the values of a,...,v for all species, the corresponding equilibrium constants K_j, the total concentrations of each metal ion ($M_{T,i}$) and each ligand ($L_{T,i}$), the pH values for which the distributions of the species are to be computed, and the initial estimates of the free metal-ion and ligand concentrations. [The latter is a convenience to reduce the number of iterations required; the program will work successfully even with the initial assumptions (M_i) = ($M_{T,i}$) and (L_i) = ($L_{T,i}$)]. The quantity on the right side of Equation 12 is computed for each kind of metal-ion and, similarly, for each ligand, to give the first-approximation quantities that can be designated ($M_{T,i}$)$_{calc.}$ and ($L_{T,i}$)$_{calc.}$. The initial estimates of the free metals and ligands, (M_i) and (L_i), are then replaced by (M_i) [($M_{T,i}$)/($M_{T,i}$)$_{calc.}$]$^{1/2}$ and (L_i) [($L_{T,i}$)/($L_{T,i}$)$_{calc.}$]$^{1/2}$. With these new values the calculations are repeated to obtain better estimates of (M_i) and (L_i). The process is repeated until all values of $(M_{T,i})_{calc.}$ and $(L_{T,i})_{calc.}$ differ from the corresponding values of $(M_{T,i})$ and $(L_{T,i})$ by less than a specified quantity (usually 0.001%). Within this accuracy, the final values of (M_i) and (L_i) satisfy all the equations for metal-ion and ligand concentrations. These final values are used to compute the equilibrium concentrations of all species, and the results are printed out in tabular form.

EQUILIBRIA IN A MODEL FRESH WATER

Organic compounds that can be used as substitutes for penta-sodium tri-polyphosphate in detergents are the type of compounds complexing metal ions in solution and attaining a widespread occurrence through human use. In this section we consider the effects of two such complexing agents, nitrilotriacetate and citrate, on the distribution of ionic species in a model fresh water. The two complexing agents are known to be feasible substitutes of phosphates in detergents, although the use of nitrilotriacetate has been effectively banned owing to detection of some possibly harmful biological effects.

The composition of a model fresh water shown in Table 43 is similar to that of Lake Ontario water,

Table 43

Composition of Model Fresh Water^a

Metal-ions	Conc.	Ligands	Conc.
Calcium(II)	1×10^{-3}	Carbonate	1×10^{-3}
Sodium(I)	5×10^{-4}	Chloride	7.5×10^{-4}
Magnesium(II)	2.5×10^{-4}	Sulfate	3×10^{-4}
Copper(II)	2×10^{-6}	Orthophosphate	1×10^{-6}
Manganese(II)	2×10^{-6}	Nitrilotriacetate	1×10^{-8} to
Strontium(II)	2×10^{-6}		2×10^{-4}
Iron(III)	2×10^{-6}	Citrate	1×10^{-8} to
Zinc(II)	1.5×10^{-6}		1×10^{-4}
Lead(II)	3×10^{-7}		
Barium(II)	1.5×10^{-7}		
Nickel(II)	1×10^{-7}		
Cobalt(II)	3×10^{-8}		
Cadmium(II)	2×10^{-8}		
Mercury(II)	1×10^{-9}		
Iron(II)	1×10^{-12}		

^{a}Total concentrations in moles/liter, in order of decreasing abundance.

estimated from the data in the Report to the International Joint Commission.[11]   As far as the major
components are concerned, the composition of Lake
Ontario water is similar to the mean composition of
the rivers of the world.[12]   Among the major ionic
constituents of fresh waters, chloride and potassium
do not form strong complexes with other cations and
anions, respectively, and these two ions were therefore omitted in the computation of the concentrations
of metal-organic complexes in the model fresh water.
The water of composition shown in Table 43 is assumed
to be aerobic, and essentially all the iron in it is
present in the ferric form, Fe^{III}. The reported
concentrations[11] of such trace metals as Cu, Cd, Zn,
Fe, Ni, and Co vary considerably, and the values in
Table 43 are means. As these metals form particularly
strong complexes with nitrilotriacetate and citrate,
and possibly with other organic complexing agents,
the distribution of their complexes in the model
fresh water should be regarded as of orientational
value, where the order of abundance of the complexes
depends on the concentration of the metal ion and
the stability constant of its complex in solution.
     Insofar as the chloride ion is a very weak
complexing agent for all the metal ions except
mercury(II), its effect as a competitor with
nitrilotriacetate and citrate would be negligible
for the conditions assumed in the calculations
(model fresh water, pH between 6 and 9). More important may be the omission of other organic complexing agents from the composition of the model
fresh water. For example, humates and amino acids
can bind metals reasonably strongly, but there is
very little quantitative information available on
the individual soluble organic components of fresh
waters. It should also be noted that equilibria
between metal-ions and long chain ligands, such as
humates, cannot be described readily in a quantitative way compatible with these equilibrium calculations for fresh water. However, nitrilotriacetate
and citrate, as well as other man-made substances,
are likely to be stronger complexing agents than
most naturally occurring ligands. This suggests
that, at least for the lake water containing
nitrilotriacetate and citrate, the results of
computations may reasonably approximate a real lake
water in which the concentrations of other soluble
organic components are not unusually high. With
regard to other inorganic ligands, as the calculations progressed it became clear that complex

formation by orthophosphate was negligible compared with complex formation by nitrilotriacetate and citrate, and that complex formation by carbonate was only of minor importance. (Consequently, any errors in the values of the equilibrium constants of metal-orthophosphate and metal-carbonate complexes, tabulated in Reference 10, would have no significant effect on the computed concentration of metal-organic complexes in the model fresh water.) The nitrilotriacetate and citrate equilibria are listed in Table 44. Other equilibria involving naturally occurring inorganic ligands used in the computation have been tabulated in Reference 10.

Table 44

Equilibrium Constants of Nitrilotriacetate and Citrate Complexes

Equilibrium	log K	Conditions[a]	Ref.

Nitrilotriacetate ($C_6H_6O_6N^{3-}$)

$$NTA^{3-} \equiv \left(N \begin{array}{l} CH_2COO^- \\ CH_2COO^- \\ CH_2COO^- \end{array} \right)$$

Equilibrium	log K	Conditions[a]	Ref.
$H^+ + NTA^{3-} \rightleftarrows HNTA^{2-}$	10.3	20°, 0.1 KCl	4
$2H^+ + NTA^{3-} \rightleftarrows H_2NTA^-$	13.3	20°, 0.1 KCl	4
$3H^+ + NTA^{3-} \rightleftarrows H_3NTA^0$	14.9	20°, 0.1 KCl	4
$Cu^{2+} + NTA^{3-} \rightleftarrows CuNTA^-$	13.0	20°, 0.1 KCl	4
$Cd^{2+} + NTA^{3-} \rightleftarrows CdNTA^-$	9.5	20°, 0.1 KCl	4
$Ca^{2+} + NTA^{3-} \rightleftarrows CaNTA^-$	6.4	20°, 0.1 KCl	4
$Ca^{2+} + 2NTA^{3-} \rightleftarrows Ca(NTA)_2^{4-}$	9.8	20°, 0	4
$Co^{2+} + NTA^{3-} \rightleftarrows CoNTA^-$	10.6	20°, 0.1 KCl	4
$Fe^{3+} + NTA^{3-} \rightleftarrows FeNTA^0$	15.9	20°, 0.1 KCl	4
$Fe^{3+} + NTA^{3-} + H_2O \rightleftarrows$ $Fe(OH)NTA^- + H^+$	10.9	25°, 1 KCl	4
$2Fe^{3+} + 2NTA^{3-} + 2H_2O \rightleftarrows$ $(Fe(OH)NTA)_2^{2-} + 2H^+$	25.8	25°, 1 KCl	4
$Fe^{3+} + NTA^{3-} + 2H_2O \rightleftarrows$ $Fe(OH)_2NTA^{2-} + 2H^+$	3.1	20°, 0.1 KCl	4
$Pb^{2+} + NTA^{3-} \rightleftarrows PbNTA^-$	11.8	20°, 0.1 KCl	4

Table 44, continued

Equilibrium	log K	Conditions[a]	Ref.
$Ni^{2+} + NTA^{3-} \rightleftarrows NiNTA^-$	11.3	20°, 0.1 KCl	4
$Zn^{2+} + NTA^{3-} \rightleftarrows ZnNTA^-$	10.4	20°, 0.1 KCl	4
$Zn^{2+} + 2NTA^{3-} \rightleftarrows Zn(NTA)_2^{4-}$	13.4	20°, 0.1 KCl	4
$Mn^{2+} + NTA^{3-} \rightleftarrows MnNTA^-$	7.4	20°, 0.1 KCl	4
$Ba^{2+} + NTA^{3-} \rightleftarrows BaNTA^-$	4.8	20°, 0.1 KCl	4
$Fe^{2+} + NTA^{3-} \rightleftarrows FeNTA^-$	8.8	20°, 0.1 KCl	4
$Mg^{2+} + NTA^{3-} \rightleftarrows MgNTA^-$	5.4	20°, 0.1 KCl	4
$Mg^{2+} + 2NTA^{3-} \rightleftarrows Mg(NTA)_2^{4-}$	8.6	20°, 0.1 KCl	4
$Na^+ + NTA^{3-} \rightleftarrows NaNTA^{2-}$	2.2	20°, 0	4
$Sr^{2+} + NTA^{3-} \rightleftarrows SrNTA^-$	5.0	20°, 0.1 KCl	4

Citrate $(C_6H_5O_7^{3-})$

$$L^{3-} \equiv \left(HO - \begin{matrix} CH_2COO^- \\ C-COO^- \\ CH_2COO^- \end{matrix} \right)$$

$3H^+ + L^{3-} = H_3L$	12.9	20°, 0.1 NaClO_4	18
$2H^+ + L^{3-} = H_2L^-$	10.0	20°, 0.1 NaClO_4	18
$H^+ + L^{3-} = HL^{2-}$	5.7	20°, 0.1 NaClO_4	18
$Cu^{2+} + L^{3-} \rightleftarrows CuL^-$	5.9	20°, 0.1 NaClO_4	18
$Cu^{2+} + 2L^{3-} \rightleftarrows CuL_2^{4-}$	8.4	20°, 4 NaClO_4	4
$2Cu^{2+} + 2L^{3-} \rightleftarrows Cu_2L_2^{2-}$	13.2	25°, 1 KNO_3	19
$2Cu^{2+} + L^{3-} \rightleftarrows Cu_2L^+$	8.1	10°, 0.1 NaClO_4	18
$Cu^{2+} + L^{3-} + H^+ \rightleftarrows CuHL$	9.8	20°, 0.1 NaClO_4	18
$Cu^{2+} + L^{3-} + H_2O \rightleftarrows$ $Cu(OH)L^{2-} + H^+$	1.6	20°, 0.1 NaClO_4	18
$2Cu^{2+} + 2L^{3-} + 2H_2O \rightleftarrows$ $Cu_2(OH)_2L_2^{4-} + 2H^+$	5.1	25°, 1 KNO_3	18
$2Cu^{2+} + L^- + H_2O$ $Cu_2(OH)L + H^+$	5.0	20°, 0.1 NaClO_4	10

Table 44, continued

Equilibrium	log K	Conditions	Ref.
$Fe^{3+} + L^{3-} \rightleftarrows FeL$	11.4	20°, 0.1 NaClO₄	20
$Fe^{3+} + L^{3-} + H \rightleftarrows FeHL^+$	12.7	- , 1 NaNO₃	4
$Fe^{3+} + L^{3-} + H_2O \rightleftarrows$ $Fe(OH)L^- + H^+$	9.5	25°, 0.1 NaNO₃	4
$Fe^{3+} + L^{3-} + H_2O \rightleftarrows$ $Fe(OH)L^- + H^+$	(7)	estimated and used	
$Fe^{3+} + L^{3-} + 2H_2O \rightleftarrows Fe(OH_2L^{2-}$	(1)	estimated	
$2Fe^{3+} + 2L^{3-} + 2H_2O \rightleftarrows$ $Fe_2(OH)_2L_2^{2-} + 2H^+$	21.2	20°, 0.1 NaClO₄	20
$Cd^{2+} + L^{3-} = CdL^-$	3.8	20°, 0.1 NaClO₄	18
$Cd^{2+} + L^{3-} + H^+ = CaHL$	8.5	20°, 0.1 NaClO₄	18
$Co^{2+} + L^{3-} = CoL^-$	5.0	20°, 0.1 NaClO₄	18
$Co^{2+} + L^{3-} + H^+ = CoHL$	9.4	20°, 0.1 NaClO₄	18
$Pb^{2+} + L^{3-} = PbL^-$	6.5	30°, → 0	4
$Pb^{2+} + L^{3-} + H^+ = PbHL$	12.1	25°, -	4
$Ni^{2+} + L^{3-} = NiL^-$	5.4	25°, 0.1 NaClO₄	18
$Ni^{2+} + L^{3-} + H^+ = NiHL$	9.7	25°, 0.1 NaClO₄	18
$Zn^{2+} + L^{3-} = ZnL^-$	5.0	25°, 0.1 NaClO₄	18
$Zn^{2+} + L^{3-} + H^+ = ZnHL$	9.4	25°, 0.1 NaClO₄	18
$Mn^{2+} + L^{3-} = MnL^-$	3.7	25°, 0.15	4
$Mn^{2+} + H^+ + L^{3-} = MnHL$	8.5	25°, 0.15	4
$Ba^{2+} + L^{3-} = BaL^-$	2.9	25°, 0.1 NaClO₄	18
$Ba^{2+} + L^{3-} + H^+ = BaHL$	8.2	25°, 0.1 NaClO₄	18

Table 44, continued

Equilibrium	log K	Conditions	Ref.
$Fe^{2+} + L^{3-} = FeL^{-}$	3.1	25°, 1 NaClO$_4$	4
$Fe^{2+} + L^{3-} + H^{+} = FeHL$	8.5	25°, 1 NaClO$_4$	4
$Mg^{2+} + L^{3-} = MgL^{-}$	3.4	25°, 0.1 NaClO$_4$	18
$Mg^{2+} + L^{3-} + H^{+} = MgHL$	8.2	25°, 0.1 NaClO$_4$	18
$Sr^{2+} + L^{3-} = SrL^{-}$	2.9	25°, 0.15	4
$Hg^{2+} + L^{3-} = HgL^{-}$	10.9	25°, 0.1 NaClO$_4$	10,21
$Hg^{2+} + L^{3-} + H_2O = Hg(OH)L^{2-} + H^{+}$	(5)	estimated	
$Hg^{2+} + L^{3-} + 2H_2O = Hg(OH)_2L^{3-} + 2H^{+}$	(1)	estimated	
$Hg^{2+} + L^{3-} + H^{+} = HgHL$	(12)	estimated	

[a]Temperature in °C and ionic strength of the medium in moles/
liter used in the determination. Presence of ionic constituents in the medium, other than the main reacting species is
indicated next to the values of the ionic strength.

Fresh Water with Nitrilotriacetate (NTA)

For a model fresh water (Table 43), concentration
of metal-NTA complexes were computed for several total
nitrilotriacetate concentrations in the range from
1×10^{-8} to 2×10^{-4} moles/liter, and for the pH
values 6, 7, 8, and 9. Figures 33 and 34 show the
results for pH = 8, which is close to the values
observed in the Great Lakes, and other lakes the
waters of which are in contact with carbonate rocks.
In Figure 33 the vertical scale shows the concentration of the metal-NTA complex as a percentage of the
total NTA concentration in the water, and the horizontal scale shows the total NTA concentration.
Figure 34 gives the concentrations of NTA-containing
species in moles/liter, as a function of the total

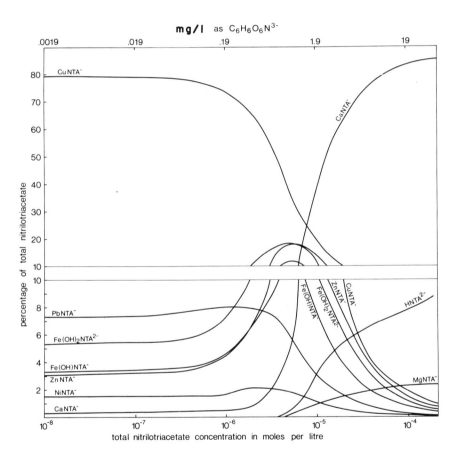

Figure 33.   Metal-nitrilotriacetate complexes in a model
fresh water. Fresh water composition given in
Table 43; pH = 8. Concentrations of metal-NTA
complexes shown as percentage of the total NTA
concentration. Note difference in scale between
0-10% and 10-80%.

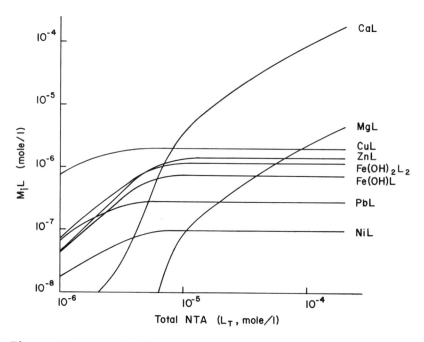

Figure 34.   *Metal-nitrilotriacetate complexes in a model*
fresh water, pH = 8. Concentrations of metal-
NTA complexes shown in moles/liter, as a function
of the total NTA concentration.

nitrilotriacetate concentration. Table 45 shows how
the concentrations of the most important metal-NTA
species vary with the pH for three given total NTA
concentrations.

The results in Figures 33 and 34 and Table 45
show the following:

(1) For a total nitrilotriacetate concentration
below 2×10^{-6} moles/liter most of the nitrilotriacetate is in the form $CuNTA^-$, whereas $PbNTA^-$,
$Fe(OH)_2NTA^{2-}$, $Fe(OH)NTA^-$, $ZnNTA^-$, and $NiNTA^-$ account
for most of the remainder.

(2) For a total nitrilotriacetate concentration
above 2×10^{-5} moles/liter, most is in the form
$CaNTA^-$ (pH of 7 and higher), whereas $MnNTA^-$, $MgNTA^-$,
and $HNTA^{2-}$ account for most of the remainder.

It should be noted that the conditions in (1)
and (2) correspond to a shortage and excess, respectively, of the total nitrilotriacetate concentration
in solution relative to the sum of the concentrations
of the trace metal-ions.

Table 45

Effect of pH on the Concentration of
Nitrilotriacetate Complexes[a]

Species	pH6	pH7	pH8	pH9
(a) Total nitrilotriacetate = 1×10^{-7} moles/liter				
$CuNTA^-$	85	91	79	51
$CaNTA^-$	0	0	0	2
$FeNTA^0$	1	0	0	0
$Fe(OH)NTA^-$	12	5	3	2
$Fe(OH)_2NTA^{2-}$	0	1	5	31
$PbNTA^-$	1	3	7	6
$NiNTA^-$	0	0	1	2
$ZnNTA^-$	0	1	3	5
(b) Total nitrilotriacetate = 3×10^{-6} moles/liter				
$HNTA^{2-}$	1	0	0	0
$CuNTA^-$	59	61	55	43
$CaNTA^-$	0	0	1	4
$FeNTA^0$	3	0	0	0
$Fe(OH)NTA^-$	32	22	9	2
$Fe(OH)_2NTA^{2-}$	1	3	14	34
$(Fe(OH)NTA)_2^{2-}$	1	0	0	0
$PbNTA^-$	3	7	8	6
$NiNTA^-$	0	1	2	2
$ZnNTA^-$	1	4	10	9
(c) Total nitrilotriacetate = 2×10^{-4} moles/liter				
$HNTA^{2-}$	86	46	9	1
$CuNTA^-$	1	1	1	1
$CaNTA^-$	10	49	85	92
$Fe(OH)NTA^-$	1	1	0	0
$Fe(OH)_2NTA^{2-}$	0	0	1	1
$ZnNTA^-$	1	1	1	1
$MnNTA^-$	0	1	1	1
$MgNTA^-$	0	1	3	3

[a]Concentrations as percentages of the total nitrilotriacetate
concentration, rounded to nearest 1%.

(3) In the pH range from 6 to 9, FeNTA^0 is virtually nonexistent in solution, and the important iron (III) species are Fe(OH)NTA$^-$ and Fe(OH)$_2$NTA^{2-}.

(4) None of the 1:2 (metal-ion to NTA^{3-}) complexes included in calculations appear in significant concentrations.

(5) The main effects of increasing the pH from 6 to 9 are:

    (a) increase in the concentrations of Fe(OH)$_2$NTA^{2-} relative to Fe(OH)NTA$^-$

    (b) increase in the concentration of CaNTA$^-$

    (c) decrease in the concentration of the free nitrilotriacetate HNTA^{2-}.

Fresh Water with Citrate

Concentrations of metal-organic complexes in a model fresh water were done for the case of citrate ligand present in solution (Table 43). Figure 35 shows the distribution of metal-citrate species

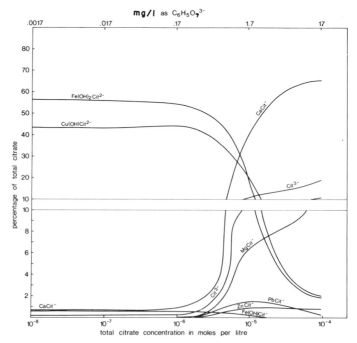

Figure 35.   *Metal-citrate complexes in a model fresh water (Table 46, pH = 8). Concentrations of metal-citrate complexes shown as percentage of the total citrate concentration. Note difference in scale between 0-10% and values over 10%.*

at pH = 8, and the dependence of the concentrations on the pH is shown in Table 46, for two values of the total citrate concentration.

Table 46

Effect of pH on the Concentration of Citrate Complexes[a]

Species	pH6	pH7	pH8	pH9
(a) Total citrate = 1×10^{-7} moles/liter				
$(FeOHCit)_2{}^{2-}$	30	1	–	–
$Fe(OH)_2Cit^{2-}$	31	55	55	54
$FeOHCit^-$	31	6	1	–
$CuOHCit^{2-}$	7	37	43	45
$CaCit^-$	–	1	1	1
Cit^{3-}	–	–	–	–
(b) Total citrate = 1×10^{-5} moles/liter.				
$(FeOHCit)_2{}^{2-}$	16	5	–	–
$Fe(OH)_2Cit^{2-}$	2	13	20	20
$FeOHCit^-$	2	1	–	–
$CuOHCit^{2-}$	19	20	20	20
$CaCit^-$	34	39	40	41
$(CaHCit)^0$	3	–	–	–
Cit^{3-}	9	10	11	12
$HCit^{2-}$	5	1	–	–
$MgCit^-$	5	6	6	6
$PbCit^-$	2	2	1	–
$ZnCit^-$	1	1	1	–

[a] Concentrations as percentage of the total citrate concentration, rounded to nearest 1%.

The main species which account for the citrate at low total citrate concentrations (less than 2×10^{-6} moles/liter) are hydrolyzed Cu^{II} and Fe^{III} species. At higher citrate concentrations, the calcium-citrate complex predominates. (The distribution of the Fe^{III} complexes depends heavily on the estimated values of the equilibrium constants for $Fe(OH)_2Cit^{2-}$ and $Fe(OH)Cit^-$, and these may be in error. Further information on equilibria in Fe^{III}-citrate solutions would resolve this uncertainty.) It is of interest to note that at the relatively low pH of 6, the dimeric species $(FeOHCit)_2^{2-}$ is an important complex in the model fresh water. In general, however, the complexes of citrate with Cu, Pb, Ni, and Zn are less important than their complexes with nitrilotriacetate.

Discussion

The results of the calculations shown in Figures 33-35, and Tables 45 and 46 can only be as good as the data used to make them. Among the background complexation equilibria, no allowance has been made for such possible interactions as the formation of complexes between hydrolyzed metal-ions and sulfate, nor for the complexes between carbonate and such metal-ions as Cd^{2+}, Mn^{2+}, Sr^{2+}, and Ba^{2+}. These equilibria, however, probably have no significant effect on the distribution of metal-organic complexes formed by the addition of a strong complexing agent, such as nitrilotriacetate or citrate, to fresh water. Somewhat more serious might be some minor gaps in information on the metal-NTA complexes. For example, the $HNTA^{2-}$ anion has two deprotonated carboxylate groups, and it is likely to be an effective complexing agent for Ca^{2+} and Mg^{2+}, probably about as effective as citrate in near neutral solution. Studies of such complex formation are absent from the literature. Another possibility which has not been studied is the formation of mixed metal-ion complexes by NTA^{3-} which has four coordination sites available for cations. In particular, for the model fresh water situation, a significant (though probably not serious) omission could be mixed metal-ion complexes of the type $Ca(NTA)Cu^+$.

In both sets of calculations for the model lake water with nitrilotriacetate or citrate no attempt has been made to adjust the equilibrium constant values to some common ionic strength. This was not done because the nature of the ionic medium, the experimental methods, and the prejudices of the investigators are often of as equal importance as the ionic strength of the medium in determining the value obtained for an equilibrium constant.

In this section, only the complexation equilibria in solution have been treated for the model fresh water. In a real situation, in a heterogeneous system consisting of solution and solids (*i.e.*, water and sediments), some of the metal-organic complexes in solution may be involved in uptake by or exchange on solid phases. Such reactions may modify the concentrations of some of the complexes in solution when solid phases in the sediment act as strong suppliers or sinks of certain metals to the organic ligands in the water. Although the nature and extent of such reactions has not been studied, the general problem of changing concentrations of metal-organic complexes in the presence of uptake by sediment in a natural body of water will be discussed in a later section.

KINETIC RELATIONSHIPS

The fact that organic substances are unstable with respect to their oxidation products under the conditions prevailing near the surface of the earth suggests that organic ligands in solution should sooner or later break down and decay. Decay of organic ligands in solution may be controlled by combinations of such processes as bacterial degradation, oxidation, and photodissociation. Little is known on the rates of decay or decomposition of dissolved organic substances in natural waters. The rates of oxidation of particulate and dissolved organic matter in natural waters (measured through changes in the oxygen content of water), decomposition and breakdown in a solid state or on solid grain surfaces,[13] and racemization of amino acids in oceanic and fresh water sediments[14] may be very different from the rates of oxidation and decomposition of organic complexing agents in natural waters. For metal-NTA complexes in solution, evidence has been presented[15] suggesting that the rates of decay of the nitrilotriacetate ligand depend to some extent on the nature of the metal-ion in the complex, and the decay half-lives are of the order of days and weeks.

If the rate of decay of an organic ligand depends on the nature of a metal-ion it complexes,

then in a natural water containing different metal-
ions, the rate of decay may be bracketed by two
decay models:

    (1) decay without equilibrium, where the
organic ligand decays in each complex at some
characteristic rate but an equilibrium is not
reestablished between the complexes;

    (2) decay with equilibrium, where the organic
ligand in each complex decays at some characteristic rate and an equilibrium is "instantaneously"
reestablished between the metal-organic complexes
in solution.

    The two decay schemes are considered in more
detail below.

Decay Without Equilibrium

    In a scheme of decay without equilibrium, an
organic ligand L in each metal-organic complex (M_iL)
and free L in solution decay according to their rate
constants, and no equilibrium is reestablished between the remaining amounts of the free and complexed
L. Assuming that L in each metal-L species decays
according to the first-order rate law, the rate of
decay of L in each M_iL complex is

$$- \frac{d(M_iL)}{dt} = \lambda_i (M_iL),$$
(13)

where parentheses denote concentrations and λ is
the first-order reaction rate constant, of dimensions
time^{-1}. The total rate of decay of the complexing
agent, $-d(L_T)/dt$, is the sum of the terms given by
Equation 13 and a similar term for the free L:

$$- \frac{d(L_T)}{dt} = - \left[\frac{d(M_iL)}{dt} + \ldots + \frac{d(M_{n-1}L)}{dt} + \frac{d(L)}{dt} \right].$$
(14)

    If the initial concentrations of the complexes
(M_iL) and free L can be expressed as simple fractions
f_i of the total ligand concentration (L_T), as in
Equation 7, then the solution of Equation 14 that
gives (L_T) as a function of time is

$$(L_T) = (L_T)_{t=0} \left[f_1 e^{-\lambda_1 t} + \ldots + f_n e^{-\lambda_n t} \right],$$
(15)

where the initial total concentration of the complexing
agent is $(L_T)_{t=0}$.

Decay With Equilibrium

In the scheme of decay with equilibrium, it is assumed that equilibria between metal ions and the complexing agent are attained much faster than the decay of the free and complexed ligand L, as shown schematically in the following reaction:

$$
\begin{array}{ccc}
& \text{slow} & \\
& \text{decay} & \\
M_2 + M_1L & \xrightarrow{\hspace{1.5cm}} & (M_1 + \text{products}) \\
\text{fast} \quad \Big\updownarrow & & \\
\text{equil.} \qquad & \text{slow} & \\
& \text{decay} & \\
M_1 + M_2L & \xrightarrow{\hspace{1.5cm}} & (M_2 + \text{products})
\end{array}
\tag{16}
$$

In reaction 16, the ligand L in each L-containing species decays at its characteristic rate, but the concentrations of all the species relative to each other are maintained in the ratios given by the equilibrium relationships, as in Equations 2 and 7. Then the total ligand concentration (L_T) as a function of time is, by reasoning similar to the preceding section,

$$
(L_T) = (L_T)_{t=0} \left[e^{-(f_1\lambda_1 + \ldots + f_n\lambda_n)t} \right] \equiv (L_T)_{t=0}\, e^{-\Lambda t}, \tag{17}
$$

where $(L_T)_{t=0}$ is the initial total ligand concentration and Λ (time^{-1}) is the sum of $f_i\lambda_i$.

The main difference between the two postulated decay models is that in the latter case--decay with equilibrium between complexes, Equation 17,--the total concentration of the complexing agent in the system, (L_T), decreases faster than in the case of decay without equilibrium, as can be verified by comparing Equations 17 and 15. A corollary of a faster decay of L_T is that in a water system with a continuous constant input, a steady-state concentration would be attained faster if equilibrium is maintained between the metal-ligand complexes. in solution.

PHYSICAL AND CHEMICAL FACTORS: A SYNTHESIS

In the preceding sections the thermodynamic and kinetic relationships between metal-organic complexes were discussed only with reference to a homogeneous solution. Natural water systems, however, are heterogeneous systems, where mineral phases and biogenic particles may affect the behavior of chemical species in solution. Natural water systems may also be open systems in the sense that they receive input of a chemical species, and they may be relieved of it by water outflow, decay, and/or uptake by sediments.

In such natural water systems as rivers, lakes, and coastal ocean waters, concentrations of a newly introduced chemical species are controlled by the physical characteristics of the system no less than by the chemical reactions the new species is involved in.

One of the most important questions that are usually asked about man-made chemical substances in natural waters is what would be the concentrations in a given water system at different times and under different conditions of input. Accurate answers to this question require detailed information on the chemical and physical characteristics of the system which is, as a rule, not available at the time when the question needs to be answered. Approximate predictions, however, may be made for simplified model systems, aimed at providing estimates bracketed by some lower and upper "reasonable" limits of the relevant chemical and physical characteristics.

In considering events that take place over a long period of time, a useful approximation is to regard a water system as a well-mixed body of water in which the chemical species introduced by input are dispersed "instantaneously." The assumption that a body of water is well-mixed is valid if some characteristic time of mixing or homogenization (t_{mix}) is shorter than the lifetime of a reacting species and is also shorter than the residence or renewal time of water in the system. A characteristic lifetime of a chemical species decaying according to a first-order rate law may be taken as $1/\lambda$ (λ is the decay rate constant of dimensions time^{-1}), and the residence time of water in the system may be taken as τ = (volume of system)/(rate of outflow, vol.time^{-1}). Then an approximation to a well-mixed state is valid of $t_{mix} < 1/\lambda$ and $t_{mix} < \tau$. Many bodies of water characterized by linear dimensions of a few kilometers or less may be considered well mixed on a time scale of months or years.

The dependence of the concentrations of reacting species in water on their interactions with sediment and on some physical characteristics of water systems is discussed below.

Interactions with Sediment

In solution, metal-organic compounds may be taken up by biogenic particles, as well as by mineral grains settling through the water column and accumulating on the bottom. The reactions between metal-organic complexes in solution and solid phases may involve exchange of metal-ions in the complexes for other metal-ions held on mineral grain surfaces, as well as adsorption of the complexes from solution on surfaces of clay and other mineral particles. Among different adsorption models, Freundlich's adsorption isotherm[16],[17] is fairly common for cations and anions on sediments suspended in water. The Freundlich adsorption equation relates the concentration of a species on substrate (C_S) to its concentration in solution (C) by the following relationship:

$$C_S = KC^n , \tag{18}$$

where K is a constant and the power exponent n has a value in the range $0 < n < 1$.

The effect of adsorption of metal-organic complexes from solution by sediment particles, according to Equation 18, will be considered for one complexed species *or* for the total ligand in solution, provided its uptake by solids can be approximated by two constants, K and n, of Equation 18.

We consider a well-mixed body of water characterized by some residence time of water with respect to outflow, $\tau = $ (volume)/(rate of outflow) years. The body of water receives a constant input of an organic ligand at a range J (in units of mass per unit volume of water per unit time). The ligand or one of its complexes is adsorbed by sediment particles according to the relationship given in Equation 18. Then the rate of change of the ligand concentration in water, dC/dt, may be expressed as

$$\frac{dC}{dt} = J - \lambda C - C/\tau - \frac{dC_S}{dt} . \tag{19}$$

In the model described by Equation 19, the organic ligand in adsorbed state on sediments does not decay (or it decays much slower than in solution). Such an assumption adds weight to the role of sediments as a storage sink of the organic species introduced into the system. In this scenario the newly introduced species can reside in the system for a much longer period of time than in the case of decay taking place both in water and in sediments.

In Equation 19 we denote

$$\alpha = \lambda + 1/\tau \ (\text{time}^{-1}), \tag{20}$$

and substitute for C_S from Equation 18, obtaining*

$$\frac{dC}{dt} = \frac{J - \alpha C}{1 + KnC^{n-1}}. \tag{21}$$

For the case when initial concentration $C_{t=0} = 0$, the solution of Equation 21 is

$$-\frac{1}{\alpha} \ln(1 - C\alpha/J) + Kn \sum_{j=0}^{\infty} \frac{C^{n+j}\alpha^j}{(n+j)J^{j+1}} = t. \tag{22}$$

Concentration in solution, C in Equation 22, is more conveniently expressed graphically by using dimensionless concentration $X = C\alpha/J$ and dimensionless time αt. Then Equation 22 becomes

$$- \ln(1 - X) + Kn(J/\alpha)^{n-1} \sum_{j=0}^{\infty} \frac{X^{n+j}}{n+j} = \alpha t. \tag{23}$$

It should be noted that dimensionless concentration $X = C\alpha/J$ is always less than 1: a steady-state concentration ($dC/dt = 0$ in Equation 19) is J/α; therefore transient concentrations C, increasing from nil, are lower than the steady-state concentration J/α. Thus dimensionless concentration X is a fraction of a steady-state concentration, and dimensionless time αt is a fraction or multiple of the combined rate of removal from solution by decay and outflow, as given by α in Equation 20.

Discussion of Equation 23.

Since dimensionless concentration X is < 1, the first term in Equation 23 is positive, and dimensionless time αt is the sum of two positive terms. Therefore, to attain any given value of concentration X in water, the time required for this is longer the larger the value of K. Constant K represents the efficiency of adsorption by sediment, as given in Equation 18, and stronger uptake of dissolved species by sediment causes their concentration in solution to rise more slowly.

---

*In the case of n=1, Equations 21 and 18 describe processes with chemical exchange characterized by an equilibrium constant K.

X being a fraction, the infinite sum of the powers of X in Equation 23 is a decreasing series, and for the purpose of computation only a number of terms need to be taken in the summation.

The curves in Figure 36, computed from Equation 23, show the dimensionless time, αt, required to attain different values of the fractional concentration X = Cα/J of a dissolved species in the presence of different intensities of adsorption by sediment, as expressed in the values of K between 0 and 0.04 shown in Figure 36. The curves in Figure 36 show that the time required to attain any concentration value in water strongly depends on

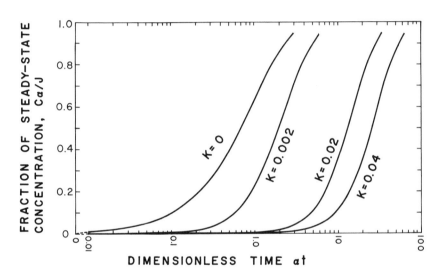

Figure 36. *Concentrations in a well-mixed body of water with adsorption by sediment; Equation 23. Curves give the fractions of a steady-state concentration (Cα/J) attained at different values of dimensionless time (αt). α is combined rate (time^{-1}) of removal of the dissolved species by decay and outflow, Equation 20; J is rate of input (mass per unit volume of the body of water per unit time); K is adsorption parameter, Equation 18. The case of no adsorption by sediment is the curve labelled K = 0. Increasing values of K correspond to stronger adsorption. For any given value of input (J) and removal (α) rate, longer time is needed to attain some concentration level in the presence of stronger adsorption by sediment.*

the intensity of adsorption.   In the case of no
adsorption (K = 0), the time required to attain
the concentration value of 95% of the steady-state
concentration (X = 0.95) is 3/α.   In the case of
K = 0.02, the 95% level of the steady-state concentration is attained in a length of time equal
to 34/α.   To translate these values into some conventional units of time, one may, for example,
consider a well-mixed lake characterized by the
residence time of water τ = 1 yr, receiving a
steady input of a reacting species of half life of
2 weeks (Λ = 18 yr^{-1}).   Then, the combined rate of
removal by outflow and decay, α from Equation 20,
is α = 1/1 + 18 = 19 yr^{-1}.   The times to 95% of
steady-state are, respectively, 3/19 = 0.158 yr
(8.2 weeks) and 34/19 = 1.8 yr.

Other parameters that enter in Equation 23 are
the steady-state concentration J/α, power exponent
n, and coefficient K.   In the computation of the
curves shown in Figure 36, J/α was taken as 2×10^{-6}
moles/liter; the choice of this value was inspired
by the distribution of metal-nitrilotriacetate
complexes in a model fresh water, at the total
nitrilotriacetate concentrations of 2×10^{-6} moles/
liter, or less (Figures 33,34).   The value of the
power exponent n was taken as n = 0.5. The choice
of the values of K, however, was somewhat less
arbitrary, and it was based on the following considerations.

Concentration of 0.1 mg/l in solution and 5000
µg/gram on solids in contact with water represent
strong uptake.   The amount of solids in contact
with water includes suspended particles and the
uppermost layer of sediment, about 1 cm thick.   The
latter usually contains about 0.5 gram of solids
per 1 cm^3 of bulk sediment (by volume, about 20%
solids and 80% water).   Concentration of suspended
matter of the order of 0.5 grams/liter is probably
near the upper limit in shallow lakes.   Thus, a
10-meters deep column of water may be in contact
with about 1 gram of solids, which corresponds to
the concentration of 1 gram/liter.   Gram-formula
weights of metal-nitrilotriacetate and metal-citrate
complexes are in the range 200-250; the value of
200 was taken to estimate K.   From Equation 18,
using the above numbers, K can be estimated:

$$C_s = \frac{5 \times 10^{-3} \dfrac{\text{grams}}{\text{gram solids}} \times 1 \dfrac{\text{grams solids}}{\text{liter solution}}}{200 \ (\text{grams/mole})}$$

$$= 2.5 \times 10^{-5} \ \text{moles/liter.}$$

$$c^n = \left(\frac{1 \times 10^{-4} \ \text{gram/liter}}{200 \ \text{grams/mole}}\right)^{\frac{1}{2}} = 7 \times 10^{-4} \ (\text{moles/liter})^{\frac{1}{2}}$$

$$K = C_s/c^n = 2.5 \times 10^{-5}/(7 \times 10^{-4}) = 0.04 \ (\text{moles/liter})^{\frac{1}{2}}.$$

The above value of $K = 0.04$, corresponding to the distribution (5000 µg/g on solids)/(0.1 mg/l in solution) was taken as an upper value in the computation of concentration-time curves shown in Figure 36. The lower values of $K = 0.02$ and $K = 0.002$ correspond to distribution ratios of 2500 µg/g and 250 µg/g on solids, to 0.1 mg/l in solution.

No Interaction With Sediment

In the absence of uptake of an organic complexing agent by sediment, a relationship for the concentration in water as a function of time is derivable from Equations 19 and 22 by setting $K = 0$. Denoting the concentration of each ligand-containing species C_i, its rate of input into a body of water J_i, and its decay rate constant λ_i, the appropriate equation is

$$\frac{dC_i}{dt} = J_i - C_i/\tau - \lambda_i C_i. \tag{24}$$

The total ligand concentration in water (L_T) is the sum of the concentrations, C_i, of metal-ligand species and the free ligand,

$$\sum_{i=1}^{n} C_i = (L_T) \equiv C. \tag{25}$$

Summation of n equations of the type of Equation 24 gives the rate of change of the total ligand concentration in water:

$$\frac{dC}{dt} = J - C/\tau - \sum_{i=1}^{n} \lambda_i C_i . \qquad (26)$$

It has been shown in Figures 33 and 35 that for such complexing agents as nitrilotriacetate and citrate the fractions of the total concentration taken up in metal-organic complexes are nearly constant, when the total ligand concentration is less than about 2×10^{-6} moles/liter. Under such conditions the concentration of each ligand-containing species, C_i, may be expressed as $C_i = f_i C$ (f_i is a fraction, by analogy with Equation 7), and Equation 26 may be rewritten as

$$\frac{dC}{dt} = J - C/\tau - \Lambda C$$

$$= J - \alpha C \qquad (27)$$

where

$$\Lambda = \sum_{i=1}^{n} f_i \lambda_i, \quad \text{and} \quad \alpha = 1/\tau + \Lambda. \qquad (28)$$

The solution of Equation 27, for the case of initial concentration in water $C_{t=0} = 0$, is

$$C = \frac{J}{\alpha} (1 - e^{-\alpha t}) . \qquad (29)$$

Equation 29 does not apply to a body of water the mixing time of which is significantly longer than the decay half life of a ligand, $1/\lambda$, or, in general, if the mixing time is longer than $1/\alpha$. Similarly, Equations 23 and 29 cannot be used strictly to compute transient concentrations at the values of time t smaller than the characteristic mixing time of water in the system.

In large bodies of water, or for short time intervals, dispersal of input may be treated by mathematical methods involving anisotropic eddy diffusivity and flow, as commonly done in meteorological, oceanographic and limnological problems. A simplified alternative may be to view a water system as made of a series of cells or boxes, each box having its characteristic mixing time shorter than $1/\lambda$. If a system can be so likened to a series

of consecutive reservoirs, each characterized by
the same residence time of water (τ_r), with the
first reservoir receiving external input at a rate
J, then the concentration in any reservoir (C_m) is
given by

$$C_m = \frac{J}{\alpha_r^m \tau_r^{m-1}} \left(1 - e^{-\alpha_r t} \sum_{j=0}^{m-1} \frac{(\alpha_r t)^j}{j!} \right) , \tag{30}$$

where $\alpha_r = 1/\tau_r + \Lambda$, and m is the number of the cell
in a lake, or reservoir in a chain.

Equation 30 for one reservoir only (*i.e.*, m=)
reduces to the form given in Equation 29.

Steady-state concentrations of reacting species
in one body of water (C_{ss}) or in a chain ($C_{m,ss}$) are
derivable from Equations 29 and 30 by letting $t \to \infty$.
Steady-state concentrations are

$$C_{ss} = \frac{J\tau}{1 + \Lambda\tau} , \tag{31}$$

$$C_{m,ss} = \frac{J\tau_r}{(1 + \Lambda\tau_r)^m} \tag{32}$$

Equation 32 shows that in a chain of cells or
reservoirs the concentration falls off rapidly with
distance (m) from the point of input. If the
physical conditions justify, the simple model of
one string of well-mixed reservoirs, given by
Equations 30 and 32, can be modified to represent
a more elaborate geometry of cells, such as, for
example, two or three tiers of cells (vertical
stratification of the water column) and more than
one horizontal row (two-dimensional flow pattern),
in which the pattern of water transport from one
cell to another must be stipulated.

Discussion

To give an example of the magnitudes of
steady-state concentrations of a reacting species
in one well-mixed reservoir, and in a reservoir
subdivided into a string of cells, we consider a
reacting species with decay half life of 2 weeks

($\Lambda = 18$ yr^{-1}). In a lake with the residence time of water with respect to outflow $\tau = 1$ yr, a steady-state concentration (C_{ss}), expressed as a fraction of input concentration ($J\tau$) from Equation 31, is

$$\frac{C_{ss}}{J\tau} = \frac{1}{1 + 18x1} = 0.053 \, ,$$

or 5.3% of input concentration (Figure 37). This is a large dilution of input.

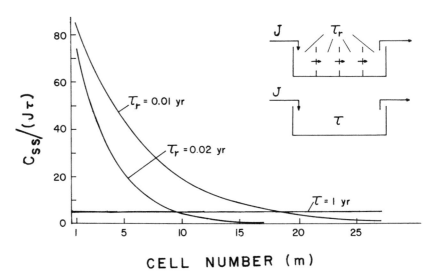

CELL   NUMBER   (m)

Figure 37.   Steady-state concentrations in a well-mixed body of water (water residence time $\tau = 1$ yr) shown by straight line, and steady-state concentrations in the same body of water subdivided into 100 and 50   well-mixed "cells," each of residence time $\tau_r = 0.01$ yr and $\tau_r = 0.02$ yr, respectively. Equations 31 and 32. Vertical scale denotes concentration expressed as percentage of input concentration (J is rate of input, in units of mass per unit volume of the body of water, or cell, per unit time). The smooth curves in the figure connect discrete concentration points representing each "cell."

If the lake of $\tau = 1$ yr is subdivided into 100
or 50 "cells," each characterized by residence time
of water $\tau_r = 0.01$ yr or $\tau_r = 0.02$ yr, respectively,
then, as shown in Figure 37, a strong stationary
concentration gradient is established in the system,
and concentrations fall below 5% of input concentration ($J\tau_r$) at a distance that is fairly short
compared with the length of the lake (near the 10-th
and 20-th cell, or 1/5 of the lake length subdivided
into 50 and 100 cells, respectively).

In a well-mixed lake receiving input at a
constant rate, the time required to attain a steady-
state concentration of a new reacting species is
controlled by the combined rate of removal by outflow and decay, given by parameter α in Equation 28.
In lakes of long residence time of water, the term
$1/\tau$ is a small number, and it may be much smaller
than the reaction or decay rate λ. For species
that decompose relatively fast (decay half lives
of the order of days or a few weeks) it is unlikely
that large bodies of water, such as the larger
Great Lakes, can be considered well mixed. But for
more stable substances, whose behavior is considered
over longer periods of time, the size of a body of
water and its water residence time may not limit
application of the models discussed in the preceding
sections.

One of the points that deserve consideration
in the study of the behavior of degradable substances
in lakes is the length of time required to attain a
steady-state concentration (under the conditions of
input at a constant rate). With reference to a
well-mixed body of water, Equation 29 defines concentration as a function of time. The rate at which
concentration (C) approaches a steady-state value
(J/α) is controlled by the combined rate of removal
α. A concentration equal to 95% of a steady-state
concentration (J/α) is attained at time

$$t_{ss} \simeq 3/\alpha \ . \qquad (33)$$

Continuing with the example used in the preceding
sections, in a lake of $\tau = 1$ yr, receiving steady
input of a species characterized by $\Lambda = 18$ yr^{-1}
(half life = 2 weeks), the time to steady-state is

$$t_{ss} \simeq 3/(1 + 18) = 0.158 \text{ yr} = 8.2 \text{ weeks}$$

But if the species introduced is stable ($\Lambda = 0$),
the combined rate of removal is $\alpha = 1/1$, and the
time to steady-state is much longer,

$$t_{ss} \simeq 3/1 = 3 \text{ years.}$$

As far as the time to steady-state is concerned,
greater stability of a reacting species has an effect
comparable to stronger adsorption from water by
sediment (Figure 36). Adsorption by sediment particles, however, has no effect on the value of a
steady-state concentration, insofar as the steady-
state value is determined only by the rates of
input (J) and removal (α).

SUMMARY AND CONCLUSIONS

Two organic compounds, nitrilotriacetate (NTA)
and citrate, which may conceivably appear in natural
waters affected by industrial societies, have a
potential to modify the existing distributions of
ionic species in waters. The two compounds form
strong complexes with such metal-ions as Cu, Pb, Fe,
Ni, Co, and Zn, and somewhat weaker complexes with
such major constituents of natural waters as Ca and
Mg.
A study of thermodynamic equilibria in a model
fresh water (similar in composition to an average
river and lake water) shows that NTA and citrate
complex virtually all available Cu, Fe, Pb, Ni, Co,
and Zn. When no more of these metal-ions are available for complexation but the ligand concentration
continues to rise owing to input from outside, then
NTA and citrate form complexes with Ca and, to a
lesser extent, Mg. Complexing of Ca becomes significant at the total ligand concentration in the range
from 2 mg/l to 15 mg/l (Figures 33-35, and Tables
45 and 46). In fresh waters, such concentrations
of NTA or citrate can significantly affect the
concentration of Ca in ionic form, and they may
therefore be potentially damaging to many fresh
water environments.
These results may be generalized by stating
that any complexing agent with strong affinities to
the major and minor metal-ions in solution can per-
turb the existing distribution of ionic species in
natural waters. In the case of NTA, the equilibrium
constant of the CaNTA$^-$ complex is of the order of
10^6 (Table 44). If for some other complexing agent

the equilibrium constant were higher, its effect on the concentration of Ca-ion in solution may have been pronounced at concentrations lower than those of NTA in the NTA-fresh water system, shown in Figure 33.

If an organic complexing agent is removed from solution by such processes as decay, oxidation or biodegradation, the preexisting distribution of metal-ions may be restored. In large bodies of water characterized by longer residence time (of the order of years to decades), a relatively fast decay (half lives of organic ligands of the order of weeks to months) assures that the total concentration of the complexing agent is low by comparison with its input concentration. On the other end of the half life scale, input of a slowly decaying or stable complexing agent into the same system would result in relatively much higher concentrations in water.

Uptake of the ligand-containing species from solution by solid mineral phases reduces, analogously to decay, the ligand concentration in water. Such uptake mechanisms as equilibrium exchange and adsorption by sediment particles may very strongly retard the rise in concentration of a new complexing agent introduced into the water system by input from outside. The immediate effects of equilibrium exchange or adsorption in a transient state are lower concentrations in solution. The final steady-state concentration, however, is not affected by exchange and adsorption processes: provided the conditions of input to, and removal from the water system are such that a steady-state can exist, chemical exchange and adsorption from solution only slow down the rate of approach to steady-state, without affecting the steady-state concentration value (Equation 21).

ACKNOWLEDGMENTS

Part of the work reported in this paper began in 1970, when both authors were affiliated with the Canada Centre for Inland Waters, Burlington, Ontario. Continuation of the work was supported by Oceanography Section, National Science Foundation, NSF Grant GA-30769 (A. Lerman, Northwestern University). The support received from C.C.I.W. and N.S.F. is gratefully acknowledged.

REFERENCES

1. Duursma, E. K. "The Dissolved Organic Constituents of Sea Water," *Chemical Oceanography*, Riley, J. P. and G. Skirrow, eds., Vol. 1 (New York: Academic Press, 1965).

2.   Degens, E. T. *Geochemistry of Sediments* (Englewood Cliffs, N.J.: Prentice-Hall, 1965).

3.   Stumm, W. and J. J. Morgan. *Aquatic Chemistry* (New York: Wiley-Interscience, 1970).

4.   Sillen, L. G. and A. E. Martell. *Stability Constants of Metal-Ion Complexes*. Spec. Pub. 17, (London: Chemical Society, 1964), Supplement No. 1 (1971).

5.   Bailar, J. C. Jr. "Coordination Compounds," *Encyclopaedia Britannica*, Vol. 6 (1964).

6.   Goldberg, E. D. "Minor Elements in Sea Water," *Chemical Oceanography*, Riley, J. P. and G. Skirrow, eds., Vol. 1 (New York: Academic Press, 1965).

7.   Perrin, D. D. and I. G. Sayce. "Computer Calculation of Equilibrium Concentrations in Mixtures of Metal-Ions and Complexing Species," *Talanta 14*, 833 (1967).

8.   Childs, C. W., P. S. Hallman, and D. D. Perrin. "The Application of Digital Computers in Analytical Chemistry," *Talanta 16*, 1119 (1969).

9.   Morel, F. and J. J. Morgan. "A Numerical Method for Solution of Chemical Equilibria in Aqueous Systems," *Aquatic Chemistry*, Stumm, W. and J. J. Morgan, eds. (New York: Wiley-Interscience, 1970).

10.   Childs, C. W. "Chemical Equilibrium Models for Lake Water Which Contains Nitrilotriacetate and for 'Normal' Lake Water," *Proc. 14th Conf. Great Lakes Res*. 198 (1971).

11.   International Lake Erie Water Pollution Board and International Lake Ontario and St. Lawrence River Water Pollution Board, "Report to the International Joint Commission on the Pollution of Lake Erie, Lake Ontario, and the International Section of the St. Lawrence River," Vol. 3 (1969).

12.   Livingstone, D. A. "Chemical Composition of Rivers and Lakes," *Data of Geochemistry*, Fleischer, M., ed. U.S. Geol. Surv. Prof. Pap. 440, Chap. G. (1963).

13.   Abelson, P. H. "Geochemistry of Organic Substances," *Researches in Geochemistry*, Abelson, P. H., ed. (New York: Wiley, 1959).

14.   Bada, J. L. "Kinetics of the Nonbiological Decomposition and Racemization of Amino Acids in Natural Waters," *Nonequilibrium Systems in Natural Water Chemistry*. Hem, J. D., Symp. Chair. Adv. Chem. Ser. 106 (Washington, D.C.: Americal Chemical Society, 1971).

15.   Personal communications from H. Bouveng (Swedish Water and Air Pollution Research Laboratory, Stockholm, Sweden, 1970), Y. K. Chau (Canada Centre for Inland Waters, Burlington, Ontario, 1970), and I. D. Hill (Monsanto Company, St. Louis, Missouri, 1970).

16.   Adamson, A. W. *Physical Chemistry of Surfaces* (New York: Interscience, 1960).

17.   Lerman, A. "Boron in Clays and Estimation of Paleo-salinities," *Sedimentology 6*, 267 (1966).

18.   Campi, E., G. Ostacoli, M. Meirone, and G. Saini. "Stability of the Complexes of Tricarballylic and Citric Acids with Bivalent Metal-Ions in Aqueous Solution," *J. Inorg. Nucl. Chem. 26(4)*, 553 (1964).

19.   Rajan, K. S. and A. E. Martell. "Polymeric Copper(II) Complexes of Hydroxy Acids," *J. Inorg. Nucl. Chem. 29*, 463 (1967).

20.   Timberlake, C. F. "Iron-Malate and Iron-Citrate Complexes," *J. Chem. Soc.*, 5078 (1964).

21.   Skorik, N. A., V. N. Kumok, and V. V. Serebrennikov. "Compounds of Mercury(II) with Nitrilotriacetic Acid and Citric Acids," *Russ. J. Inorg. Chem. 12*, 1429 (1967), Engl. edit.

8. BIODEGRADATION OF NTA METAL CHELATES IN RIVER WATER

R. D. Swisher, T. A. Taulli, and E. J. Malec.
Monsanto Company, St. Louis, Missouri.

INTRODUCTION

The projected use of trisodium nitrilotriacetate (NTA) in detergent formulations could eventually lead to concentrations in the range of 10-20 mg/liter in domestic wastewaters. If biodegradation did not occur, this would mean 1-2 mg/liter in those surface waters receiving 10% sewage, or proportionally higher or lower concentrations for other circumstances of dilution. However, NTA is biodegradable in the receiving waters as well as in sewage treatment processes,[1] so the final steady state concentrations in surface waters would most likely be another two orders of magnitude lower, 0.1-0.01 mg/liter.

Even though NTA is in itself not significantly toxic to either mammalian[2] or aquatic[1] life, nevertheless questions have been raised about possible environmental consequences. Some of those questions deal with NTA-metal interactions because NTA is a moderately strong complexing agent for bivalent and polyvalent metals.

For use as a detergent raw material, NTA is produced as the trisodium salt monohydrate (I).

$$N \diagdown \begin{matrix} CH_2CO_2Na \\ CH_2CO_2Na \\ CH_2CO_2Na \end{matrix} \cdot H_2O \qquad (I)$$

In aqueous solution the state which NTA assumes is an equilibrium determined in part by the pH of the system, with species ranging from the trianion back

through the di- and monoanion to the undissociated
free acid (Equation 1). The trianion also reacts

$$N(CH_2CO_2H)_3 \overset{OH^-}{\underset{H^+}{\rightleftarrows}} NTA^- \rightleftarrows NTA^= \rightleftarrows NTA^\equiv \qquad (1)$$

with appropriate metal ions to produce the metal
chelate, as illustrated in Equation 2 for the case

$$NTA^\equiv + M^{++} \rightleftarrows NTAM^- \qquad (2)$$

of a bivalent metal. This reaction is also rever-
sible, the chelate dissociating back into the free
metal ion and the ligand as shown. The equilibrium
constant K, as written in Equation 3, is the stability

$$K = [NTAM^-]/[NTA^\equiv][M^{++}] = 10^5 \text{ to } 10^{16} \qquad (3)$$

constant for the chelate, and its magnitude depends
upon the specific metal involved. In natural waters
at the concentrations of interest to us, below 100
mg/liter, the NTA will exist mainly in the chelated
form since the magnesium and calcium ions usually
present will give complexes with stability constants
of 10^5 or 10^6. The heavy metals of interest in this
study give still more stable complexes, with constants in the range of 10^{10} to 10^{13} and up to 10^{16}
for ferric iron.

In systems containing several chelants and
several metals, the equilibrium distribution of all
free and complex species can be calculated from the
stability constants of the various chelant-metal
pairs. Due attention must also be given to pH
effects since protons compete with the metals for
the NTA; the amount of metal chelated diminishes
markedly with pH decreasing below neutrality.
Furthermore, increasing pH may also decrease the
amount of metal chelated because of the competition
of hydroxyl ions with the NTA for the metal ions,
giving hydroxides or oxides for example. Likewise
phosphate, carbonate, silicate, sulfide and other
ions found in natural waters will compete with the
NTA to an extent depending on the solubility product
constants of the corresponding metal salts.

Natural chelants, both inorganic and organic,
occur everywhere in the environment at concentrations
comparable to those contemplated for NTA.[3],[4] As a
first approximation, then, no spectacular effects
on the environment are to be anticipated from the

addition of another chelant, NTA. A more precise estimate of possible effects requires more detailed information, and one very important factor is the biodegradation of the NTA chelates. If they are not biodegradable, accumulation in the environment might occur, and evaluation of the eventual environmental acceptability might be rather complex. On the other hand, if the chelates do degrade the problem becomes much simpler.

Recent publications have indicated the ready biodegradability of many NTA metal chelates, particularly in activated sludge and in soil (Table 47).

Table 47

Biodegradation of NTA Metal Chelates (or of NTA in Presence of the Metal Ions)

	Ca^{++a}	Cd^{++}	Cu^{++}	Fe^{+++}	Hg^{++}	Mn^{++}	Ni^{++}	Pb^{++}	Zn^{++}
Water									
Björndal et al.[4]		o	o+	+	o		o		o
Chau et al.[15c]	+	+	o+	+	o	+	o+	+	+
Activated Sludge									
Björndal et al.[4]		+	+				o		
Swisher et al.[5]				+					
Gudernatsch[6]			o	+[b]		+	o		
Huber et al.[7]	+		o+						
Soil									
Tiedje et al.[8]	+	+	+	+	+	+	+	+	+

+ = Extensive degradation, approaching 90-100%.

o = Poor biodegradation, approaching zero.

[a]Biodegradation of the calcium and magnesium chelates is implicit in many other studies on NTA biodegradation[1] wherever natural waters or synthetic media containing calcium or magnesium salts were used.

[b]Fe^{++}.

[c]Also found that the Al^{+++}, Cr^{+++} and Mg^{++} chelates degraded.

Such biodegradation might occur by either of two mechanisms: (1) attack on the chelate molecule itself, or (2) attack on the free NTA in equilibrium with it, which would be continuously replenished by reequilibration at the expense of the remaining chelate (Equation 2). The first mechanism appears to be operative in the experiments of Bouveng's group[4] and Huber and Popp,[7] who were able to bring about the biodegradation of the Cu^{++} and Cd^{++} chelates by increasing the ratio of calcium ion in their systems. Such a change would decrease the concentration of free NTA, so if mechanism 2 were the only one available the biodegradation should have diminished instead of improved.

The present work was undertaken to examine the biodegradability of a range of NTA metal chelates in river water, as an aid in the evaluation of environmental consequences if NTA were to be used extensively by the detergent industry. This work was begun in 1970, before publication of the references cited in Table 47 (except for one), and interim results have been transmitted to interested scientists and governmental agencies from time to time since then.

TEST CONDITIONS

The water was collected from the Meramec River at Valley Park, Missouri. This stream is not pristine but is relatively unpolluted. The presence of small communities and numerous single dwellings upstream ensures that it is seeded with and reasonably well acclimated to household wastewaters. We have used water from this source since 1957 for biodegradation studies on surfactants and other materials. These waters have been quite consistent in their biodegradation properties since that time, and our results on LAS in the present work are quite in line with those obtained in the past. Wide variations in chemical analysis and bacterial count have been observed in past experiments (Table 48), but these have had no detectable effect upon biodegradation capability.

The waters used in the present work were not analyzed. Each lot was settled for at least two days and then decanted before use; occasionally it was also run through a coarse paper filter. Holding for a week or two before use had no noticeable effect upon its biodegradation capabilities.

One-liter portions of the water were placed in two-liter glass jars and dosed with the test compounds

Table 48
Analyses of Meramec River Water[a]

	Range mg/liter	Average mg/liter	No. of Samples
Total dissolved solids	158-292	220	34
Silica	3-80	19	34
Iron and aluminum oxides	1-27	9	34
Calcium	13-60	76	34
Magnesium	9-33	20	34
Sulfate	13-47	22	34
Chloride	Tr-7	3	34
Alkalinity--carbonate	0-22	11	18
Alkalinity--bicarbonate	90-290	200	34
Sodium	3-19	6	33
Potassium	0-5	2	30
Bacterial count	200-200,000[b]	20,000[b]	36

[a] Samples taken during 1958-1963; a 1972 sample showed Ca 27, Mg 19, Na 6 by atomic absorption. The tabulated chemical analyses were by St. Louis Testing Laboratories, Inc. and bacterial plate counts by Scientific Associates, Inc.

[b] Cells/ml

The NTA was added as the trisodium salt monohydrate (commercial product, 99% pure), but the amounts were calculated as the free acid, mw 191. Five mg/liter was the usual amount. In the first series of experiments, analyzed by the zinc-zincon method, the selected metals were added separately as the chlorides at three levels, the highest being equivalent to the 5 mg/liter of NTA present. The lower concentrations were 0.3 and 0.1 times that amount, leaving most of the NTA unchelated except by the hardness ions present. Along with these three, a fourth sample was set up with NTA and no added metal. To serve as an internal indicator of viability, 5 mg/liter of linear alkylbenzene sulfonate (C_{12} LAS, Standard #2, Soap & Detergent Association) was added to each. An additional set of four samples was made up without

NTA, with only LAS and the metals. The jars were all
stored in the dark and analyzed for NTA and LAS at
intervals to determine the rate and extent of
biodegradation.

In the second series of experiments, analyzed
for NTA by both the zinc-zincon and gas chromatographic methods, six of the metals were added as a
mixture of chlorides in the proportions reported for
their occurrence in US surface waters (Table 49).
As in the earlier series, these also were set up at

Table 49
Six Metals in US Waters, 1962-1967[a]

Metal	Frequency %	Maximum mg/liter	Mean mg/liter	NTA Equiv., mg/liter[b]
Pb	19	0.14	0.02	0.02
Cd	3	0.12	0.01	0.02
Cu	74	0.28	0.02	0.06
Ni	16	0.13	0.02	0.07
Zn	77	1.18	0.06	0.18
Fe	76	4.60	0.05	0.17
				0.52

[a] Data from Kopp and Kroner.[9]

[b] Basis: one molecule of NTA (mw 191) for one atom of metal.

three metal levels, corresponding to equivalence
ratios of 0.1, 0.3 and 1.0 relative to the 5 mg of
NTA present. A second set of mixed metal runs was
made with the iron levels ten times those in the
first set.

The third series of experiments, analyzed using
the specific cupric ion electrode, was made at NTA
concentrations as low as 0.15 mg/liter, generally
in the presence of equivalent amounts of $CuCl_2$.
Polyethylene jars were used instead of glass; later
experiments suggested that glass would have served
equally well.

ANALYTICAL

Total NTA By Zinc-Zincon

The manual procedure of Thompson and Duthie[10] (pp. 317-318) was used with minor modifications. Zinc-zincon reagent is a complex of zinc with a dye, zincon, the complex having a characteristic absorption band. If NTA is present in the sample, it abstracts a corresponding amount of zinc from the zincon and the decrease in absorbance of the band is thus a measure of the amount of NTA present. Hardness ions or other metals interfere since they compete with the zinc for the NTA and for the zincon. In the Thompson-Duthie method the interfering metals are removed by a preliminary treatment of the analytical sample with a cation exchange resin, sodium form.

The sensitivity of the method was increased to about 0.1 mg/liter NTA by using a 25-ml sample containing up to 50 µg, diluting to 50 ml for resin treatment, mixing 40 ml of resin filtrate with 10 ml of zinc-zincon reagent (made with 20 instead of 15 ml of zinc sulfate solution) and measuring absorbance immediately in a 5 cm cell (Cary model 14M spectrophotometer). Under these conditions each microgram of NTA reduces the absorbance by about 0.01 unit, so that the uncertainty should be no more than about 0.1 mg/liter in laboratory samples with known baseline absorbance.

The original Thompson-Duthie procedure does not detect NTA in its more stable chelates such as those with Cu^{++} and Ni^{++} since the fresh sodium form resin does not remove the metal from the complex. However, removal is much improved by using old, recycled, unregenerated resin which has already been used on natural water samples preserved with acid. After a few such uses the resin, partly in the magnesium, calcium, hydrogen form, will liberate and make detectable some 80% or more of copper-chelated NTA. In the absence of adequate characterization and quality control for the old unregenerate resin, knowns containing the metals as well as NTA are included in the analyses. Use of sodium form resin at pH 2-4 gave some response with NTA Cu, but was much less effective.

NTA by Gas Chromatography

The method developed by Warren and Malec[11] was used, as modified for the presence of heavy metals,[12]

wherein the NTA is converted to the tributyl ester.
Knowns containing 0.1 to 5 mg/liter of NTA and the
highest levels of mixed metals used in the biodegradation experiments gave recoveries from 0.86 to
1.38 of the added NTA, comparable to those in the
absence of the metals.

Cupric Ion by Specific Ion Electrode.

The Cu^{++} electrode (Orion Solid State, model
94-29A) was used in conjunction with a single junction electrode (Orion model 90-01, filling solution
Orion 90-00-01), measured with a Beckman Research
pH meter, reproducibility 0.05 millivolt, accuracy
0.1 mv. The sample was agitated by a magnetic
stirrer in a blackened Teflon beaker with black lid
to minimize fluorescent light effect. Approximately
10-15 minutes were required for the millivolt reading
to drop to its final value. Four calibration runs
are shown in Figure 38, three of them on one sample

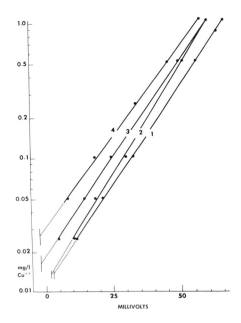

Figure 38. Calibration of cupric ion electrode. Electrode
potentials in river water containing added $CuCl_2$;
vertical lines at left show initial potential of
the water without $CuCl_2$. Curves 1, 2 and 3 come
from a single sample of water (series BO) on three
different days, curve 4 from a different sample
(Series BP).

of water on three different days. The millivolt
readings for the water before addition of any $CuCl_2$
would seem to indicate that somewhat over 0.01
mg/liter of Cu^{++} was already present, but atomic
absorption analysis indicated less than 0.01. In
any case, 0.05 mg/liter of Cu^{++} was readily detected
by the electrode.

LAS by Methylene Blue

The Hellige modification of the methylene blue
procedure[13] was used. A 50-ml sample containing up
to 100 µg of LAS was shaken with 15 ml of methylene
blue reagent and 10 ml of chloroform. LAS (and any
other anionic surfactants present) passed into the
chloroform as the methylene blue salt and its amount
was estimated by comparison of the blue color with
a set of glass standards calibrated from 0.1 to 2
mg/liter.

RESULTS AND DISCUSSION

Individual Metals

The results with lead are shown in Figure 39.
The upper section gives the decrease in NTA content
of the river water samples as time passed. Sample A
had no lead, and its 5 mg/liter of NTA was gone by
the eighth day. Sample B contained 0.5 mg/liter of
lead, and it was definitely slower. The dotted line
crossing line B is at 0.5, the amount of NTA corresponding to the added lead. If the chelate had not
degraded, curve B would have leveled off there. The
fact that it penetrates on down to zero shows that
the chelate did indeed degrade. The portion of curve
B above the dotted line corresponds to NTA unchelated
by lead. It, too, is slower than curve A, indicating
that the lead retards the degradation of that portion
of the NTA as well.

Curve C shows the next higher level of lead,
1.5 mg/liter. We can offer no explanation for the
rapid disappearance of the NTA in this case. What-
ever the reason, a second 5 mg/liter of NTA added
to this sample took between 11 and 14 days to disappear (not shown in the figure), quite in harmony
with runs B and D.

In Sample D, 5 mg/liter of lead was present,
sufficient to chelate all of the NTA. Degradation
was somewhat slower, but unquestionably it did
degrade.

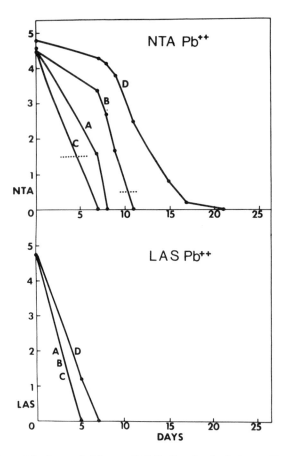

Figure 39.   *Biodegradation of NTA lead chelate. Upper: NTA*
content of river water initially containing 5
mg/liter NTA, 5 mg/liter LAS and (A,B,C,D) 0,
0.54, 1.62, 5.40 mg/liter Pb⁺⁺. Lower: LAS
content of parallel runs without NTA.

    The lower section of Figure 39 gives the parallel
set of runs using LAS without NTA. Here the lead had
little if any delaying action, indicating that the
LAS-degrading enzyme systems, or the LAS-degrading
bacteria, were less sensitive to lead.
    LAS biodegradation data in the presence of the
metals are further shown in Table 50. Here the LAS
degradation time in the absence of NTA (which is
also shown in Figure 39 and the succeeding ones) is
compared with that in the presence of NTA. In most
cases where LAS degradation was delayed by the

Table 50

LAS Biodegradation in Presence of Metals[a]

Metal	Initial NTA mg/liter	LAS Degradation Time, Days[b]			
		A	B	C	D
Pb^{++}	0	5	5	5	7
	5	5	5	5	5
Hg^{++}	0	5	15	15	>31
	5	5	8	21	37
Cd^{++}	0	4	6	12	10
	5	4	6	6	6
Cu^{++}	0	4	6	7	39
	5	4	4	4	6
Ni^{++}	0	4	4	6	11
	5	4	4	4	4
Zn^{++}	0	4	4	4	11
	5	4	4	6	4
Fe^{+++}	0	4	4	4	4
	5	4	4	4	4
6 Metals	0	5	5	5	7
	5	5	5	5	5
6 Metals with excess Fe	0	4		4	4
	5	4		4	4

[a] Initial LAS concentration 5 mg/liter in all cases.

[b] Metal concentrations in A, B, C, D are equivalent to 0, 0.5, 1.5, 5 mg/liter NTA; see Figures 39-47. LAS degradation measured by methylene blue analysis; degradation times may have been shorter in many cases since analyses were seldom made daily.

presence of metal without NTA, the delay was significantly shorter when NTA was present along with the metal. Chelation of the metal by NTA thus reduces its interference with the LAS biodegradation process significantly. In the case of lead, only at the highest level was any interference noticeable; here too the protective action of the NTA was quite evident.

Mercury gave a somewhat different pattern (Figure 40). Here the increased toxicity with

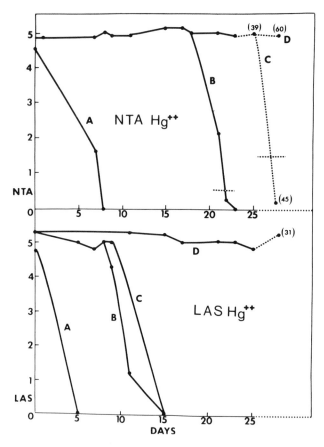

Figure 40. *Biodegradation of NTA mercury chelate. As in Figure 39 except that A, B, C, D are 0, 0.53, 1.58, 5.25 mg/liter Hg^{++}.*

increased mercury content is quite evident on both LAS and NTA. But at the two lower concentrations the bacteria eventually acclimated, and once they did, disappearance of both mercury-free and mercury-chelated NTA was rapid. Thus at the middle level, 1.5 mg/liter, curve C, acclimation took some 39 days, but the NTA, including mercury-chelated, was all gone 6 days later. LAS in the presence of the NTA did not degrade consistently faster than in its absence (Table 50), so there is no strong indication that the NTA reduced the toxicity of the mercury by chelating it.

Cadmium showed yet another pattern (Figure 41). LAS degradation was only moderately retarded, even at the highest cadmium level, whereas the NTA chelate was degraded only at the lowest cadmium level. LAS degradation in the presence of NTA was more rapid than in its absence, at the two higher metal levels (Table 50), again indicating a protective effect of the NTA.

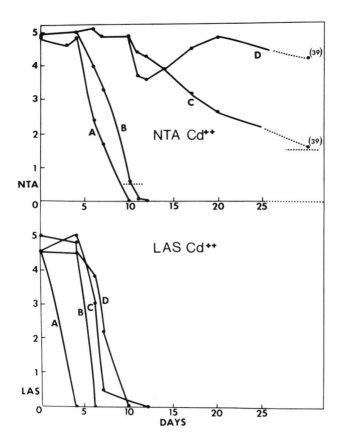

Figure 41. *Biodegradation of NTA cadmium chelate. As in Figure 39 except that A, B, C, D are 0, 0.29, 0.88, 2.94 mg/liter Cd⁺⁺.*

Copper, in contrast (Figure 42), seems to have affected the LAS relatively more than it did the NTA. At the two lower levels the NTA disappeared rather

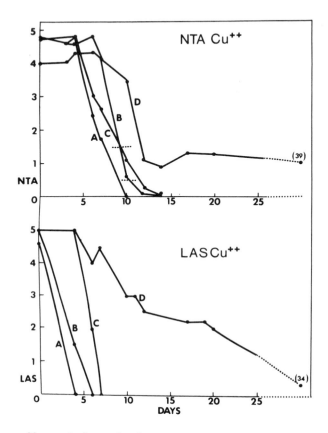

Figure 42.   Biodegradation of NTA copper chelate. As in
              Figure 39 except that A, B, C, D are 0, 0.17,
              0.50, 1.66 mg/liter Cu++.

quickly, including that chelated by the copper. At
the highest copper level, where all the NTA was
chelated, the degradation stopped short of completion.   We might speculate that by that time enough
free copper ion had been liberated to exert a toxic
action.   In the presence of NTA, LAS degraded
within 4 to 6 days at all copper levels, indicating
that NTA-chelated copper was much less toxic for
LAS biodegradation (Table 50).

Nickel and zinc (Figures 43 and 44) likewise interfered with the NTA degradation at the highest metal concentration, whereas the iron chelate degraded readily at all three levels (Figure 45). Nickel and zinc inhibited LAS degradation, but NTA protected against this action (Table 50).

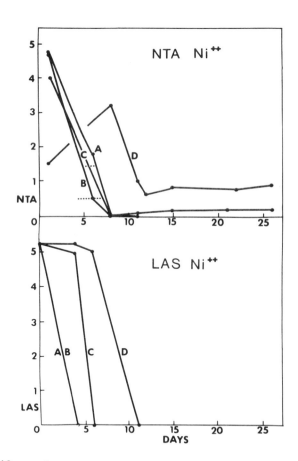

Figure 43.   *Biodegradation of NTA nickel chelate. As in Figure 39 except that A, B, C, D are 0, 0.15, 0.46, 1.54 mg/liter Ni++.*

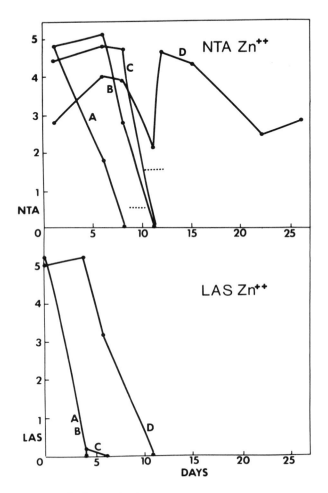

Figure 44.   Biodegradation of NTA zinc chelate. As in
              Figure 39 except that A, B, C, D are 0, 0.17,
              0.51, 1.71 mg/liter Zn^{++}.

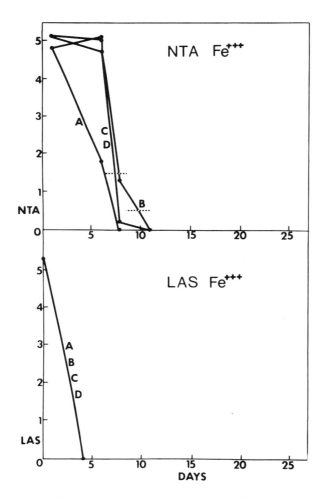

Figure 45.   *Biodegradation of NTA iron chelate.   As in*
             Figure 39 except that A, B, C, D are 0, 0.15,
             0.44, 1.46 mg/liter Fe^{+++}.

The foregoing results must remain largely
descriptive for the present; interpretations in
terms of the chemical and biological mechanisms
involved can be little more than speculations at
this time in view of the great complexity of these
river water systems and the scanty information on
them. The toxic and inhibitory effects shown by
the metals may well result from their interference
with various bacterial enzyme systems, perhaps with
some of those directly involved in initiating the
NTA or LAS biodegradation processes, or perhaps with
one or more of those necessary for the basic life
processes themselves. Likewise, the means by which
the cells eventually overcome this interference have
not been clarified. And in the one case where interference with degradation was not observed (the case
of Fe^{+++}), it was not obvious whether this was simply
because of lack of inherent toxicity of the ion
itself, or because of the extremely tight binding
of the ion in the form of the NTA chelate or of
hydrated oxide or other species formed in equilibrium
with the various components of the river water.

Mixed Metals

     As pointed out above, if NTA enters the natural
environment it will exist as an equilibrium mixture
of different chelates. In simulation of this, some
runs were made using metals in a mixture instead of
individually. The proportions were based on recent
analyses of rivers and lakes (Table 49). Our lowest
level contained the six metal ions at the mean concentrations found in the survey, as listed in the
fourth column. Mercury was not included because it
occurs only rarely--in 7% of the samples--and then
at very low concentrations, averaging only 0.004
mg/liter in those where it was found.[14] If included
in the mixture at higher levels, comparable to the
other metals, its toxic effects (Figure 40) would
probably have obscured any biodegradation information.
     In addition to the tests at the normal mean
levels, the six metal mixture was also run at 3
times and 10 times those concentrations. At 10
times, the metals are present in amounts near the
maxima reported in the survey (Table 49, column 3).
     The mixed metal results (Figure 46) resemble
those from the individual metals. At the two lower
levels NTA degradation proceeded normally, but at
10 times, curve D, degradation was slower and
stopped short of completion. The increase in NTA

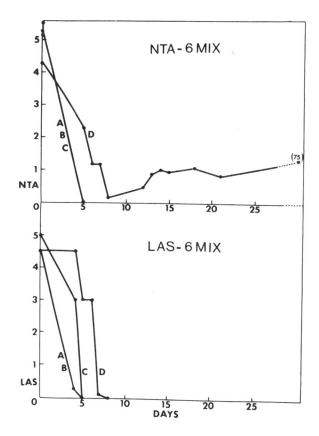

Figure 46. *Biodegradation of NTA mixed chelates. As in Figure 39 except that A, B, C, D are 0, 1, 3, 10 times the metals shown in Table 49, mean mg/liter.*

in curve D after 8 days is of course only apparent, since resynthesis of NTA in the system is most unlikely. It must be an analytical artifact, resulting most likely from outright errors in analyzing the samples at days 8 and 12, or perhaps from transient interferences with the analyses, either negative ones during the earlier part of the run or positive ones developing in the later stages.

The same general patterns evident in Figure 46 were found again in later confirmatory runs made at the two higher metal levels. The circles in the upper section of Figure 47 show the 3 times level

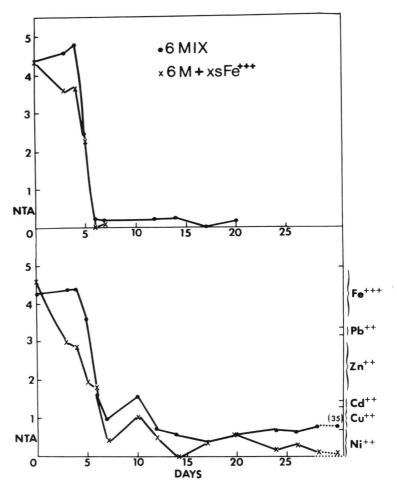

*Figure 47.   Biodegradation of NTA mixed chelates.   Upper: •,
NTA content of river water initially containing
5 mg/liter NTA, 5 mg/liter LAS and 3 times the
metals shown in Table 49, mean mg/liter.   x,
same except with 10 times as much Fe[+++].   Lower:
same except with 10 times the metals shown in
Table 49.   Scale at lower right shows NTA
equivalences of the metals present.   See Table
51 for gas chromatographic analyses.*

and in the lower section the 10 times, where again
NTA degradation fell noticeably short of completion.
The residual NTA corresponds roughly to the amount
of nickel present, or perhaps a little more, judging
by comparison with the scale at the lower right.
    The crosses in Figure 47 mark two companion
runs in which an excess of ferric iron was added
along with the other metals, to the extent of 10
times the amount present in the base runs. The idea
here was to force a higher proportion of the NTA
into the iron chelate form, a more readily biodegradable one. This hope was realized to a considerable extent, the NTA running significantly
below that of the companion experiment throughout.
    All of the preceding results have been based
on the zinc-zincon analytical method. As mentioned
above, this has an uncertainty of around 0.1 mg/
liter under favorable conditions, and possibly
several tenths when complicating factors such as
metals are present. Thus it was important that
these results be verified using other, independent
analytical methods. That has been done in two ways,
using gas chromatography and by the specific cupric
ion electrode.

Gas Chromatography

    Certain key samples in the 6-metal runs were
analyzed for NTA content using the GC method (Table
51). The zero day analyses, when the 5 mg/liter of

Table 51

Biodegradation of NTA Mixed Chelates (Figure 47)

Mixed Metal Conc.[a]	Analytical Method[b]	Apparent NTA, mg/liter			
		0 d.	7 d.	35 d.	39 d.
3x Natural	ZZ	4.4	0.2		
	GC	3.1	0.15		
3x + excess Fe^{+++}	ZZ	4.4	0.05		
	GC	3.6	0.06		
10x Natural	ZZ	4.3	1.0	0.8	
	GC	2.2	0.6		0.5
10x + excess Fe^{+++}	ZZ	4.6	0.4	0.05	
	GC	2.9	0.7		0.05

[a] Based on Table 49 , mean mg/liter.

[b] Zinc-zincon or gas chromatography.

NTA had first been added, show zinc-zincon results
ranging from 4.3 to 4.6 and the GC from 2.2 to 3.6.
We are at a loss to explain the poor recovery in
these GC analyses, but even so the later analyses
agree reasonably well and the substantial disappearance of the NTA in the runs with excess iron was
confirmed.

These GC results give reassurance that the
conclusions drawn from the zinc-zincon analyses are
valid and that the NTA chelates are indeed being
degraded in these systems.

Cupric Ion Electrode

Still another independent means of verification
is offered by the specific cupric ion sensing electrode. It does not respond to NTA-chelated copper,
so that if the electrode does show free Cu^{++} ion
appearing we can be rather sure that the chelate is
being degraded.

One potential difficulty in these river water
systems is that the free copper ion may precipitate
out, or react with natural chelants, and thus evade
detection. To correct for this we set up comparison
samples with copper but without NTA. Three such
comparisons are illustrated in Figure 48. In the
lower part the crosses show the Cu^{++} content in that
control to which 0.5 mg/liter has been added. By
weight, copper ion corresponds to 3 times its amount
of NTA and for convenience the scale at the left
is graduated in terms of 3 times the Cu^{++} content.
Thus the dotted line at 1.5 signifies the 0.5 mg/
liter of Cu^{++} added and the crosses just below it
show that a substantial amount of it remained in
solution in the Cu^{++} form throughout the run.

The circles show the parallel sample which
along with the copper received the equivalent 1.5
mg/liter of NTA. At the beginning only a small
fraction of this copper was detectable by the
electrode, but after 5 days this sample had the
same amount of free cupric ion as did the control.
The NTA copper chelate had degraded. The middle
and upper sections of Figure 48 show parallel runs
with 0.15 and 0.05 mg/liter copper, one of each
pair containing also an equivalent amount of NTA.
The same pattern of biodegradation shows up in
each.

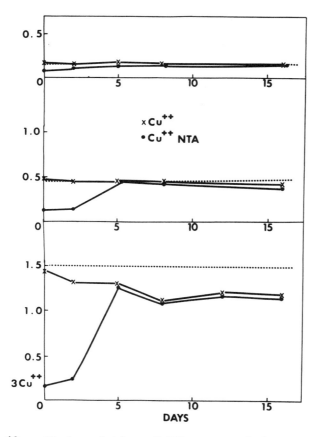

Figure 48. *Biodegradation of NTA copper chelate. Upper:*
•, *Cu^{++} content of river water (Series BP)*
initially containing 0.05 mg/liter Cu^{++}, 0.15
mg/liter NTA. x, *same except without NTA.*
Middle: same except with 0.15 mg/liter Cu^{++},
0.45 mg/liter NTA. Lower: same except with
0.5 mg/liter Cu^{++}, 1.5 mg/liter NTA.

Sometimes the results are not quite so clear
cut, as seen in the earlier run presented in Figure
49. Here the initial water was perceptibly more
turbid, and the free cupric ion precipitated out (or
disappeared in some other manner) quite extensively

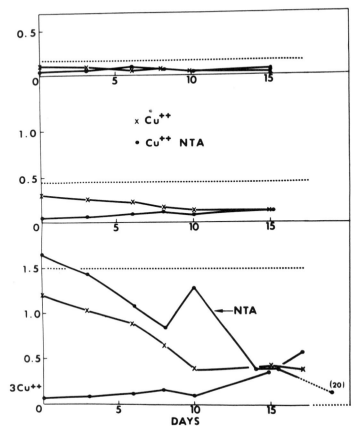

*Figure 49.   Biodegradation of NTA copper chelate.   Same as
Figure 48 except for different river water
(Series BO).   Curve marked "NTA" shows NTA
content in mg/liter by zinc-zincon analysis.*

as shown by the steady decline of the crosses in
each case.   But here too the NTA chelate degraded--
the cupric ion content of the NTA systems rose to
eventually match the unprecipitated cupric ion in
the control systems.

In this set the Cu^{++} measurements were supple-
mented in the 1.5 mg/liter NTA system by zinc-zincon
analyses (the other two levels, 0.45 and 0.15 mg/liter
NTA, were too close to the uncertainty limit of the
zinc-zincon method).   Disappearance of the NTA was
reasonably synchronous with appearance of cupric
ion, as seen in the lower part of Figure 49.

   The upper section of Figure 50 illustrates an
experiment wherein 5 mg/liter of glucose was present
along with 5 mg/liter of NTA and 0.5 mg/liter of
Cu^{++}, parallel with the glucose-free run at the
bottom of Figure 48.   Chelate degradation was delayed to about 15 days compared to the 5 days in
Figure 48.   Such delays in biodegradation upon
addition of glucose to another substrate are not
uncommon.

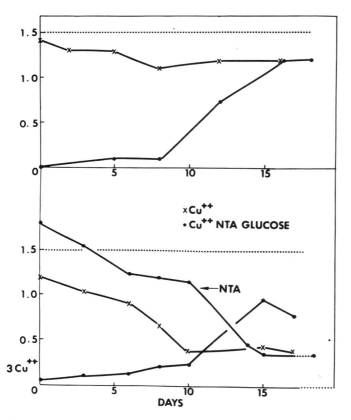

Figure 50.   *Biodegradation of NTA copper chelate. Upper:* •*,*
            Cu^{++} content of river water (Series BP) initially
            containing 0.5 mg/liter Cu^{++}, 5 mg/liter NTA, 5
            mg/liter glucose. x, same except without NTA
            and glucose. Lower: •*, Cu^{++} content of river*
            water (Series BO) initially containing 0.5 mg/liter
            Cu^{++}, 1.5 mg/liter NTA, 8.5 mg/liter glucose; curve
            marked "NTA" shows NTA content in mg/liter by zinc-
            zincon analysis. x, same except without NTA and
            glucose.

At the bottom of Figure 50 is another glucose run, parallel with the one at the bottom of Figure 49. This was made in the water which had the copper precipitating property. Here, at 15 days, the cupric ion liberated from the chelate temporarily exceeded the amount remaining detectable in the parallel control. But it too appears to be precipitating and heading downward at the end. Here again zinc-zincon analyses showed NTA disappearance at about the same time as cupric ion appearance.

CONCLUSION

The present experiments show that a variety of NTA metal chelates are biodegradable in river water at environmentally realistic concentrations, confirming and supplementing the results of other workers. Thus there is no reason to expect that NTA chelates would accumulate in the environment even if extensive amounts of NTA were used in detergent formulations.

ACKNOWLEDGMENT

We are grateful to Dr. W. M. Haynes for advice and preliminary experiments on the suitability of the cupric ion electrode as used in this work. Many of the zinc-zincon analyses were made by G. L. Mikules.

REFERENCES

1.   Thom, N. S. "Nitrilotriacetic Acid: A Literature Survey," *Water Research 5*, 391 (1971).
2.   Nixon, G. A., E. V. Buehler, and R. J. Niewenhuis. "Two-Year Rat Feeding Study with Trisodium Nitrilotriacetate and its Calcium Chelate," *Toxicol. Appl. Pharmacol. 21*, 244 (1972).
3.   Stiff, M. J. "The Chemical States of Copper in Polluted Fresh Water and a Scheme of Analysis to Differentiate Them," *Water Research 5*, 585 (1971).
4.   Björndal, H., H. O. Bouveng, P. Solyom, and J. Werner. "NTA In Sewage Treatment. 3. Biochemical Stability of Some Metal Chelates," *Vatten 28*, 5 (1972).
5.   Swisher, R. D., M. M. Crutchfield, and D. W. Caldwell. "Biodegradation of Nitrilotriacetate in Activated Sludge," *Environ. Sci. Technol. 1*, 820 (1967).
6.   Gudernatsch, H. "Verhalten von Nitrilotriessigsäure im Klarprozess und im Abwasser," *Gas-Wasserfach 111*, 511 (1970).

7. Huber, W. and K. H. Popp. "Der biologische Abbau der Nitrilotriessigsäure in Gegenwart von Cadmium-Ionen," *Fette-Seifen-Anstrichmittel 74*, 166 (1972).

8. Tiedje, J. M. and B. B. Mason. "Biodegradation of Nitrilotriacetic Acid (NTA) in Soils," *Biological Proceedings 1971*, Abstract A2; 71st Annual Meeting, American Society for Microbiology, Minneapolis, May, 1971.

9. Kopp, J. F., and R. C. Kroner. "A Five Year Summary of Trace Metals in Rivers and Lakes of the United States (1962-1967)," FWPCA, Division of Pollution Surveillance, Cincinnati.

10. Thompson, J. E. and J. R. Duthie. "The Biodegradability and Treatability of NTA," *J. Water Pollution Control Federation 40*, 306 (1968).

11. Warren, C. B. and E. J. Malec. "Quantitative Determination of Nitrilotriacetic Acid and Related Aminopolycarboxylic Acids in Inland Waters," *J. Chromatography 64*, 219 (1972).

12. Warren, C. B. and E. J. Malec. "Biodegradation of Nitrilotriacetic Acid and Related Imino and Amino Acids in River Water," *Science 176*, 277 (1972), footnote 15.

13. Swisher, R. D., J. T. O'Rourke, and H. D. Tomlinson. "Fish Bioassays of LAS and Intermediate Biodegradation Products," *J. Am. Oil Chemists' Soc. 41*, 746 (1964).

14. Durum, W. H., J. D. Hem, and S. G. Heidel. "Reconnaissance of Selected Minor Elements in Surface Waters of the United States, October 1970," *Geological Survey Circular 643*, Washington, 1971.

15. Chau, Y. K. and M. T. Shiomi. "Complexing Properties of Nitrilotriacetic Acid in the Lake Environment," *Water, Air, Soil Pollution 1*, 149 (1972).

9. THE CHEMISTRY OF FULVIC ACID AND ITS REACTIONS WITH METAL IONS

Donald S. Gamble and Morris Schnitzer. Soil Research Institute, Research Branch, Canada Agriculture, Ottawa, Ontario K1A OC6 Canada

INTRODUCTION

Humic materials are widely distributed in nature, and arise from the chemical and biological degradation of plant materials and from synthetic activities of microbes. Those found in soils are polyelectrolytes having ranges of molecular weights and chemical properties quite comparable to those of the synthetic polyelectrolyte systems, such as polyacrylic and polymethacrylic acid.[1-4] Christman and Minear have stated that organic materials appear to be present to some extent in most natural waters and in some cases are sufficiently concentrated to make the water yellow or brown in color.[5] Christman and Ghassemi[6] and Sarkanen[7] have suggested that these organic materials are dissolved from living woody tissues, decaying wood, soil organic matter, or a combination of these. Stumm and Morgan have indicated that the range of concentrations of organic materials in surface waters is 0.1 to 10 mg/liter.[8] The practical importance of organic matter in surface waters has been recognized by the publication of a colorimetric method of analysis for what are described as "tannin, lignin, tannin-like, lignin-like compounds, or hydroxylated aromatic compounds."[9]

Black and Christman have verified[10] an earlier report by Saville[11] that the organic matter found in water is negatively charged. This is consistent with the view that it is a humic polyelectrolyte. Christman and Ghassemi have carried out chemical degradation experiments from which they concluded

265

that the colored organic material found in water is
chemically very similar to soil humic materials.[6],[12]
However, Christman and Minear have concluded from
published reports that samples of organic matter
from different surface waters have significantly
different molecular size distributions.[5],[13-16]

Stumm and Morgan have expressed doubts about
the existence of humic-metal ion complexes in
natural waters.[8] There is, however, convincing evidence in the literature to show that such complexes
exist just as they do with soil humic materials.
Shapiro found that Fe^{III} was retained in solution
by dissolved humic material,[17] while Barsdate reported that a considerable part of the cobalt,
manganese, and zinc dissolved in strongly colored
lake water could not be removed by dialysis.[18]
Further confirmation comes from the work of Barsdate
and Matson,[19] who found that humic materials recovered
from surface waters react strongly with various metal
ions. Their metal ion complexing was qualitatively
comparable to that reported by Basu *et al.*,[20] for
soil humic acids. Finally, the reaction between
these materials and Fe^{III} alters both fluorescence
and Sephadex column elution behavior.[16],[21]

Duursma has been quoted as follows:[22],[23] ". . .
the problem of chelation in the sea and also in fresh
waters between trace metals and dissolved organic
matter cannot be solved in a simple way and many
efforts are necessary to study, on the one hand,
the complete chemical composition of the dissolved
organic matter, and on the other hand, their
chelating stability constants in relation to all
cations present."

Since this is equally true for soil chemistry,
and considering the evidence quoted above to show
that comparable sorts of humic material-metal-ion
complexing occur in both soils and natural waters,
the soil chemistry progress that has been achieved
will be described in detail. In particular, fulvic
acid is being used as a model humic material because
of its solubility and because of its extensive
characterization.

CHEMICAL COMPOSITION AND
CHEMICAL STRUCTURES

Description of the Fulvic Acid Sample

A fulvic acid extracted from the Bh horizon of
an imperfectly drained Podzol soil in Prince Edward

Island, Canada, has been under continuous investigation in the Soil Research Institute of the Canada Department of Agriculture for more than 15 years. This fulvic acid is relatively stable chemically and biologically, and vapor pressure osmometry measurements indicate that it is a low polymer with a number-average molecular weight of 951.[24-26] The percent elemental analysis is 50.90, C; 3.35, H; 0.75, N; 0.25, S; and O, 44.75.[27]

Functional Groups

Schnitzer and co-workers[2-6] used a number of experimental methods to determine the types of functional groups present in fulvic acid. Their techniques included functional group analysis, conductivity titrations, back titrations, and infrared spectrophotometry. The titration methods revealed acidic functional groups, and infrared spectra confirm the preponderance of functional groups containing oxygen. The absorption bands are broad because of extensive overlapping of individual absorptions. The results of this work are summarized in Tables 52 and 53.[27] The number of carboxyl groups

Table 52

Infrared Absorption Bands of Fulvic Acid

Absorption band cm^{-1}	structural feature
3400	hydrogen-bonded OH
2900	aliphatic C-H stretch
1725	C=O of COOH; C=O stretch of ketonic carbonyl
1630	aromatic C=C; hydrogen-bonded C=O of carbonyl; COO⁻
1400	COO⁻
1200	C-O stretch; OH deformation of COOH

Table 53

Fulvic Acid Functional Groups

Functional Group	Amount $\left(\dfrac{m\ equiv.}{g\ fulvic\ acid}\right)$
COOH	7.71 ± 0.25
Phenolic OH	3.3
Alcoholic OH	3.6
Ketonic C=O	2.5
Quinoid C=O	0.6

given in Table 53 was obtained from the second equivalence point of potentiometric titrations.[28] The chemical formula calculated from these data is $C_{28}H_{16}(CO_2H)_8(OH)_7(CO)_3$.

Chemical Structure

Schnitzer and various co-workers have used a number of different experimental methods for determining the chemical structure of fulvic acid. Each of these will now be outlined in turn.

Oxidation with 7.5 N HNO₃[20]

This yields 0.7% aliphatic dicarboxylic acids (succinic, glutaric, adipic and primelic acids), 5.5% nitrophenols (mainly picric acid) and 1.5% benzene-polycarboxylic acids.

Reductive degradation[30]

Reductive degradation by Zn dust distillation and fusion yields polycyclic aromatics, the most prominent of which are naphthalene (0.1%), phenanthrene (0.1%), anthracene (0.2%), pyrene (0.1%), and perylene (0.2%).

Exhaustive methylation and permanganate oxidation

In 1963, Barton and Schnitzer[31] proposed a new approach to unraveling the chemical structure of

fulvic acid. They described the exhaustive methylation of fulvic acid, which can serve at least three
different purposes. First, methylation makes most
of the fulvic acid soluble in benzene. Second, it
makes oxidation products sufficiently volatile for
gas chromatographic and mass spectrometric analyses.
In both cases, the methylation facilitates separation
of sample components. Finally, methylation prior
to alkaline permanganate oxidation protects phenolic
OH groups from electrophilic attack by the $KMnO_4$.
These advantages have been exploited with three
types of experiments.

The first type requires methylation but no
oxidation. The fulvic acid was first exhaustively
methylated in order to make most of it soluble in
benzene. The benzene-soluble methyl esters and
ethers were separated over Al_2O_3 with the aid of
solvents of increasing polarity into chromatograph-
ically homogeneous fractions which differ in molecular
weights, oxygen-containing functional groups and
spectroscopic properties. Ogner and Schnitzer[32]
and Khan and Schnitzer[33] have recently extended and
modified this approach. Following preliminary
separation over Al_2O_3, each fraction is further
separated by TLC and preparative gas chromatography
into relatively pure components which were identified by matching their mass and micro-IR spectra
with those of standards of known structures.

The second and third columns of Table 54
indicate the amounts of each of the various compounds
isolated and identified. In toto, 1.03 g of compounds
were isolated from 100 g of fulvic acid. Since preliminary experiments with known fully methylated
phenolic and benzenecarboxylic acids showed that at
least 50% of the starting material was lost during
the lengthy fractionation and purification procedure,
it is likely that the compounds listed in Table 54
may account for up to 2% of the initial fulvic acid.

It should be borne in mind that these compounds
were isolated from the fulvic acid without degradation by chemical methods in the laboratory. Thus,
about 28% of the material identified consists of
fully methylated phenolic acids, 19% of benzenecarboxylic acid methyl esters, 40% of dialkyl
phthalates and 13% of alkanes and fatty acids.
Methoxy (hydroxy) benzenepentacarboxylic acid and
benzenehexacarboxylic acid account for about 45% of
the fully methylated phenolic and benzenecarboxylic
acids, respectively.

The next type of experiment consisted of permanganate oxidation of unmethylated fulvic acid.

Table 54

Investigation of Fulvic Acid Composition and Structure by Methylation and Permanganate Oxidation

Item No.	Compounds Identified	Amounts recovered from separations after various treatments: mg/g fulvic acid		
		methylated & unoxidized	oxidized & unmethylated	methylated & oxidized
1	Dimethoxy-benzoic acid methyl esters	0.075		
2	Trimethoxy-benzoic acid methyl esters	0.032		
3	Methoxy-benzenedicarboxylic acid dimethyl esters	0.077		
4	Dimethoxy-benzenedicarboxylic acid dimethyl esters	0.231		
5	Methoxy-benzenetricarboxylic acid trimethyl esters	0.541		11.91
6	Dimethoxy-benzenetricarboxylic acid trimethyl esters	0.144		
7	Methyl-dimethoxy-benzenetricarboxylic acid trimethyl esters	0.046		
8	Methoxy-benzenetetracarboxylic acid tetramethyl esters	0.649	0.35	47.55
9	Dimethoxy-benzenetetracarboxylic acid tetramethyl esters	0.076		
10	Methoxy-benzenepentacarboxylic acid pentamethyl esters	1.00	11.57	24.55
11.	Dehydrodiveratric acid dimethyl esters	0.105	0.16	5.46
12	Benzenedicarboxylic acid dimethyl esters	0.060	3.93	2.56
13	Benzenetricarboxylic acid trimethyl esters	0.105	6.21	3.30
14	Benzenetetracarboxylic acid tetramethyl esters	0.259	37.34	31.70
15	Benzenepentacarboxylic acid pentamethyl esters	0.347	27.27	44.65
16	Benzenehexacarboxylic acid hexamethyl esters	1.17	5.46	32.68
17	Dialkyl phthalates	4.10		
18	C_{14}-C_{36} alkanes	0.78		
19	C_{14}-C_{36} fatty acids	0.50		
	Total identified	10.30	92.49	204.36
	Not identified		5.12	20.43

The procedure entailed oxidation with aqueous 4% (w/v) KMnO$_4$ at pH 10, extractions of the oxidation products into ethyl acetate, methylation and separation of the esters and ethers by preparative gas chromatography into relatively pure components which were then analyzed by mass spectrometry and micro-infrared spectrophotometry. A matching of the mass and IR spectra and gas chromatographic retention times of the isolated components with those of authentic specimens led to the identification of 1.5% aliphatic monocarboxylic acids (ranging from acetic to caprylic acids), 3.0% oxalic acid and 1.7% benzenecarboxylic acids (ranging from benzene di- to hexa-carboxylic acids).

The permanganate oxidation of exhaustively methylated fulvic acid was the third type of experiment. This degradative technique has been widely used for the elucidation of the main structural components of complex organic substances such as coal, wood, and lignin, but so far has seldom been applied to humic materials. The oxidation, separation and identification techniques already described above were used here.

Compounds produced by the permanganate oxidation of unmethylated and methylated fulvic acid are shown in Table 54. Methylation prior to oxidation increased the yield of total products by more than 200% and also allowed for the isolation of substantial amounts of phenolic compounds. As shown in Table 54, benzenecarboxylic acids constitute about 65% of the oxidation mixture, and phenolic acids the remaining 35%. The major product isolated from both the unmethylated and methylated fulvic acid is benzenepentacarboxylic acid. Other compounds produced by the oxidation of the methylated fulvic acid in relatively large amounts are benzenehexacarboxylic acid, 5-methoxy-1,2,3,4-benzene-tetracarboxylic acid and methoxy(hydroxy)-benzenepentacarboxylic acid. Only very small amounts of aliphatic dicarboxylic acids were identified. The origin of the phenolic and benzenecarboxylic acids is difficult to assess. Condensed lignin structures and/or complex organic structures of microbial origin are likely precursors of the compounds isolated and identified.

X-ray Diffraction

X-ray analysis indicates that the carbon skeleton of fulvic acid consists of a broken network of poorly condensed rings with appreciable numbers of disordered

aliphatic chains or alicyclic structures around the
edges of the aromatic layers.[34]

Most of the alkanes, fatty acids and dialkyl
phthalates isolated from fulvic acid are water-
insoluble. In the presence of fulvic acid, however,
these substances become solubilized in water.[35-37]
For example, Matsuda and Schnitzer have recently
demonstrated that fulvic acid retains in solution
relatively large amounts of certain dialkyl
phthalates.[38] In addition to this, these same
alkanes, fatty acids and dialkyl phthalates are
quite difficult to extract from fulvic acid with
organic solvents. Only about 3% of the total
alkanes present could be extracted this way. Less
than 10% of the fatty acids and only traces of the
dialkyl phthalates have been extracted. Some sort
of interaction, either physical or chemical, evidently occurs between these hydrophobic molecules
and the fulvic acid.

The complete separation of these hydrophobic
compounds from fulvic acid has been achieved, however, by Schnitzer and his co-workers during their
fulvic acid fractionation work described above.
About one-third of the total alkanes recovered after
methylation and separation consisted of n-alkanes
which ranged from C_{14} to C_{36} with a C-odd to C-even
ratio of 1.0.[35] The remaining alkanes were
branched-cyclic. The n-fatty acids ranged from
C_{16} to C_{36}, with an overall C-even to C-odd ratio
of 2.3. The dialkyl phthalates found in this way
included dibutyl phthalate, dicyclohexyl phthalate,
bis(2-ethylhexyl) phthalate, and dioctyl phthalate.
The exhaustive methylation of fulvic acid probably
facilitates the separation of the alkanes, fatty
acids and dialkyl phthalates in the same way as it
does the separation of the other fulvic acid components. Schnitzer has concluded from these
solubility and separation results that a substantial
amount of hydrogen bonding exists in fulvic acid.
The infrared data in Table 52 are consistent with
this view.

The conclusions drawn about fulvic acid structure from the available evidence are that the
chemical degradation and separation experiments
prove the existence of aromatic rings, and the
X-ray diffraction results indicate that these rings
form an open structure in a two- or three-dimensional
array. This open structure is held together at least
partly by hydrogen bonds among the functional groups.

Schnitzer and his co-workers have concluded that the fulvic acid structure may typically be represented in part by Figure 51.

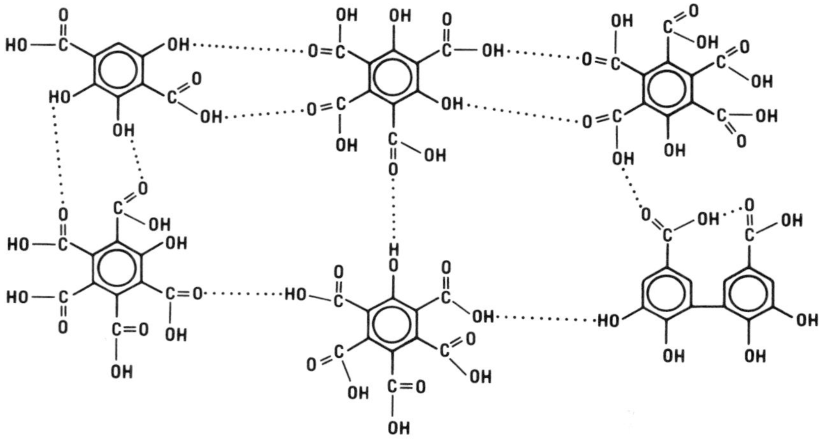

Figure 51. *A possible example of fulvic acid structure.*

Christman's chemical degradation work referred to in the Introduction consisted of two types of experiments. In one, Christman and Ghassemi carried out mild oxidative degradation with alkaline copper oxide in a sealed bomb, under nitrogen at 180°C.[6] The other was the reductive degradation with sodium amalgam under total reflux at atmospheric pressure.[12] Christman and Minear have put more emphasis on the importance of ether and ester linkages than have Schnitzer and co-workers. Their conclusions about fulvic acid structure are, however, essentially the same for samples from different geographical locations.[5]

REACTIONS WITH CATIONS

Weak Acid Properties:
Formulation of the Titration Problem

Because of its structure and functional groups, fulvic acid constitutes a weak acid polyelectrolyte in aqueous solution. Its acid dissociation proper-ties are best investigated by potentiometric titration.[28,39] For this purpose the following points about the polyelectrolyte chemistry of fulvic acid are pertinent:

(a) Because of the very irregular polymer structure which contains no identifiable monomer units,[34] one must assume, as the most general case, that no two carboxyl groups are inherently chemically identical.

(b) Since fulvic acid is a low molecular weight polymer fraction, the properties of the carboxyl groups will, in the most general case, depend on molecular weight.

(c) As neutralization advances to more weakly acidic carboxyl groups, the electrostatic charge accumulating on the polymer molecules will make the carboxyl groups even more weakly acidic.

(d) Only two acid titration end points can be found experimentally for fulvic acid.

(e) Each fulvic acid sample taken from a given geographic location will be a unique mixture of the same molecular weight components.

A practical conceptual framework for this system is obtained by adapting the "titration constant" defined originally by Simms.[40] This was used for titrations of polyvalent acids having only one end point. Simms described it as being calculated in the same way as would the acid dissociation constant of a monovalent, monomeric weak acid. As an approximation, Simms neglected activity coefficients. Because activity coefficients can deviate from unity by 20 to 25% in 0.1 m electrolyte, mass action quotients are employed here to avoid this approximation

Symbols

m_H = = molality of H^+

m_A = molality of all Type A functional groups that are ionized

m_{AH} = molality of all Type A functional groups that are unionized

m_L = molality of ionized Type I carboxyl groups

m_{LH} = molality of unionized Type I carboxyl groups

m_{SH} = molality of the bidentate chelating sites, with the first proton removed

m_{SH_2} = molality of the unionized bidentate chelating sites

m_C = molality of site-bound, bidentate M^{++} chelate

a = molal activity

γ = molal activity coefficient

K = thermodynamic equilibrium constant
K = mass action quotient corresponding to K
α = degree of ionization
C = total molality of acidic functional groups, both ionized and unionized, accumulated in the order of decreasing acid strengths up to some point in the titration
ΔG^e = electrostatic free energy

Assume that a polymer has a number of carboxyl groups, Type A, which are chemically non-identical but whose acid strengths are so similar as to prevent the appearance of separate titration end points. In this case the Type A functional groups are all of those titrated up to the first equivalence point. Following Simms, Equation 1 is written for the H^+ dissociation of all Type A functional groups:

$$AH \underset{\longleftarrow}{\overset{\overline{K}_A}{\longrightarrow}} A^- + H^+$$

$$\overline{K}_A(\alpha_A) = \frac{m_{A^-} m_H}{m_{AH}} = \overline{K}_A \frac{\gamma_{AH}}{\gamma_A \gamma_H} \tag{1}$$

This must now be related to the components of the mixture. Let C_{OH} be the molality of total base added by titration, when the degree of ionization is α_A. During the titration of fulvic acid with standard base, the most strongly acidic functional groups ionize first. Neutralization and ionization are, of course, not identical in the present case. Imagine therefore, the process of accumulating Type A acidic functional groups, both ionized and unionized, in the order of decreasing acid strength up to some value C molal. A graph may be imagined for this accumulated total, $C < C_{OH}$, plotted as a function of the acid dissociation equilibrium function. Now consider the ith infinitesimal increment of functional groups at the location C on the C vs. acid strength curve,

$$c_i = m_{A_i^-} + m_{A_i H}$$

This i^{th} infinitesimal increment of functional groups is equilibrated between ionized and unionized forms:

$$A_iH \rightleftharpoons A_i^- + H^+$$

$$K_i(\alpha_A, C) = \frac{m_{A_i} m_H}{m_{A_iH}} = K_i \left(\frac{\gamma_{A_iH}}{\gamma_{A_i} \gamma_H} \right) \tag{2}$$

The mass action quotient, K_i corresponds roughly to the dissociation constant which Katchalsky and Spitnik[41] have used for the carboxyl groups of the νth stage of ionization of polymethacrylic acid. Two important differences should be noted, however. The Katchalsky-Spitnik function was not defined for an infinitesimal increment of functional groups, as is the K_i used here. Unlike the Katchalsky-Spitnik function, the K_i is not tied by definition to particular polymer molecules.

Following the practice of Katchalsky and Gillis[42] and of Arnold and Overbeek,[43] Bjerrum's electrostatic field effect is here assumed to apply to the ith infinitesimal increment of Type A functional groups. This gives Equation 3.

$$K_i = K_i^\circ \left(\frac{\gamma_{A_i}^\circ \gamma_H}{\gamma_{A_iH}^\circ} \right) e^{-\Delta G_i^e / RT} \tag{3}$$

The superscript "$\circ$" refers to the uncharged state of the polymer. In terms of mass action quotients, Equation 3 becomes

$$K_i(\alpha_A, C) = K_i^\circ \Gamma_i e^{-\Delta G_i^e / RT}$$

$$\Gamma_i = (\gamma_{A_i}^\circ \gamma_{A_iH} / \gamma_{A_i} \gamma_{A_iH}^\circ) \tag{4}$$

While $K_i^\circ = K_i^\circ (\gamma_{A_i}^\circ \gamma_H / \gamma_{A_iH}^\circ)$ is a function only of C, that is of the particular increment c_i of functional groups being considered, K_i is a function of both C and α_A. The chief reason for this dependence on α_A is the electrostatic term ΔG_i^e in Equation 4.

The total numbers of ionized and unionized carboxyl groups A are given respectively by

$$m_A = \sum_{i=0}^{k} m_{A_i} \tag{5}$$

$$m_{AH} = \sum_{i=0}^{k} m_{A_i H} \tag{6}$$

where k = total number of c_i increments. The experimental $\overline{K}$ may now be related to the mass action quotients, K_i, by combining Equations 1, 2, 5, and 6. The result is Equation 7.

$$\overline{K}_A = \frac{\sum_{i=0}^{k} m_{A_i H} K_i}{\sum_{i=0}^{k} m_{A_i H}} \tag{7}$$

According to Equation 7, the experimentally observed $\overline{K}$ is a weighted average of the K_i's for all of the k infinitesimal increments, c_i, of Type A functional groups. Fulvic acid samples obtained from the same source should have the same K_i values at a given degree of ionization. They will differ generally, however, in their relative values of the $m_{A_i H}$ factors. Different $\overline{K}$ vs. α_A curves will therefore be observed experimentally.

The ith infinitesimal mole fraction increment of unionized Type A functional groups is $[\Delta(1-\alpha_A)]_i = m_{A_i H}/c_A$, where $c_A = m_A + m_{AH}$. Dividing the numerator and denominator of Equation 7 by c_A and substituting from Equation 4,

$$\overline{K}_A = \frac{1}{(1-\alpha_A)} \sum_{i=0}^{k} K_i^\circ \, \Gamma_i(\alpha_A) \, e^{-\Delta G_i^e/RT} [\Delta(1-\alpha_A)]_i \tag{8}$$

It is postulated, on the basis of titration results, that the $(K_i^\circ - K_{i+1}^\circ)$ intervals are small. The titration process also imposed the experimental constraint $\alpha_A = f(C)$, so that each K_i is a function of only one independent variable instead of two as indicated in Equation 4. The multitude of mass action quotients, K_i, may therefore be replaced by a continuous function

$$K_A(\alpha_A) = K_A^\circ \, \Gamma \, e^{-\Delta G^e/RT}$$

$$= m_H \delta m_A / \delta m_{AH} \qquad (9)$$

δm_A and δm_{AH} are the infinitesimal concentration increments of functional groups Type A which occur at mole fraction $(1-\alpha_A)$ of unionized functional groups. The summation of Equation 8 can now be replaced by an integration;

$$\overline{K}_A(\alpha_A) = \frac{1}{(1-\alpha_A)} \int_o^{(1-\alpha_A)} K_A^\circ(\alpha_A) \, \Gamma(\alpha_A) \, e^{-\Delta G^e/RT} d(1-\alpha_A) \qquad (10)$$

The chemically nonidentical acidic functional groups have a central importance to the chemical properties of fulvic acid. It is therefore desirable to characterize them in more detail than is possible with the weighted average function $\overline{K}_A$. This may be achieved by calculating the differential function $K(\alpha_A)$, since it is based on the K_i mass action law quotients. For this purpose the familiar relationship

$$\overline{K}_A = \frac{\alpha_A m_H}{(1-\alpha_A)}$$

is introduced into Equation 10, after which rearrangement and differentiation give the equation required for calculating K_A. The result is Equation 11.

$$- \frac{d(\alpha_A m_H)}{d\alpha_A} = K_A \qquad (11)$$

The K_A which characterizes any particular infinitesimal increment of Type A functional groups may therefore be calculated from the experimental $(\alpha_A m_H)$ vs. α_A data, by means of Equation 11.

As the most general case, these arguments and equations developed for the Type A functional groups also apply to the Type B functional groups which correspond to the second titration equivalence point.

Experimental Titration Results

Fulvic acid potentiometric titration curves plotted as in Figure 52 exhibit only the second of two titration equivalence points. When, however,

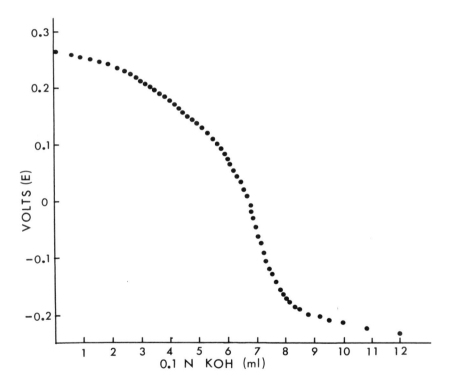

Figure 52.   *Potentiometric titration of fulvic acid Batch No. 4.*
0.1 g/(100 g 0.1 m KCl); 0.1 N standard KOH at
25.0°C.   (From Reference 39).

the corresponding H^+ molalities are plotted, a sharp discontinuity reveals the first equivalence point. This may represent a thousand-fold drop in concentration.   Dissolved KCl displaces the first

equivalence point to a higher value. Figure 53
illustrates the first equivalence point, and the
effect of KCl on it. No such salt effect has been
found for the second equivalence point.

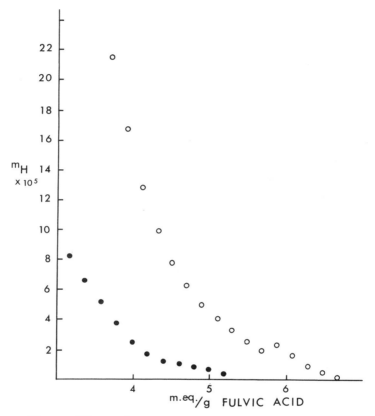

*Figure 53.   Effect of KCl on fulvic acid titration at 25.0°C.
m_H as a function of the amount of standard base
added. Typical examples with (0.1 g fulvic acid/
100 g of solution). o, 0.1 m KCl.   •, No KCl.
(From Reference 28).*

Conductivity titration curves like those of Figure
54 confirm the existence of two equivalence points
for fulvic acid, but only the second one has been
successfully measured in this way. Numerical values
for these equivalence points are given in Table 55
The potentiometric titration results were calculated
with a modified version of Gran's method.[28,44,45]

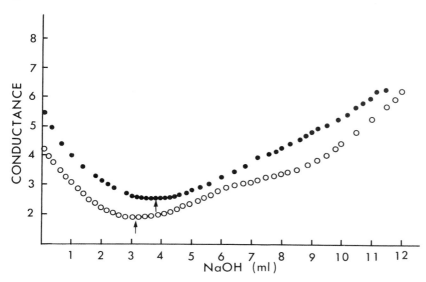

Figure 54. *Conductivity titration of fulvic acid Batch No. 4.*
●, 0.01 g/100 g H₂O; mhos X 10⁴ vs. ml 0.01 N
standard NaOH; o, 0.10 g/100 g H₂O; mhos X 10³
vs. ml 0.1 N standard NaOH (from reference 39).

Table 55

Fulvic Acid Titration Equivalence Points

Fulvic Acid Batch No.	*(m. eq. acid groups / g fulvic acid)*			
	1st Equivalence Point		*2nd Equivalence Point*	*Reference No.*
	no KCl	*0.1 m KCl*		
FA 1	4.99 ± 0.25	6.54 ± 0.06	7.71 ± 0.25	28
4			7.72 ± 0.22	39

This same calculation technique, which is an iterative procedure, also produced numerical results for $\overline{K}_A$, α_A, $\overline{K}_B$ and α_B. That is, the weighted average acid strengths of the Type A and Type B functional groups were determined. Tables 56 and 57 summarize the numerical results, while Figure 55 shows the

Table 56

Weak Acid Dissociation Equilibrium of Type A Functional Groups at 25.0°C
Fulvic Acid Batch FA1

$$\overline{K}_A \times 10^3 = \sum_{i=0}^{m} C_i \alpha_A^i$$

	g fulvic acid / 100 g solution	KCl molality	Number of titrations	Number of data points	α_A	C_0
Nominal concentrations						
1.	0.1	0.1	6	85	$0.38 \leqslant \alpha_A < 0.50$	5.508
2. *	0.1	0.1	6	58	$0.50 \leqslant \alpha_A < 0.70$	1.40794×10^3
3.	0.01	0.1	2	19	$0.70 \leqslant \alpha_A \leqslant 0.90$	0.170
4.	0.1	0.0	5	73	$0.40 \leqslant \alpha_A \leqslant 0.60$	9.009
5.	0.1	0.0	5	35	$0.60 \leqslant \alpha_A \leqslant 0.84$	1.398

Table 56, continued

	C_1	C_2	C_3	C_4	C_5	Standard Deviation ±	Standard Error ±
1.	-9.757					0.16	0.017
2.	-1.15687×10^4	3.79769×10^4	-6.22090×10^4	5.08299×10^4	-1.65700×10^4	0.060	0.0078
3.						0.006	0.0014
4.	-26.79	20.21				0.22	0.026
5.	- 2.866	1.556				0.039	0.007

*Extra significant figures are carried to avoid rounding off errors. Answers should be rounded off to two decimal places. From Reference 28.

Table 57

Acid Dissociation Equilibrium of Type B Functional Groups at 25.0°C
Fulvic Acid Batch FA1

KCl Molality	Region of the Titration Curve m.eq. g fulvic acid	Number of Titrations	Number of Data Points	$\bar{K}_B$ X 10 average	Standard Deviation X 10^{10} ±	Standard Error X 10^{10} ±
0.1	6.6 to 7.7	3	23	5.8	0.9	0.2
0.0	5.6 to 7.7	6	64	6.6	2.1	0.3
0.05 to 0.2	6.6 to 7.7	9	136	*7.	3.	

*Rough results from preliminary work.
From Reference 28.

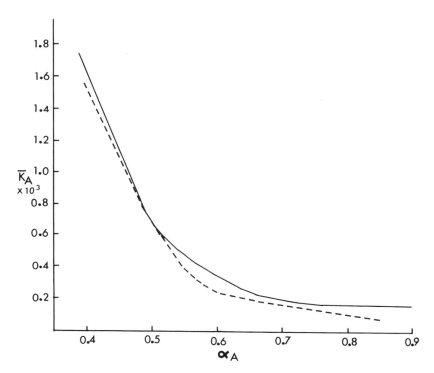

Figure 55.   The curves of Table V.
_____ 0.1 m KCl
------------- no KCl

shapes of $\overline{K}_A$ curves in Table 56. This strong dependence of $\overline{K}_A$ on α_A probably reflects the non-
identical chemical nature of the Type A functional
groups, since the curves with and without KCl have
very similar shapes. The $\overline{K}_B$ values summarized in
Table 57 show no dependence on α_B outside the experimental errors. It should be noted that the
0.0 m KCl value belongs to acid functional groups
including the 1.5 m eq./g of functional groups
strongly affected by KCl, while the 0.1 m KCl value
does not (see Figure 53 and Table 56). Considering
the sizes of the experimental errors, these two $\overline{K}_B$
values cannot be assumed to be different.

    K_A, calculated according to Equation 11 and
plotted in Figure 56, gives a detailed picture of
the acidic functional groups up to the first

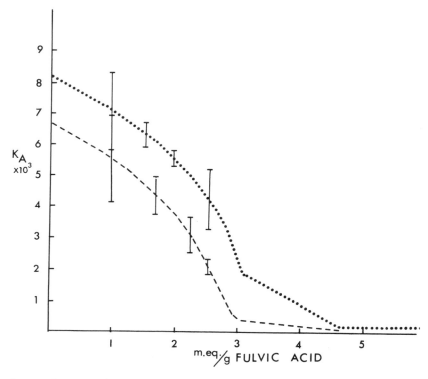

Figure 56.   *Acid dissociation equilibrium function for an infinitesimal increment of functional groups at 25.0°C as a function of its location in the titration curve. Dots: 0.1 m KCl, calculated from six titrations. Dashes: No KCl, calculated from five titrations (from Reference 28).*

equivalence point. The 0.1 m KCl and 0.0 m KCl curves in Figure 56 were calculated with data from six and five titration curves, respectively. In the former curve, the horizontal region above 5.0 $\frac{(m.eq.)}{g}$ is equal to $\overline{K}_A = 0.170 \times 10^{-3} \pm 0.006 \times 10^{-3}$ since for constant $\overline{K}_A$,

$$K_A = -\frac{d[(1-\alpha_A)\overline{K}_A]}{d\alpha_A} = \overline{K}_A.$$

The very distinct break at $3.0 \pm 0.1 \frac{(m. \ eq.)}{g}$
corresponds to the first sudden slope changes in
the curves of Figure 55. These first $3.0 \frac{(m. \ eq.)}{g}$
of acidic functional groups are evidently the
Type I carboxyl groups originally identified by
Schnitzer and co-workers.[24,25,29,39,46,47] There are
two other distinct types of functional groups,
occupying the regions $3.0 < \frac{(m. \ eq.)}{g} < 5.$ and
$5. < \frac{(m. \ eq.)}{g} < 6.54$, respectively. Not only do
they respond very differently to the presence of
KCl, but also one is a constant at constant KCl
concentration while the other is a decreasing
function of α_A.

Reactions with Metal Ions:
Cu++ Chelation

As is the case with some other polyelectrolytes,
for example the polyacrylic acids,[48,49] both soluble
and crosslinked humic acid fractions complex metal
ions.[50,51] Gamble, Schnitzer and Hoffman have
re-evaluated the evidence for the nature of the
complexing sites and of the Cu++ complex. Of the
functional groups listed in Table 53, the three
general types of functional groups that influence
any metal ion-fulvic acid complexing study are:
(a) phenolic OH groups, which remain unneutral-
ized except by prolonged reaction with concentrated
strong base;[24]
(b) Type I carboxyl groups, defined as those
ortho to the phenolic groups;
(c) all of the other ionizable functional
groups, probably including some carboxyl groups
meta to the phenolic OH groups, which contribute
to a buffering effect.
The familiar Cu++ chelation reactions of
salicylic acid and phthalic acid[52] suggest that
the following two reactions are both possible with
humic acids:

By heavily loading fulvic acid with Fe^{+++} and Al^{+++}, Schnitzer and Skinner[25] evidently complexed more metal ion per gram of fulvic acid than could be accounted for by Reaction I. Reaction II must therefore have occurred either alone or together with I. The trace Cu^{++} concentrations generally found in nature will not cause such loading, however. The available evidence indicates that under natural conditions Reaction I will occur. This evidence comes from two types of experiments:

*Selective blocking of metal
complexing functional groups*

In 1957 Himes and Barber[53] found that the blocking of noncarboxylic acid functional groups reduced Zn^{++} retention just as effectively as did the blocking of carboxyl groups. From this they concluded that the carboxyl groups did not constitute independent complexing sites. Himes and Barber suspected the existence of bidentate chelating sites as seen in Reaction I.

Schnitzer and Skinner[46] exploited this technique more fully in 1965, using Cu^{++}, Fe^{+++} and Al^{+++}. The Type I and other carboxyl groups, and the phenolic and alcoholic OH groups were blocked in various combinations. This was followed by metal retention measurements effected with cation exchanger in an aqueous acetone solvent. Their results were as follows:

(a) The largest reduction in metal ion retention by the fulvic acid occurred when carboxyl groups and phenolic OH groups were both blocked.

(b) The second largest effect was obtained when either carboxyl groups or phenolic groups were blocked.

(c) No effect resulted from the blocking of alcoholic OH groups.

There was a residual metal ion retention, even for case (a). The probable reason for this was that the aqueous solvent was required for fulvic acid with blocked functional groups. The resulting loss of metal ion solubility limited the amount of complexed metal ion to about 1 mole % of the total carboxyl groups of the unmodified fulvic acid. If the blocking of the functional groups was, for example, only 99.5% effective, then the observed residual metal ion retentions could be explained. Therefore, the residual metal ion retentions of cases (a) and (b) are compatible with Reaction I. The results obtained by Schnitzer and Skinner, therefore, lead to two conclusions. First, at least 50% of the retained Cu^{++} was chelated according to Reaction I. Second, because of the small amounts of metal bound per gram of fulvic acid and the possibility of incomplete blocking, the residual metal ion retentions cannot be interpreted as evidence for complexing at other sites. For example, Reaction II cannot be assumed to apply.

The Effect of added Cu^{++} on
the titration of humic materials

In 1959, Beckwith examined the effect of added Cu^{++} on the potentiometric titration of a humic material,[54] after first testing this procedure with salicylic and citric acids. All three samples gave the same result. That is, the Cu^{++} released an otherwise untitratable H^+ from the fulvic acid. Within the limits of what was likely their experimental error, Khanna and Stevenson[55] found the same effect of Cu^{++} on the titration curves of fulvic acid in 1962. Although referring to "nitrogenous materials" in their humic polymer fractions, Khanna and Stevenson did not consider that α-amino groups had made any significant contribution to complexing.

With further confirmation of these titration results by Schnitzer and Skinner in 1963,[25] Reaction I is assumed to be the correct description of the Cu^{++}-fulvic acid chelation reaction. From this it follows by definition that the number of bidentate chelating sites is identical to the number of Type I carboxyl groups.

The chelation equilibrium of Reaction I is best formulated in terms of the bidentate chelating sites, with their carboxyl groups ionized. The chemical postulates used for the Type A functional groups also

apply here.   That is, it is assumed that no two of
these chelating sites are chemically identical but
that they exhibit a continuity of chemical properties.   The same kind of mathematical treatment is
therefore used.   Equation 12 is written accordingly
as follows for all of the chelating sites:

$$SH^- + Cu^{++} \underset{\longleftarrow}{\overset{\overline{K}_A}{\longrightarrow}} SCu + H^+$$

$$\overline{K}_4 = \frac{m_c m_H}{m_{SH} m_M} \tag{12}$$

Likewise for the ith infinitesimal increment of
chelating sites,

$$S_i H^- + Cu^{++} \underset{\longleftarrow}{\overset{K_{4i}}{\longrightarrow}} S_i Cu + H^+$$

$$K_{4i} = \frac{m_{ci} m_H}{m_{S_i H} m_M} \tag{13}$$

For k such infinitesimal increments one has the
totals given by Equations 14 and 15.

$$m_c = \sum_{i=0}^{k} m_{c_i} \tag{14}$$

$$m_{SH} = \sum_{i=0}^{k} m_{S_i H} \tag{15}$$

In the same way as for $\overline{K}_A$, Equations 12 to 15 give a
weighted average expression for $\overline{K}_4$.   This is Equation
16.

$$\overline{K}_4 = \frac{\displaystyle\sum_{i=0}^{k} m_{S_i H} K_{4i}}{\displaystyle\sum_{i=0}^{k} m_{S_i H}} \tag{16}$$

Since by definition the number of bidentate chelating sites equals the number of Type I carboxyl groups,

$$m_T = m_L + m_{LH} = m_{SH} + m_{SH_2}$$

$$\alpha_1 = \frac{m_L}{m_T} = \frac{m_{SH}}{m_T} \, .$$

Correspondingly,

$$[\Delta\alpha_1]_i = \frac{m_{L_i}}{m_T} = \frac{m_{S_iH}}{m_T} \, .$$

The results finally obtained in this way are Equations 17 and 18.

$$\overline{K}_4 = \frac{1}{\alpha_1} \int_o^{\alpha_1} K_4(\alpha_1) \, d\alpha_1 \qquad (17)$$

$$K_4(\alpha_1) = \frac{m_H \delta m_c}{m_M \delta m_{SH}} = \frac{d(\alpha_1 \overline{K}_4)}{d\alpha_1} \qquad (18)$$

A detailed characterization of the chelation sites is achieved by evaluating the function $K_4 = f(\alpha_1)$. This requires that α_1 be estimated from the experimental α_A values. The weighted average equations required for this purpose will now be developed.

The degree of ionization of the ith increment of Type A functional groups is

$$\beta_i = \frac{m_{A_i}}{c_i} \qquad (19)$$

With substitution into Equations 5 and 19 and division by C_A, α_A is found to be weighted average of the β_i values.

$$\alpha_A = \frac{1}{C_A} \sum_{i=0}^{k} \beta_i c_i \qquad (20)$$

Assume that the increments become very small, so that

$$\lim_{c_i \to 0} c_i = \delta C = (\delta m_A + \delta m_{AH})$$

The β_i values may then be replaced by a continuous fraction,

$$\beta_A = \frac{\delta m_A}{C} = \frac{K_A}{K_A + m_H} = \beta_A(C, m_H) \tag{21}$$

β_A is a function of two variables for the same reason that K_A is. The summation of Equation 20 may therefore be replaced by an integration at a constant H^+ molality.

$$\alpha_A = \frac{1}{C_A} \int_0^{C_A} \beta_A dC = \alpha_A(m_H) = \alpha(C_A, m_H) \tag{22}$$

Differentiating Equation 22 at this same H^+ molality,

$$\left(\frac{\partial \alpha_A}{\partial C} \right)_{m_H} = \frac{\beta_A}{C_A}$$

$$\partial m_A = \beta_A(C, m_H) \partial C.$$

For the carboxyl groups in the region $C \leqslant C_L$, that is, the Type I carboxyl groups,

$$\partial m_A \equiv \partial m_L = C_L \partial \alpha_1$$

$$\left(\frac{\partial \alpha_1}{\partial C} \right)_{m_H} = \frac{\beta_A}{C_L} \tag{23}$$

The degree of ionization of the Type I carboxyl groups as a function of H^+ molality is therefore calculated by integrating between the limits 0 and C_L.

$$\alpha_1 = \frac{1}{C_L} \int_0^{C_L} \beta_A(C, m_H) dC = \alpha_1(m_H) = \alpha(C_L, m_H) \tag{24}$$

The rigorous calculation of α_1 according to Equation 24 requires an experimental measurement of C_L, the total molality of Type I carboxyl groups. It also requires an empirical relationship for $\beta_A(C, m_H)$, with m_H being the experimental parameter of Equations 22 to 24. This empirical relationship for β_A is not experimentally obtainable, however, because the titration procedure imposes the extra constraint

$$\acute{m}_H = f(C).$$

$\acute{m}_H$ is therefore an experimental variable, in contrast to the parameter m_H of Equations 22 to 24. If K_A can be shown to be a sufficiently insensitive function of $\acute{m}_H$ over the range $0 < C < C_L$, then Equation 24 may be used with the approximation

$$\beta_A(C, m_H) \simeq \beta_A(C, \acute{m}_H).$$

The degree of ionization of Type I carboxyl groups, α_1, has been estimated in this way for fulvic acid solutions both with and without KCl. These are given in Tables 58 and 59, together with corresponding α_A values for comparison.

Gamble, Schnitzer and Hoffman have used an ion exchange method for measuring $\overline{K}_4$ for the Cu^{++} chelation Reaction I.[55] The measurements were obtained for two ranges of α_1 values, and are given in Table 60. These α_1 values in Table 60 were calculated with the assumption that the first titration equivalence point corresponds to the Type I carboxyl groups.[39] Although this assumption has since been found to be incorrect,[28] the resulting α_1 values are valid. Considering that Tables 58 and 60 represent different batches of fulvic acid (see Table 55), and taking into account the estimated experimental errors, reasonable agreement has been found between these two sets of α_1 values.

On the basis of Reaction I, $\overline{K}_4$ might be expected to be an increasing function of α_1. If any such trend exists, it has been obscured in this case by other effects. For example, 0.1 m KCl, which was used to reduce the risk of ligand sorption by the ion exchange beads, may have resulted in some K^+-polyelectrolyte ion pairing, and some cupric chloride complex formation. There is a more likely explanation for the observed shape of the $\overline{K}_4$ vs. α_1 curve. Assuming as a general case,

Table 58

Ionization of Fulvic Acid in 0.1 m KCl at 25.0°C
Fulvic Acid Batch No. FAl

Type I Carboxyl Groups, 3.0 ± 0.1 $\dfrac{\text{m. eq.}}{\text{g fulvic acid}}$

Type A Acid Groups, 6.54 ± 0.06 $\dfrac{\text{m. eq.}}{\text{g fulvic acid}}$

H^+ Molality	α_A $\pm 2\%$	α_1 $\pm 2\%$
2.60×10^{-3}	0.392	0.72
2.21×10^{-3}	0.406	0.72
1.50×10^{-3}	0.442	0.77
1.30×10^{-3}	0.454	0.82
1.00×10^{-3}	0.472	0.85
5.08×10^{-4}	0.52	0.92
1.97×10^{-4}	0.61	0.97
1.00×10^{-4}	0.68	0.98
1.00×10^{-5}	0.94	1.00
1.00×10^{-6}	0.99	

From Reference 28.

Table 59

Ionization of Fulvic Acid in Aqueous Solution at 25.0°C
Fulvic Acid Batch No. FA1

Type I Carboxyl Groups, 3.0 ± 0.1 $\dfrac{\text{m. eq.}}{\text{g fulvic acid}}$

Type A Acid Groups, 4.99 ± 0.25 $\dfrac{\text{m. eq.}}{\text{g fulvic acid}}$

H^+ Molality	α_A ± 4.0%	α_1 ± 4.0%
2.00 X 10^{-3}	0.41	0.66
1.80 X 10^{-3}	0.42	0.67
1.60 X 10^{-3}	0.43	0.69
1.40 X 10^{-3}	0.44	0.71
1.20 X 10^{-3}	0.46	0.74
1.00 X 10^{-3}	0.47	0.78
8.00 X 10^{-4}	0.49	0.81
6.00 X 10^{-4}	0.51	0.85
4.00 X 10^{-4}	0.53	0.88
2.00 X 10^{-4}	0.58	0.92
1.00 X 10^{-4}	0.66	0.96
6.00 X 10^{-5}	0.71	0.99
2.00 X 10^{-5}	0.83	1.00
5.00 X 10^{-6}	0.92	

From Reference 28.

Table 60

Cu^{++}-Fulvic Acid Chelate Formation Equilibrium
Fulvic Acid Batch No. 4

$\overline{K}_4$	α_1	0.1 m KCl 25.00°C Total Cu^{++} $m \times 10^5$ ± 3%	H$^+$ $m \times 10^4$ ± 1%	Total Chelating Sites $m \times 10^4$ ± 0.7%
5.0 ± 17%	0.871 ± 0.2%	2.59	4.47	
4.8	0.865	2.75	4.77	
4.7	0.861	2.86	4.97	
4.7	0.861	2.88	5.01	
4.6	0.856	3.03	5.26	
4.6	0.853	3.15	5.43	
4.6	0.851	3.22	5.53	4.26
4.6	0.849	3.32	5.66	
4.6	0.846	3.4	5.79	
4.6	0.843	3.6	5.96	
4.6	0.840	3.8	6.16	
4.6	0.838	3.9	6.25	
4.5	0.836	4.1	6.37	
4.5 ± 20%	0.834 ± 0.2%	4.3	6.50	
13. ± 28%	0.67 ± 5%	1.41	23.9	
13.	0.67	1.54	24.8	
14.	0.66	1.62	25.3	
14.	0.66	1.80	26.4	
15.	0.65	1.88	26.9	
15.	0.65	2.01	27.7	
15.	0.65	2.09	28.1	
15.	0.64	2.26	28.9	4.49
15.	0.64	2.50	30.0	
14.	0.63	2.73	30.7	
14.	0.63	3.07	31.4	
14.	0.63	3.18	31.6	
14.	0.63	3.4	32.0	
14.	0.63	3.5	32.1	
15. ± 31%	0.63 ± 5%	3.6	32.3	

From Reference 55.

$$K_4 = K_4^{\circ}(C) e^{-\Delta G^e (C, \alpha_1)/RT}$$

Equation 17 becomes

$$\overline{K}_4(\alpha_1) = \frac{1}{\alpha_1} \int_0^{\alpha_1} K_4^{\circ}(C) e^{-\Delta G^e (C, \alpha_1)/RT)} d\alpha_1 \quad (25)$$

If all of the polymer molecules of the fulvic acid are chemically very similar, then as C increases the electrostatic energy term must increase, while K_4° will change little if at all. If on the other hand chemically different polymer molecules come into the chelation reaction as C increases, then each of the two factors K_4° and $e^{-\Delta G^e/RT}$ might separately increase, decrease, or remain constant. From this it is concluded that the shape of the $\overline{K}_4$ curve in Table 60 reflects chemically different fulvic acid fractions. The calculation of K_4 as a function of α_1 according to Equation 18 should therefore provide a very interesting picture of the chelation properties of the fulvic acid. Unfortunately the only relevant chelation data presently available are the ion exchange measurements of Table 59. Ion exchange is a low precision method for measuring chelation equilibria, and differentiation magnifies the experimental errors. The calculation in Equation 18 has therefore not been carried out. Chelation measurements with specific ion electrodes should make the calculation of K_4 possible.

The Relative Stabilities of
Various Metal Ion Chelates

Schnitzer and Hansen have measured fulvic acid chelation equilibria for a number of metal ions, using both ion exchange and Job's continuous variations techniques.[56] The continuous variations experiments were done photometrically. The two methods gave results in reasonable agreement. They presented their results in the form of an equilibrium constant, K', defined in Equation 26.

$$K' = \frac{M_c}{M_M M_p} \quad (26)$$

M_p is the molarity of number-average polymer molecules. Most of the equilibrium measurements of
Schnitzer and Hansen have been recalculated for the
purpose of relating the equilibrium functions to
the chelating sites and to the degree of ionization
of the Type I carboxyl groups. The resulting $\overline{K}_4$
estimates are summarized in Table 61.

Table 61

Fulvic Acid-Metal Ion Chelates
K_4 *Values Recalculated From the Work of*
Schnitzer and Hansen[56]

Metal Ion	$\overline{K}_4$		
	0.1 M KCl pH = 3 $\alpha_1 = 0.85$	0.1 M KCl pH = 5 $\alpha_1 = 1.00$	no KCl pH = 3 $\alpha_1 = 0.78$
Cu^{++}	1.04	0.044	23
Ni^{++}	0.73	0.070	15
Co^{++}	0.37	0.062	9.3
Pb^{++}	0.23	0.049	1.8
Ca^{++}	0.23	0.0099	1.8
Zn^{++}	0.10	0.020	0.74
Mn^{++}	0.072	0.022	0.37
Mg^{++}	0.041	0.70	0.23

In addition to information for the cations
listed in Table 61, Schnitzer and Hansen obtained
results for Fe^{+++} and Al^{+++} at somewhat lower pH's.
Fe^{+++} is much more strongly chelating than the
other metal ions investigated, while Al^{+++} is next.
Schnitzer and Hansen observed a strong ionic
strength effect on the chelation equilibria. The
same effect is seen here on H^+ dissociation, in
Figures 53, 55, and 56. The cause is assumed to
be the electrostatic binding of cations by the
fulvic acid polyelectrolyte molecules.

DISCUSSION

It must be remembered that fulvic acid is
primarily a polyelectrolyte, and that it is even

more a mixture than are most polymers. Since, in fact, one cannot assume that any two of the ionizable functional groups are chemically identical, it follows necessarily that an experimentally observed equilibrium function is a weighted average function. This has certain practical consequences for the study of water chemistry.

(a) The experimentally measured equilibrium function for the reaction of fulvic acid with a cation is a variable. This variable is a function of the weighting factors of the participating functional groups, and of the electrostatic charge accumulated on the polymer molecules. Any attempt to describe such an equilibrium with a simple numerical constant is therefore a fundamental flaw in logic. This mistake may lead to wrong conclusions about the chemistry of natural waters. Figures 55 and 56 graphically illustrate the point.

(b) It cannot be assumed that the alkali, alkaline earth, and heavy metal ions all react with the same functional groups, or with the same molecular weight fractions.

(c) Any phase change or heterogeneous reaction could cause fractionation of the fulvic acid. An equilibrium function measured for unfractionated fulvic acid would then not be applicable.

Water quality standards for heavy metals should be related in some practical way to the concentrations of humic materials such as fulvic acid, by taking into account the stabilities and solubilities of the various chemical species. Natural waters may often contain higher molecular weight humic materials, some of which could be colloidal, undissolved, or sorbed onto clay particles. Consequently, a homogeneous fulvic acid solution might not be a fully adequate model for particular natural systems. In addition, the available $\bar{K}_4$ data in Tables 60 and 61 are not fully relevant to the conditions found in natural waters. A few sample calculations of metal ion chelation are none the less instructive.

Stumm and Morgan give the range of organic matter concentrations in surface waters as 0.1 to 10. mg/liter.[8] They also report that the average river in the United States has, among other ions, 3.8×10^{-4} M Ca^{++}. Consider two hypothetical cases having 1.0 $\dfrac{\text{mg. fulvic acid}}{\text{liter}}$, pH 5., and 5.0×10^{-6} M metal ion. At this pH for a dilute solution, one may take $m_H \simeq 1.0 \times 10^{-5}$, and $\alpha_1 \simeq 1$. Using $\bar{K}_4$ values of 23. and 0.37 for Cu^{++} and Mn^{++} respectively from Table 61, Equation 12 predicts that the Cu^{++} would be about half chelated, while only about 9% of the Mn^{++} would be chelated. Next suppose that instead of just one metal ion, the solution has 5.0×10^{-6} M Cu^{++} and 4.0×10^{-4} M Ca^{++}. Equation 12 now predicts that the fulvic acid chelating sites will be almost fully loaded with Ca^{++}, and that hardly any Cu^{++} will be chelated.

While these calculations are speculative, they demonstrate the need for quantitative $\bar{K}_4$ data relevant to natural waters. Prerequisite to this are Duursma's experimental requirements.[22,23] Specifically, these include determination of the elemental composition, molecular structure, and polyelectrolyte chemistry of fulvic acid samples collected from natural waters.

Beyond this level of complexity, the heterogenous systems require that the chemical and physical properties of higher molecular weight humic materials and of clay-humic complexes be measured quantitatively. It is expected that a detailed understanding of fulvic acid chemistry will facilitate the investigation of the ion exchange and chelation properties of these solid phase and colloidal systems.

REFERENCES

1. Arnold, R., and J. Th. G. Overbeek. *Rec. Trav. Chim.* *69*, 192 (1950).

2. Gregor, Harry P. and Michael Frederick. *J. Polymer Sci.* *23*, 451 (1957).

3. Wall, Frederick T. *J. Phys. Chem. 61*, 1344 (1957).

4. Mandel, M., J. C. Leyte, and M. G. Stadhouder. *J. Phys. Chem. 71*, 603 (1967).

5. Christman, R. F. and R. A. Minear. "Organics in Lakes," In *Organic Compounds in Aquatic Environments*, Faust, Samuel J. and Joseph V. Hunter, eds. (New York: Marcel Dekker, 1971).

6. Christman, R. F. and M. Ghassemi. *J. Amer. Water Works Assoc. 58*, 723 (1966).

7. Sarkanen, K. V. *The Chemistry of Wood*, Browning, B. L., ed. (New York, J. W. Wiley, 1963), Chapter 10.

8. Stumm, Werner and James J. Morgan. *Aquatic Chemistry*. (New York: Wiley-Interscience, 1970).

9. Taras, Michael J., Arnold E. Greenberg, R. D. Hoak, and M. C. Rand, eds. *Standard Methods for the Examination of Water and Wastewater*, 13th ed. (Washington, D.C.: American Public Health Association, 1971).

10. Black, A. P. and R. F. Christman. *J. Amer. Water Works Assoc. 55*, 753 (1963).

11. Saville, A. *J. New Eng. Water Works Assoc. 31*, 78 (1917).

12. Christman, R. F. Symposium on Organic Matter in Natural Waters, University of Alaska, College, Alaska, September 1968.

13. Gjessing, E. T. *Nature 208*, 1091 (1965).

14. Shapiro, J. The Symposium of the Hungarian Hydrological Society, Budapest-Tihang, Hungary, September 1966.

15. Gjessing, E. T. and G. F. Lee. *Environ. Sci. Technol. 1*, 631 (1967).

16. Ghassemi, M. and R. F. Christman. *Limnol. Oceanog. 13*, 583 (1968).

17. Shapiro, J. *J. Amer. Water Works Assoc. 56*, 1062 (1964).

18. Barsdate, R. J. Symposium on Organic Matter in Natural Waters, University of Alaska, College, Alaska, September 1968.

19. Barsdate, R. J. and W. R. Matson. In *Radioecological Concentration Processes,* Aberg, B. and F. Hungate, eds. (Oxford: Pergamon Press).

20. Basu, A. N., D. C. Mukherjee, and S. K. Mukherjee. *J. Ind. Soc. Soil Sci. 12*, 311 (1964).

21. Ghassemi, M. Ph. D. Thesis, University of Washington, Seattle, Washington, 1967.

22. Sigel, Alvin. "Metal-Organic Interactions in the Marine Environment," In *Organic Compounds in Aquatic Environments,* Faust, Samuel J. and Joseph V. Hunter, eds. (New York: Marcel Dekker, 1971).

23. Duursma, E. K. Symposium on Organic Matter in Natural Waters, University of Alaska, College, Alaska, September 1968.

24. Schnitzer, M. and J. G. Desjardins. *Soil Sci. Soc. Amer. Proc. 26*, 362 (1962).

25. Schnitzer, M. and S. I. M. Skinner. *Soil Sci. 96*, 86 (1963).

26. Hansen, E. H. and M. Schnitzer. *Anal. Chim. Acta, 46*, 247 (1969).

27. Schnitzer, M. and H. Kodama. *Geoderma 7*, 93 (1972).

28. Gamble, Donald S. *Can. J. Chem. 50*, 2680 (1970).

29. Schnitzer, M. and J. R. Wright. *Proc. 7th Intern. Congr., Soil Sci. 2*, 112 (1960).

30. Hansen, E. H. and M. Schnitzer. *Soil Sci. Soc. Amer. Proc. 33*, 29 (1969).

31. Barton, D. H. R. and M. Schnitzer. *Nature 198*, 217 (1963).

32. Ogner, G. and M. Schnitzer. *Can. J. Chem. 49*, 1053 (1971).

33. Khan, S. U. and M. Schnitzer. *Can. J. Chem. 49*, 2302 (1971).

34. Kodama, H. and M. Schnitzer. *Fuel 46*, 87 (1967).

35. Ogner, G. and M. Schnitzer. *Geochim. et Cosmochim. Acta 34*, 921 (1970).

36. Schnitzer, M. and G. Ogner. *J. Israel Chem. 8*, 505 (1970).

37. Ogner, G. and M. Schnitzer. *Science 170*, 317 (1970).

38. Matsuda, K. and M. Schnitzer. *Bull. Environmental Contamination and Toxicology 6*, 200 (1971).

39. Gamble, Donald S. *Can. J. Chem. 48*, 2662 (1970).

40. Simms, Henry S. *J. Amer. Chem. Soc. 48*, 1239 (1926).

41. Katchalsky, Aharaon and Pnina Spitnik. *J. Polymer Sci. 2*, 432 (1947).

42. Katchalsky, A. and J. Gillis. *Rec. Trav. Chim. 68*, 879 (1949).

43. Arnold, R. and J. Th. G. Overbeek. *Rec. Trav. Chim. 69*, 192 (1950).

44. Gran, Gunnar. *Analyst 77*, 661 (1952).

45. Gran, Gunnar. *Acta Chem. Scand. 4*, 559 (1950).

46. Schnitzer, M. and S. I. M. Skinner. *Soil Sci. 99*, 278 (1965).

47. Schnitzer, M. and I. Hoffman. *Geochim. Cosmochim. Acta 29*, 859 (1965).

48. Gregor, Harry P., Lionel B. Luttinger, and Ernest M. Loebl. *J. Phys. Chem. 59*, 34 (1955).

49. Gustafson, Richard L. and Joseph A. Lirio. *J. Phys. Chem. 72*, 1502 (1968).

50. Khanna, S. S. and F. J. Stevenson. *Soil Sci. 93*, 289 (1962).

51. Coleman, N. T., A. C. McClung, and David P. Moore. *Science 123*, 330 (1956).

52. Ringbom, Anders. *Complexation in Analytical Chemistry.* (New York: Interscience Publ., 1963).

53. Himes, Frank L. and Stanley A. Barber. *Soil Sci. Soc. Amer. Proc. 21*, 368 (1957).

54. Beckwith, R. S. *Nature 184*, 745 (1959).

55. Gamble, Donald S., M. Schnitzer, and I. Hoffman. *Can. J. Chem. 48*, 3197 (1970).

56. Schnitzer, M. and E. H. Hansen. *Soil Sci. 109*, 333 (1970).

10. THE STABILIZATION OF FERROUS IRON BY ORGANIC COMPOUNDS IN NATURAL WATERS

Thomas L. Theis and Philip C. Singer. Department of Civil Engineering, University of Notre Dame, Notre Dame, Indiana

In oxygen-free aquatic environments, such as groundwaters and hypolimnetic waters of eutrophic lakes, iron exists predominantly in the ferrous state, Fe(II). For those groundwaters used for domestic and industrial purposes, removal of iron is desirable because it can form rust (iron oxide) deposits, causing staining of plumbing fixtures, laundered goods, and manufactured products, as well as imparting a "metallic" taste to the water. Conventional water treatment for the removal of iron consists of aeration of the raw water, providing for the oxidation of ferrous iron

$$Fe^{+2} + 1/4 \ O_2 + H^+ = Fe^{+3} + 1/2 \ H_2O \qquad (1)$$

The resultant ferric iron hydrolyzes to form the highly insoluble ferric hydroxide

$$Fe^{+3} + 3H_2O = Fe(OH)_3(s) + 3H^+ \qquad (2)$$

which is subsequently removed by sedimentation and filtration. The net reaction is

$$Fe^{+2} + 1/4 \ O_2 + 5/2 \ H_2O = Fe(OH)_3(s) + 2H^+ \qquad (3)$$

In natural surface waters, iron serves as a nutrient for algae, higher plants, and other forms of aquatic life. During the spring and fall overturns in eutrophic lakes, the anoxic hypolimnetic water containing ferrous iron is exposed to oxygen and Fe(II) is normally oxidized to Fe(III), as in Equation 3. The availability of iron as a nutrient is dependent upon the rate of Fe(II) oxidation and the stability of the resultant Fe(III); the

concentration of total soluble ferric iron in
equilibrium with ferric hydroxide at pH 8 is approximately 0.2 μg/l. [The solubility product of ferric
hydroxide, Kso, is approximately 10^{-36}.[1]]

Stumm and Lee[2] found the kinetics of the oxygenation of Fe(II) in simple aqueous media to be in
accordance with the following expression:

$$\frac{-d[Fe(II)]}{dt} = k \ [Fe(II)] \ (P_{O_2}) \ [OH^-]^2 \qquad (4)$$

where k = 8.0 x 10^{13} mole^{-2} atm^{-1} min^{-1} at 20.5°C.
(At 25°C, k = 1.36 x 10^{14} mole^{-2} atm^{-1} min^{-1}.) The
oxidation of ferrous iron proceeds relatively rapidly
at neutral pH values; the half-time of the reaction
is 4 minutes at pH 7 and a partial pressure of oxygen of 0.21 atm. Thus, under oxygen-rich conditions,
Fe(II) would be rapidly oxidized, making iron removal a relatively simple process and limiting iron
availability as a nutrient for aquatic growth in
natural surface waters.

However, certain natural organic compounds,
usually referred to as humic material, exert a
stabilizing effect on iron in aquatic systems, *i.e.*,
high concentrations of iron are often associated
with these organic compounds. These humic substances
are primarily products of natural vegetative decay
and, as such, are ubiquitous throughout the aquatic
environment. They impart a yellowish color to water
and are the source of true color. In fact, many
studies have shown that stabilization of iron by
humic substances increases its availability as a
nutrient for plant growth.[3-5]

Many conventional water treatment plants report
difficulties[6] in achieving satisfactory iron removal
in accordance with the U.S. Public Health Service
recommended limit of 0.3 mg/l[7] for iron in drinking
water. These difficulties have been traced to the
influence of humic-type substances[8,9] which apparently retard the oxidation of Fe(II), negating the
simple oxidation kinetics of Equation 4.

The exact mechanism or mechanisms by which
humic substances interact with iron has not been
established as yet. In general, three types of
interactions have been considered: (a) peptization
of the organic molecule by charged hydroxo-ferric
iron complexes;[10,11] (b) chelation of the metal ion
(either ferric or ferrous iron) by the humate ligand
leading to the formation of polynuclear complexes;[12,13]
and (c) reduction of ferric iron to ferrous iron by
the humic substances, thereby generating the more
soluble form of iron.[14,15]

In view of the importance of Fe(II) oxidation and iron-organic interactions in water quality management, this investigation was undertaken to discern the specific reactions between both ferrous and ferric iron and organic species of natural origin. Emphasis is placed upon the kinetics of Fe(II) oxidation and Fe(III) reduction in the presence of humic material. The study makes use of various model organic compounds which are chemically simpler than the naturally-occurring humic substances. The latter have the disadvantages of being somewhat difficult to isolate and possessing a great amount of metal impurities. Those compounds which were used in this study are phenol, resorcinol, pyrogallol, syringic acid, vanillin, vanillic acid, gallic acid, tannic acid, citric acid, tartaric acid, histidine, glutamic acid and glutamine. These are pictured in Figure 57. The results of the studies with these model compounds are then applied to natural water systems.

EXPERIMENTAL TECHNIQUES

Various concentrations of ferrous or ferric perchlorate were added to oxygenated solutions containing the appropriate organic compound in a constant temperature reactor (25°C). For the Fe(II) oxidation studies, pH was maintained constant at 6.3 with a CO_2- HCO_3^- buffer, and the system was subjected to a constant oxygen partial pressure. Under these conditions, Equation 4 can be reduced to a pseudo-first order expression:

$$\frac{-d[Fe(II)]}{dt} = k' \, [Fe(II)] \tag{5}$$

where

$$k' = k \, P_{O_2} \, [OH^-]^2 = \text{constant} \tag{6}$$

Integration of Equation 5 gives

$$\ln \, [Fe(II)]_o - \ln \, [Fe(II)] = k' \, t \tag{7}$$

which yields a straight line on a semilogarithmic plot. ($[Fe(II)_o]$ is the initial concentration of ferrous iron.) Any deviation from this simple rate expression can be interpreted as an effect of the given organic substance. Fe(III) reduction studies were also performed in buffered, oxygenated systems at a variety of pH values.

In both studies, aliquots were removed from the reactors at various times and analyzed for residual ferrous iron using the bathophenanthroline procedure

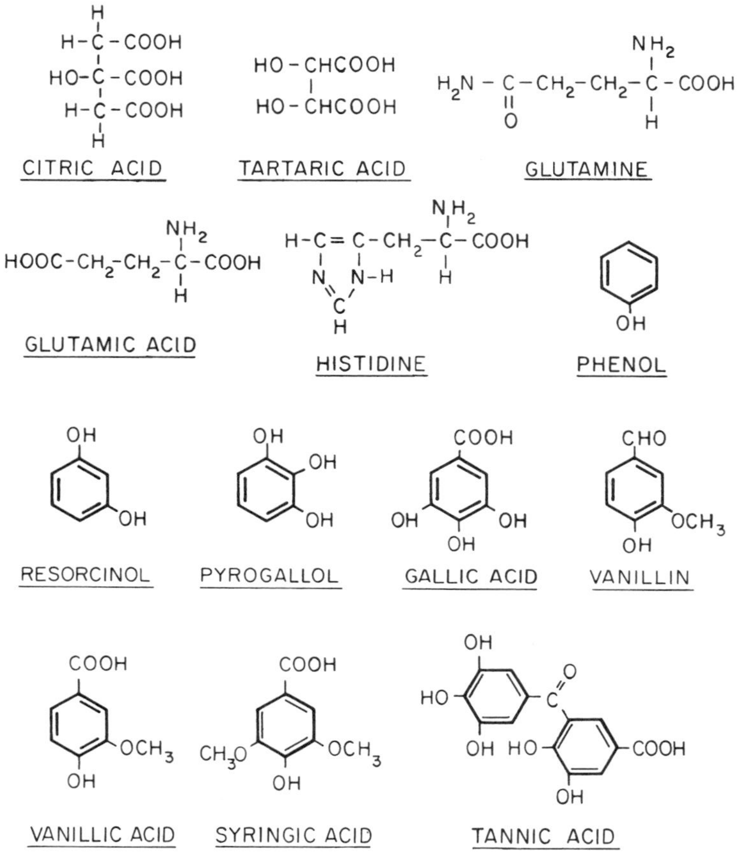

Figure 57.   *Model organic compounds used in this study.*

of Lee and Stumm.[16]   The concentration of Fe(III) was calculated as the difference between the total iron and Fe(II) concentrations.   In place of acidification to pH 1 with hydrochloric acid in order to quench the reaction, the aliquots were immediately added to a pH 4 sodium acetate-acetic acid buffer. Lower pH's were avoided, as it was found that many of the compounds were capable of reducing Fe(III) rapidly if the pH was allowed to go below 3.   This particular phenomenon will be considered later in this paper.

The oxidation of Fe(II) was also investigated in the following systems:

    a. solutions of humic material extracted from plant material (commercially available from K and K Chemical Co.)

b. anoxic hypolimnetic waters subjected to aeration

c. activated sludge from the aeration tanks of the South Bend, Indiana, Wastewater Treatment Plant

d. secondary effluent following conventional activated sludge treatment. [Rebhun and Manka[17] classified 50% of the residual organic matter in conventional secondary effluent as humic material.]

Samples of these waters were brought into the laboratory, spiked with an initial amount of ferrous iron (if necessary), and aerated.

A modification of the bathophenanthroline analysis was found necessary for the concentrated samples from *b* and *c* above because simple acidification to pH 4 did not release all the iron from hydroxo-organo complexes and the suspended material prior to filtration. Consequently, bathophenanthroline was added to each of the aliquots following pH adjustment with the acetate buffer, but before filtration, in order to "pull" the iron away from the organic matter. The samples were then filtered (0.45 μ Millipore filters), the filters rinsed with hexanol, and the analysis continued.

RESULTS AND DISCUSSION

The results of the Fe(II) oxidation studies are displayed in Figure 58; the reported rate[2] in simple aqueous media (Equations 4 and 7) is shown for purposes of comparison. It is evident that the organic compounds exhibit a wide range of effects on the rate of the oxidation reaction. The four compounds shown are representative of these different effects. Tannic acid, as well as gallic acid and pyrogallol, completely inhibited the oxygenation of ferrous iron over the 60-minute observation period. Glutamic acid and tartaric acid produced effects similar to glutamine in significantly decreasing the rate of the reaction, without causing total retardation. Vanillic acid, phenol, resorcinol, syringic acid, vanillin, and histidine had no observable effect on the oxidation reaction; the rate of Fe(II) oxidation in the presence of these compounds, as given by the slope of the line in Figure 58, is seen to be identical to the rate in simple aqueous media.[2] Citric acid was the only compound investigated which accelerated the oxidation of ferrous iron but only up to a point, beyond which the oxidation kinetics conformed to Equation 4.

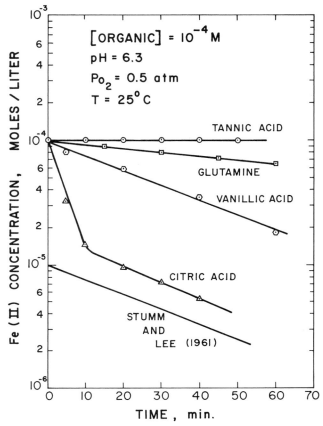

Figure 58.   *Effects of representative organic compounds on the oxidation rate of ferrous iron.*

Figure 59 illustrates, more extensively, the Fe(II) stabilization phenomenon with tannic acid. For solutions containing the same initial concentration of Fe(II) (3.5×10^{-5} M, or about 3 mg/l) but different concentrations of tannic acid (T.A.), an initial period of iron oxidation takes place after which the remainder of the Fe(II) is stabilized. The amount of Fe(II) stabilized is dependent upon the concentration of tannic acid present, the stabilized amount increasing as the tannic acid concentration increases. Furthermore, Figure 59 shows that the iron resisted oxidation for seven days; in fact, the solutions of tannic acid and ferrous iron were found to be stable for several weeks, demonstrating that ferrous iron can exist for extended periods of time in an oxidative environment. This behavior of ferrous iron in the

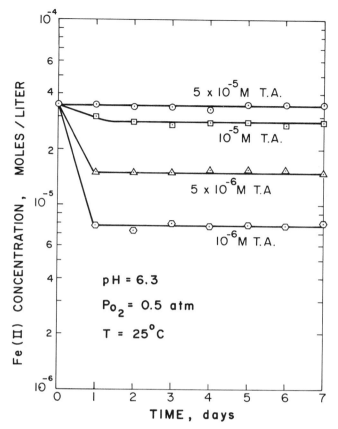

Figure 59. *Inhibition of ferrous iron oxidation and stabili-*
zation of ferrous iron by various concentrations
of tannic acid (T.A.).

presence of tannic acid (and other organic compounds)
is indicative of the formation of a kinetically stable
ferrous-organic complex. Depending upon the specific
organic compound, the oxidation can be enhanced or
retarded to varying degrees, as in Figure 58.

   It should be noted, analytically, that simple
acidification of sample aliquots to pH 4 (with the
acetic acid-sodium acetate buffer) was sufficient
to recover 100% of the initial ferrous iron added
to the systems in Figures 58 and 59. In fact, this
procedure produced no recovery difficulties when
working with the model compounds.

   Reduction of ferric iron by tannic acid at pH
6.3 is shown in Figure 60. Ferric iron was added
to the various solutions of tannic acid in the form

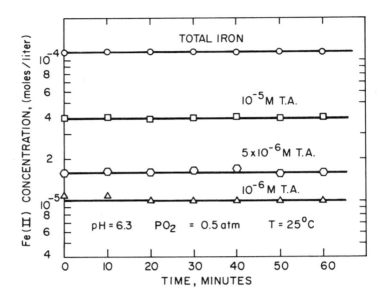

Figure 60.   *Steady state concentrations of ferrous iron following instantaneous reduction of ferric iron by tannic acid (T.A.).*

of crystals of ferric perchlorate. The reduction was virtually instantaneous; before an initial sample could be taken (time zero), the Fe(III) had been reduced and subsequently stabilized by the residual tannic acid and its oxidation products. Again, the extent of Fe(III) reduction is a function of the amount of tannic acid added; as the concentration of tannic acid is increased, a greater amount of Fe(III) is reduced and a greater concentration of Fe(II) is stabilized.

The extent of reduction of Fe(III) by other model compounds is given in Table 62. The reduction of Fe(III) by these compounds was also instantaneous. However, the resultant ferrous iron formed through the reduction of Fe(III) by the organic compounds in Table 62 underwent subsequent oxidation in accordance with the results in Figure 58. Subsequent oxidation of the Fe(II) was completely inhibited by the residual tannic acid, gallic acid, and pyrogallol (as in Figure 60), while subsequent oxidation in the presence of syringic acid and resorcinol proceeded unimpeded as in Figure 58.

Table 62
Reduction of Fe(III) by Model Organic Compounds*

Compound, 10^{-4} M	Fe(III) Reduced,[†] M/l
Tannic acid	7.9×10^{-5}
Pyrogallol	6.9×10^{-5}
Gallic acid	3.4×10^{-5}
Syringic acid	2.6×10^{-5}
Resorcinol	5.2×10^{-6}
Citric acid	4.7×10^{-6}
Tartaric acid	1.9×10^{-6}
Vanillic acid	9.0×10^{-7}
Vanillin	7.0×10^{-7}

*pH 6.3; Fe(III) added = 8.0×10^{-5} M.

[†]Reduction was virtually instantaneous.

When the order of addition of the ferric iron and organic compound to the reactor was reversed, *i.e.*, when the iron was added first and the pH was adjusted prior to the addition of the organic compound, the reduction of Fe(III) by the organic species was not nearly as rapid nor as extensive as previously observed. Figure 61 illustrates the reduction of Fe(III) by tannic acid at several pH values. The rate of Fe(III) reduction increases with increasing pH up to pH 2.3. However, as the pH is increased further, the rate of the reaction slows considerably until no reduction is measurable at pH 6.3. This is in contrast to the virtual instantaneous reduction of Fe(III) in Figure 60 and Table 62 when the organic compound was added first.

The tendency of ferric iron to hydrolyze extensively at pH values greater than 3 is undoubtedly responsible for this phenomenon, as shown in Figure 62. The left-hand ordinate is the amount of ferric iron reduced by tannic acid after one day when Fe(III) was added first and allowed to undergo hydrolysis at the pH indicated; the concentration of Fe(III) reduced is plotted as a percentage of the total Fe(III) added. The right-hand ordinate is the calculated equilibrium concentration of free aqueous ferric iron, Fe^{+3}, as a percentage of the total Fe(III) species in solution. Figure 62 indicates

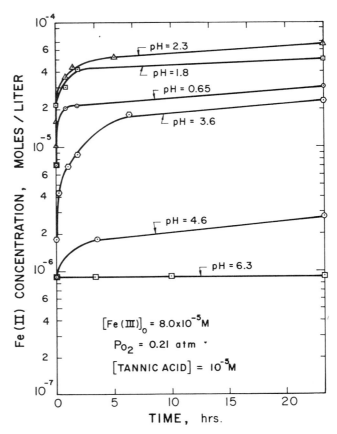

Figure 61.   *Rate of formation of ferrous iron due to reduction of ferric iron by tannic acid at various pH values.*

that, in the acidic region, the rate of Fe(III) reduction increases with pH as long as Fe^{+3} is the dominant Fe(III) species.   The rate of reduction drops significantly as the concentration of free Fe^{+3} decreases, indicating that it is the free trivalent iron which is most reactive and is most easily reduced by tannic acid.   At higher pH values, the hydroxo complexes of Fe(III) kinetically resist reduction by tannic acid.

Upon acidification of the system containing hydroxo-ferric species and tannic acid, the rate of Fe(III) reduction is limited by the rate of dissolution or conversion of the hydroxo-ferric species to free aqueous ferric iron.   It is this factor which allows for Fe(III) stability and aids Fe(II) specificity in the bathophenanthroline analysis; Fe(III)

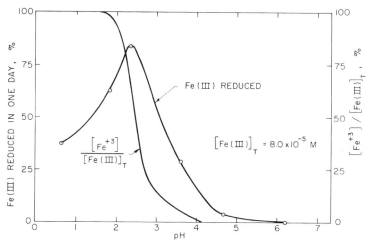

Figure 62. *Extent of ferric iron reduced compared to equilibrium concentration of free ferric iron, Fe^{+3} (both as a percentage of the total Fe(III) concentration).*

reduction by the organic species is limited during the analytical test for Fe(II) in which the pH of the samples was buffered to pH 4.

Consequently, the kinetic stability of ferrous iron in oxygen-rich environments may be due either to the formation of Fe(II)-organic complexes which resist oxidation (Figure 58) or to the reduction of the resultant Fe(III) by the organic species. The latter reaction is dependent upon the degree of hydrolysis of the ferric iron, with Figures 60 and 61 illustrating two extreme conditions. In Figure 60, the iron was added as unhydrolyzed Fe^{+3} and was instantaneously reduced; the curves in Figure 61 were obtained by allowing the iron to hydrolyze first. In reality, the occurrence of either extreme in a natural water is dubious. As ferrous iron is oxidized, one can envision a competition for the newly-formed ferric iron between hydroxide and the organic material present. As a result, some degree of hydrolysis and reduction takes place, the relative proportions of which depend upon pH and the concentration and specific type of organic matter.

Figure 62 shows results of the oxidation study of ferrous iron in the presence of 40 mg/l of natural humic acids. The humic acids contained approximately 2% by weight of metal impurities; thus it was

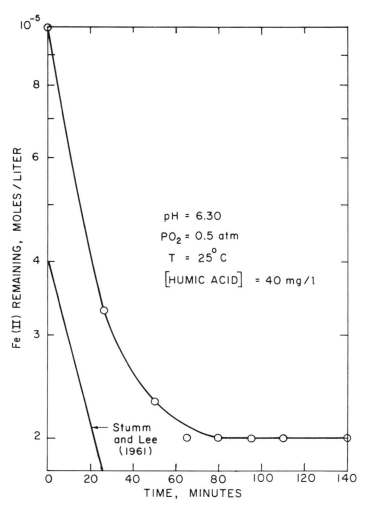

Figure 63.   *Rate of oxidation of ferrous iron in the presence*
of humic acids extracted from natural waters.

anticipated that their complexing ability would be
masked somewhat. The reaction proceeds initially
at a rate in accordance with Equation 4. After
about 60 minutes, 2×10^{-6} moles/liter (approximately 0.1 mg/l) of Fe(II) are stabilized by the
humic substances. These results are identical to
those reported by Komolrit[8] for the oxidation of
Fe(II) in Illinois groundwaters where difficulties
in iron oxidation and removal have been reported.[6]

Reduction of ferric iron by the same concentration of humic acids is shown in Figure 64. The effect is seen to be qualitatively similar to the results with tannic acid shown in Figure 61.

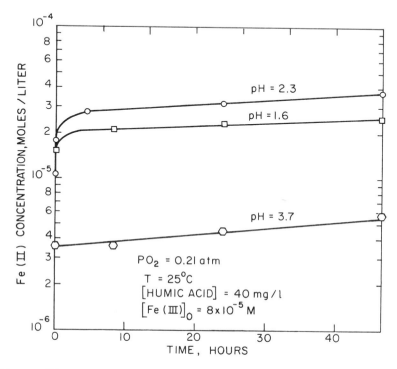

Figure 64.   *Reduction of ferric iron by humic acids.*

Oxidation of ferrous iron following the aeration of three anoxic hypolimnetic water samples is shown in Figure 65 and compared with the rate in simple aqueous media (Equation 4). (The increase in pH during the course of this study was due to the release of carbon dioxide through aeration.) The oxidation rate is significantly slower than Equation 4 would predict. A rough correlation exists between the degree of retardation and the chemical oxygen demand (COD) of the sample. (The COD serves as a measure of the organic content of the sample, but is by no means reflective of the quantity of humic-type material present.) Calculations indicate the Stone Lake and St. Joseph Lake (12.0 meters) samples were initially oversaturated with respect to ferrous

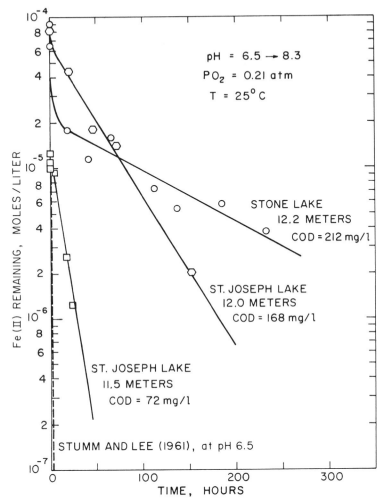

Figure 65.   *Oxidation of ferrous iron in aerated hypolimnetic waters from St. Joseph Lake, Notre Dame, Indiana, and Stone Lake, Cassopolis, Michigan.*

carbonate, $FeCO_3$ ($K_{SO} = 5.1 \times 10^{-11}$).[18]   Thus the retardation effect for these two curves in Figure 65 may be partially due to precipitation of ferrous carbonate.   St. Joseph Lake at 11.5 meters, however, was undersaturated with respect to $FeCO_3$.   In any case, Figure 65 indicates that the oxidation of ferrous iron in lakes following overturn may be much

slower than previously anticipated from simple
kinetic considerations. The resultant increased
stability of Fe(II) may be nutritionally significant
with respect to aquatic growth.

Several wastewater treatment plants have begun
utilizing ferrous iron (*e.g.*, in the form of waste
pickle liquor from the iron and steel industry[19])
in order to precipitate phosphorus as ferric phosphate, $FePO_4$, following oxidation of Fe(II) to
Fe(III) in the aeration tanks of the activated
sludge process. Figure 66 illustrates the oxidation
of ferrous iron added to conventional activated
sludge under aeration in the laboratory. The oxidation appears to be first order in Fe(II) with a
half-time of 2.8 hours, considerably longer than
anticipated; as much as 35% of the initial Fe(II)

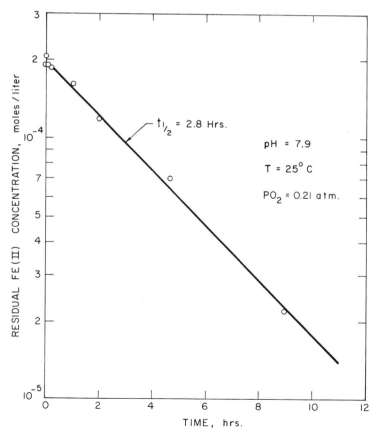

Figure 66.   *Oxidation of ferrous iron added to activated
sludge.*

remained unoxidized at the end of a detention period
of 4 hours despite the oxidizing conditions prevailin•
Figure 67 shows the relative rapidity with which
10^{-4} M ferrous iron is oxidized when added to effluen
from a conventional secondary wastewater treatment
plant. Despite the fairly rapid rate of oxidation

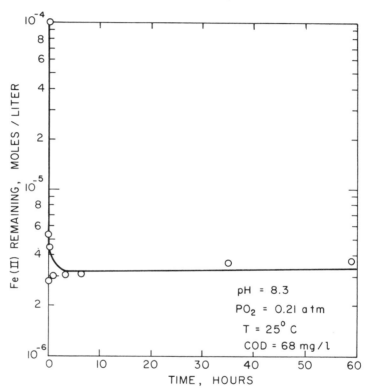

Figure 67.   *Oxidation of ferrous iron added to secondary
sewage effluent.*

due to the high pH of the water (in comparison to
Figure 66), approximately 0.18 mg/l of ferrous iron
was stabilized by the effluent. Rebhun and Manka[17]
reported that humic material comprises about 50% of
the residual organic matter in conventional secondary
effluent. The similarity of Figure 67 to Figure 63
is apparent.

CONCLUSIONS

This study has demonstrated that organic matter is capable of creating conditions whereby the oxidation of ferrous iron in well-oxygenated environments is inhibited. Humic substances, as represented by the model compounds used in this investigation, can stabilize ferrous iron through the formation of Fe(II)-organic complexes which resist oxidation or through reduction of the resultant ferric iron by the organic species. The stability of iron in many natural aquatic systems can create specific problems in water supply and wastewater treatment, and may be nutritionally significant with respect to aquatic growth.

ACKNOWLEDGMENTS

This study was supported by Grant No. 5F1-WP-26,602 of the Office of Water Programs of the Environmental Protection Agency. The authors wish to thank Dr. John T. O'Connor of the Department of Civil Engineering at the University of Illinois for his assistance in making his research results available to us.

REFERENCES

1. Sillen, L. G., and A. E. Martell. *Stability Constants of Metal-Ion Complexes*, Special Publ. No. 17 (London: The Chemical Society, 1964).
2. Stumm, W., and G. F. Lee. Ind. Eng. Chem. *53*, 143 (1961).
3. Prakash, A., and M. A. Rashid. Limnol. Oceanog. *13*, 598 (1968).
4. Davies, A. G. J. Marine Biol. Assoc., U.K. *50*, 65 (1970).
5. Martin, D. F., M. T. Doig, and R. H. Pierce. "Distribution of Naturally Occurring Chelators (Humic Acids) and Selected Trace Metals in Some West Coast Florida Streams, 1968-1969," Professional Papers Series Number 12, Florida Department of Natural Resources Marine Research Laboratory (St. Petersburg, Florida, 1971).
6. Ghosh, M., J. T. O'Connor, and R. S. Engelbrecht. Proc. Amer. Soc. Civil Eng., J. San. Eng. Div. *92*, 120 (1966).
7. *Public Health Service Drinking Water Standards*, (Washington, D.C.: U.S. Dept. Health, Education, and Welfare, 1962).
8. Komolrit, K. "Measurement of Redox Potential and Determination of Ferrous Iron in Ground Waters," M.S. Thesis, University of Illinois, Urbana, Illinois (1962).
9. Robinson, L. R. Water & Sewage Works *114*, 377 (1967).
10. Shapiro, J. J. Am. Water Works Assoc. *56*, 1062 (1964).

11. Hutchinson, G. E. *A Treatise on Limnology*, Volume 1 (New York: John Wiley and Sons, 1957).
12. Oldham, W. K., and E. F. Gloyna. J. Am. Water Works Assoc. *61*, 610 (1960).
13. Christman, R. F. Environ. Sci. Technol. *1*, 302 (1967).
14. Coulson, C. B., R. I. Davies, and D. A. Lewis. J. Soil Sci. *11*, 30 (1960).
15. Morgan, J. J., and W. Stumm. *Proc. Second International Water Pollution Research Conference* (New York: Pergamon Press, 1964).
16. Lee, G. F., and W. Stumm. J. Am. Water Works Assoc. *52*, 1567 (1960).
17. Rebhun, M., and J. Manka. Environ. Sci. Technol. *5*, 606 (1971).
18. Singer, P. C., and W. Stumm. J. Am. Water Works Assoc. *62*, 198 (1970).
19. Sewerage Commission of the City of Milwaukee, "Phosphorus Removal with Pickle Liquor in an Activated Sludge Plant," Rept. No. 11010 FLQ, Water Pollution Control Research Series (Washington, D.C.: Environmental Protection Agency, 1971).

11. ORGANIC LIGANDS AND PHYTOPLANKTON GROWTH IN NUTRIENT-RICH SEAWATER

Richard T. Barber. Duke University Marine
Laboratory, Beaufort, North Carolina

Natural waters always contain a background
concentration of dissolved organic compounds in
milligram per liter amounts. Identification of the
kinds of organic compounds present and their geographic and temporal variations have been the
subject of considerable study in the last decade.
Two recent symposium volumes, *Organic Matter in
Natural Waters*[1] and *Organic Compounds in Aquatic
Environments*,[2] contain 56 reports on the subject
of organic matter in natural waters; however, it
is clear from reading these two volumes that the
functional involvement of dissolved organic compounds and organometallic complexes in regulating
biogeochemical processes is poorly known. Descriptive work on naturally occurring organic matter is
well underway but mechanistic analysis of the
biogeochemical processes involving organic compounds
has only started.

There has been considerable speculation on the
importance of organic complexation in natural water.
Much of the concern grew out of attempts to develop
defined media for phytoplankton culturing. Provasoli,
McLaughlin, and Droop[3] in their comprehensive review
of media development have described the work[4-7]
leading to the use of synthetic organic ligands,
such as ethylenediaminetetraacetic acid (EDTA), to
replace the undefined "soil extract." Work in the
author's laboratory[8-11] has related variations in
phytoplankton growth rate with variations in
naturally occurring organic compounds and ligand/
metal interactions. This work is now converging
with the older work[3] on the development of artificial

seawater media for marine phytoplankton and is producing convincing evidence for the existence of
naturally occurring organic ligands that are functional analogues of synthetic chelators such as
EDTA. The work described in this report tests
several specific aspects of the general hypothesis
that natural organic ligands are involved in phytoplankton growth and, therefore, in biogeochemical
processes in natural waters. The ideas discussed
are (1) growth-enhancing compounds are synthesized
by phytoplankton and released into the surrounding
seawater; (2) the phytoplankton-synthesized compounds are functionally analogous to synthetic
organic ligands since the synthetic ligands can
substitute for the natural compounds and support
normal phytoplankton growth; (3) the growth-
enhancing compounds are organic and present in
some natural waters, but absent in others; (4) the
specific chemical benefit to phytoplankton involves
altering the availability or toxicity of metals.
While there is too little evidence, or too much
contradictory evidence, to prepare an overall model
of the specific mechanism of organometallic involvement in phytoplankton growth, the evidence on each
of the ideas must be taken into consideration in
the development of mechanistic models of the natural
processes. The author's goal in presenting this
information is to stimulate chemical theoreticians
to develop mechanistic models of biogeochemical
complexation phenomena that can be tested by experimental analysis on natural ecosystems. The
interplay between theoreticians and experimentalists
which has proved fruitful in other disciplines seems
to have been signally absent in this area of environmental science; further advances in understanding
biogeochemical processes will require this interplay.
Satisfactory analysis of the complex and frequently
contradictory biological and environmental information requires the logical rigor of a mechanistic
framework based on chemistry.

EXPERIMENTAL DESIGN

Growth, defined here an an increase of biomass,
is the end product of a complex series of biochemical
processes. Biologists, in their concern with
organisms *per se*, often dispute the validity of one
or another method of measuring the growth of phytoplankton. To answer this criticism, several indices
of phytoplankton growth have been used in the

experiments described here. One method was long term photosynthetic uptake of inorganic [14]carbon. During the course of these experiments the mass of newly-synthesized cell material increased several orders of magnitude in the presence of the radioactive tracer. The kind of error common to short-term ^{14}C uptake experiments results from incomplete equilibration of the tracer in the various internal pools of the phytoplankton cells present at the initial exposure to the tracer. In the experiments described here the original cell biomass was always less than 10^{-2} and usually less than 10^{-3} of the final biomass; therefore, the consequences of incomplete tracer equilibration in the original cells cannot affect the interpretation of the experiments.

Long term ^{14}C uptake has been found by other workers to give a reliable measure of the net increase in new particulate cell carbon.[12-15] Previous work[9] in the author's laboratory using a high temperature combustion method[16] to measure particulate organic carbon determined that the correlation coefficient of long term ^{14}C uptake and total cell carbon was very high ($r^2 = 0.976$). Having confirmed the validity of long term ^{14}C uptake as a measure of increase in biomass, the relationship of long term ^{14}C uptake to chlorophyll a[18] synthesis was examined in all of the experiments reported here. The results of this comparison (given below) indicate that the interpretation of the experiments is the same regardless of the index of growth that is used. The high correlations among the various independent measurements of growth (cell carbon, ^{14}C uptake, chlorophyll a) would be extremely unlikely if any one of the methods was subject to an artifact of the major experimental variables (ligand additions, ligand removal, and metal additions).

It is especially important to emphasize to nonbiologists that the ligand/metal interactions examined in this work take their place in a hierarchy of environmental factors which regulate the magnitude and rate of phytoplankton growth. The availability of nitrogen, phosphorus, silicon, and energy in the form of sunlight are primary factors regulating phytoplankton production. The potential role of chelation phenomena in natural systems is most likely a second level of regulation which functions when inorganic nutrients and light are present in amounts which will support phytoplankton

growth.[9] The experiments were designed to provide
nonlimiting amounts of major inorganic nutrients
and light, but not to exceed the nutrient and light
conditions that marine phytoplankton encounter in
the ocean. Natural nutrient-rich seawater was
provided by collecting deep water from below the
euphotic zone and the pycnocline. In the western
North Atlantic the water was "artificially upwelled"
from 800 meters deep, while off the North African
coast the water was collected from 75 meters. The
concentration of inorganic nutrients was 10 to 20
μM NO_3; 0.5 to 1.5 μM PO_4; and 10 to 15 μM SiO_4.
The North African experiments were incubated in
natural sunlight with running seawater as a coolant.
Other experiments were incubated at 24°C under
cool-white fluorescent and incandescent light (8 x
10^4 ergs/cm^2/sec) on a 12-hour light-dark cycle.
This experimental design provided the phytoplankton
with an optimal (*i.e.*, saturating but not inhibiting)
supply of inorganic nutrients and light.[20,21]
    Both natural populations of marine phytoplankton
and a pure culture of the marine diatom *Chaetoceros
socialis* were used. The natural populations were
freshly collected from surface waters before each
experiment; the North African experiments used
populations collected in the upwelling region off
Cabo Blanco, Spanish Sahara. Other work used
natural populations from the continental shelf
waters of North Carolina, U.S.A. The pure culture
of *Chaetoceros socialis* was provided by H. Kayser,
Biologische Anstadt Helgoland, F.R.G. This species
was selected for the simulated upwelling experiments
because it is a common dominant phytoplankter in
the productive upwelling regions off the west coast
of North America, Peru,[22,23] and Southwest Africa.[24]
Details of the experimental procedure have been
described[9] and further information on the chemistry
and hydrography of the North African area is
available.[19]

PHYTOPLANKTON SYNTHESIS OF
GROWTH-ENHANCING LIGANDS

    Despite optimal light and nutrient conditions,
very low density inoculum populations of phytoplankton grow poorly when nutrient-rich deep ocean
water is used as a medium for either batch or continuous culture experiments. This phenomenon is
assumed to be conceptually related to the critical
minimum inoculum effect observed in other kinds of

microorganisms. Jannasch[25] suggests that the best
interpretation of the critical size effect is that
if the initial population is large enough it can
modify the unfavorable environment by its own
metabolic products. The existence of an inoculum
size effect on the initiation of exponential growth
by marine phytoplankton has been documented by
Bunt;[26] however, it has not been established that
the specific chemical modification of the initially
unfavorable culture medium (in this case nutrient-
rich deep ocean water) involves the synthesis and
excretion of organic ligands. To test this idea
the growth rate and lag period of various density
inocula with and without the addition of the synthetic organic ligand, EDTA, were examined. The
overall growth response of the pure culture (Figure
68) of *Chaetoceros socialis* and the natural population (Figure 69) were similar; the lowest density
inoculum (0.5 cell per ml) without EDTA showed the
longest lag phase. In both cases the lowest density
inoculum with EDTA showed a lag phase only half as
long as the equivalent inoculum without EDTA. Increases in cell density to 5.0 cells per ml and
500.0 cells per ml shortened the lag phase in both
the natural population and the pure culture. At
the maximum cell density (500 cells per ml) the
growth rate and lag period with and without EDTA
were similar. The simplest explanation of these
results is that each phytoplankton cell has a
certain capacity to synthesize and excrete a natural
compound that makes the environment more favorable
to exponential growth. If there are more cells
contributing their metabolic products, the favorable
condition is achieved more rapidly and the lag period
is shortened.

FUNCTIONAL ANALOGY OF EDTA AND
NATURAL GROWTH-ENHANCING COMPOUNDS

The chemical activity of the compound produced
by phytoplankton appears to be analogous to the
artificial ligand, EDTA, because when a high density
inoculum was used no further benefit was conferred
by the addition of EDTA. Conversely, when EDTA was
provided, the low density inocula grew almost as
rapidly as the high density inocula and the lag
period was shortened considerably. The similar
slopes of all the "with EDTA" growth curves in
Figures 68 and 69 indicate that when the synthetic

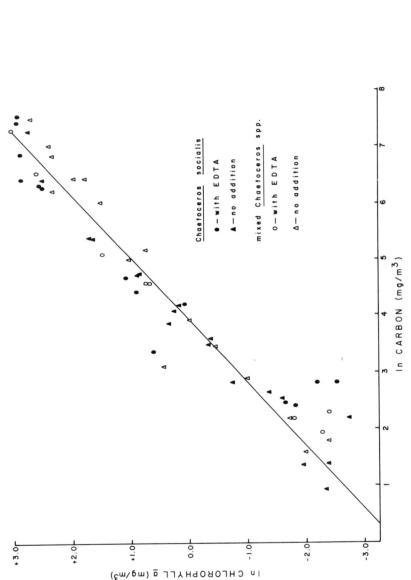

Figure 68.   The effect of initial inoculum density and the addition of an artificial ligand, EDTA, on the growth of Chaetoceros socialis in artificially upwelled seawater from 800 meters. The concentration of EDTA was 1.0 μM.

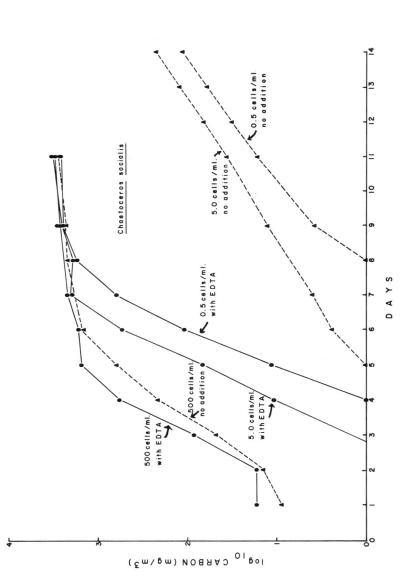

Figure 69. The effect of initial inoculum density and the addition of an artificial ligand, EDTA, on the growth of natural populations of marine phytoplankton in artificially upwelled seawater from 800 meters. The concentration of EDTA was 1.0 μM.

ligand was present the specific growth rate was about equal regardless of the initial density of the inoculum.

The pure culture and the natural populations, while similar in their response to the addition of EDTA, showed an interesting difference in the specific growth rate of the low density inoculum when no addition of EDTA was provided. Both low density inocula had long lag periods (7 and 8 days) but the natural population of mixed *Chaetoceros* spp. and *Nitzschia* spp. (Figure 69) eventually grew at a specific growth rate equal to the rate of the phytoplankton provided with EDTA. It is perhaps significant that *Nitzschia* was one of the first diatoms successfully grown in the laboratory;[27] the ability of this diatom to modify and eventually grow rapidly in initially unfavorable nutrient-rich conditions may be related to its success as a laboratory organism before synthetic ligands were routinely added to culture media.

In these experiments chlorophyll a, as well as carbon, was measured throughout the experiment. The curves of chlorophyll a versus time were similar enough to the carbon curves that there is no benefit in presenting them; however, a regression analysis of the natural logarithm of carbon concentration versus the natural logarithm of chlorophyll a concentration is useful to examine the impact of the experimental treatments on the composition of the phytoplankton cells. Figure 70 gives results of the regression analysis of carbon and chlorophyll a data from the exponential portion of the growth curves shown in Figures 68 and 69. Over several orders of magnitude there is a high correlation between chlorophyll a measured fluorometrically[18] and carbon determined by [14]carbon uptake.[28] More importantly the correlation coefficient for the phytoplankton grown in the presence of EDTA was identical to that of phytoplankton without EDTA; both were $r^2 = 0.93$ (Figure 70). The timing of exponential growth and the specific growth rates were different in the two groups but the carbon/chlorophyll a composition was identical suggesting that EDTA-enhanced growth is fundamentally similar to growth-enhancement by increased inoculum density.

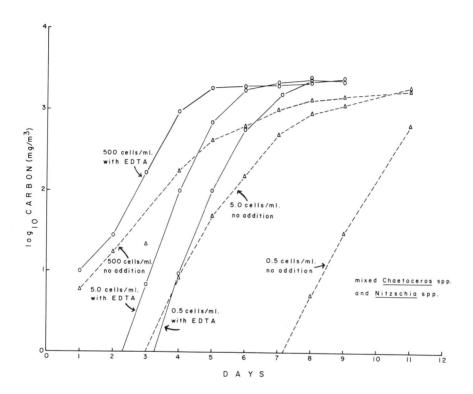

Figure 70.   The relation between chlorophyll *a* and carbon in
the experiments described in Figures 68 and 69.
The data were taken from the exponential portion
of the growth curves presented in Figures 68 and
69.   Chlorophyll *a* was determined fluorometrically;[18]
carbon was determined by **carbon-14** uptake.   The
correlation coefficient of the natural logarithms
of all data in Figure 70 is $r^2 = 0.93$; for the
Chaetoceros socialis data, $r^2 = 0.92$; for the mixed
Chaetoceros spp. data, $r^2 = 0.94$; for the "with
EDTA" data, $r^2 = 0.93$; for the "without EDTA" data,
$r^2 = 0.93$.

ORGANIC NATURE OF THE
GROWTH-ENHANCING COMPOUNDS

Ultraviolet photooxidation effectively oxidizes some portion of the organic compounds dissolved in seawater.[29-31] Ferric iron and other metals photosensitize the oxidation reaction,[29] so that organometallic complexes will be preferentially oxidized relative to other organic compounds making this method particularly effective for removing organic ligands if they are present in the water being irradiated. Photooxidation was used in a series of experiments (Figures 71 and 72) in the North African upwelling region[19,32] to determine if UV-labile, growth-promoting compounds were present in this region of high biological productivity.

The experimental strategy consisted of selecting nutrient-rich water in which phytoplankton are growing, photooxidizing the dissolved organic compounds by ultraviolet irradiation, then adding back a population of phytoplankton to determine if the irradiated water can support growth as well as control water that has not been irradiated. Strong ultraviolet light is lethal to phytoplankton; it is necessary to emphasize that the seawater, not the phytoplankton, was irradiated. The seawater in 500-ml volumes was exposed for 30 minutes to a 4-inch immersion lamp (UV Products, Inc., San Jose, California). After cooling to the *in situ* temperature (27°C), the ligand and metal additions were made according to the factorial enrichment design of the experiment (Figures 71 and 72). The final concentration was 2 µM EDTA, 1 µM Mn and 1 µM Fe. After UV exposure the water was reaerated by shaking before the phytoplankton were added; 30 minutes elapsed between the end of the UV exposure and the addition of the living phytoplankton cells.

Considerable concern has been given to the possibility that ultraviolet photooxidation may introduce a toxic artifact unrelated to the removal of organic solutes. The *in situ* pH in this region varied from 8.3 at the surface to around 7.9 at the depth of the oxygen minimum at 500 meters.[19] The pH change resulting from photooxidation of organic carbon into inorganic carbon was less than 0.1 pH units and, therefore, was not greater than the magnitude of the environmental gradients of pH in this upwelling region. Variations of pH in this range, for example from 8.2 to 8.1, have not been found to decrease the growth-supporting ability of nutrient-rich deep ocean water when the pH modification is

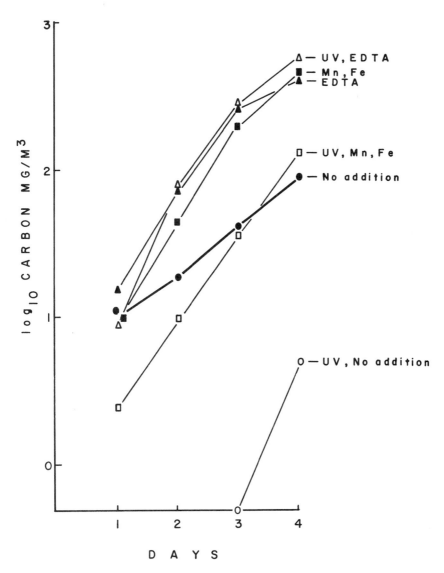

Figure 71. *The effect of organic removal via UV photooxidation and EDTA, Fe, and Mn enrichment on the growth of phytoplankton in nutrient-rich water from the North African upwelling region. The water was from 75 meters at Station 312; the inoculum was a natural population of phytoplankton from 20 meters. A description of the hydrography and biology is available.[19,32] The heavy line describes the growth of the "No addition" control.*

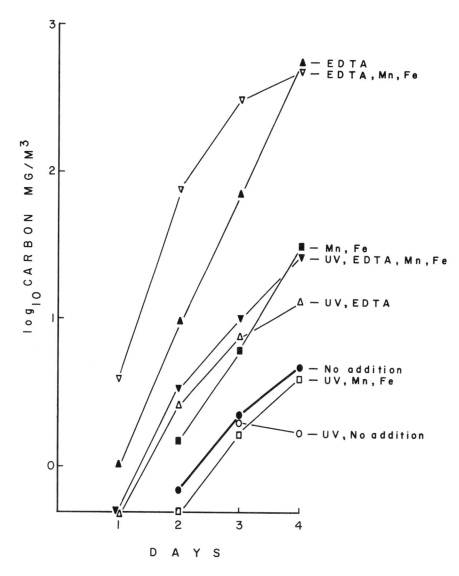

Figure 72.   *The effect of organic removal via UV photooxidation*
and EDTA, Fe, and Mn enrichment on the growth of
phytoplankton in nutrient-rich water from the
North African upwelling region. The water was
from 60 meters at Station 217; the inoculum was a
natural population of phytoplankton from 20 meters.
The heavy line describes the growth of the "No
addition" control.

made with inorganic buffers.[36] The author recognizes
the importance of pH in determining the chemical
speciation of metal ions in seawater as established
by Zirino[33-35] but at the present time it does not
appear that the growth variations presented in
Figures 71 and 72 are the result of UV-induced
changes in pH. Obviously, this subject and the
potential existence of extremely toxic UV-induced
radicals requires further analysis.

Comparison of growth in natural seawater (the
"No addition" growth curve in Figure 71) and in UV-
treated water indicates that the UV treatment did
reduce ability of the seawater to support phytoplankton growth. The results given in Figure 71
indicate that the seawater used in this experiment
did contain UV-labile, growth-enhancing compounds.
A different situation exists in Figure 72; a comparison of the growth of the "No addition" and "UV,
No addition" populations shows that there is little
difference between the two. Apparently the seawater
used in this experiment (Figure 72) did not contain
a significant concentration of UV-labile compounds.
The decreased growth of the "No addition" population
in Figure 72 as compared with the "No addition"
population in Figure 71 suggests that there was an
original difference in the ability of the two waters
to support rapid phytoplankton growth. The difference
was not due to major inorganic nutrient limitations;
the initial concentration of NO_3 in both experiments
was over $10\mu M$ and other nutrients and light were in
abundant supply. The author has observed similar
variations in the growth-supporting quality of
nutrient-rich waters in the Peru upwelling region[9]
and in equatorial upwelling.[8] The presence or
absence of labile, growth-enhancing compounds is
apparently related to the past circulation and
biological history of the seawater.[8,9]

THE CHEMICAL BENEFIT TO PHYTOPLANKTON

The most interesting information in Figures 71
and 72 concerns the additions that were able to
reverse the effect of UV-removal of organic solutes.
In both Figure 71 and 72 the addition of EDTA in 2
μM concentration (the "UV, EDTA" growth curves) was
able to completely reverse the negative effect of UV
photooxidation and restore the growth-supporting
ability of the seawater. These results indicate
that the compounds removed by UV are functionally
analogous to synthetic ligands such as EDTA and are

not specific organic micronutrients such as vitamin B-12 or plant auxins.[37] The finding that EDTA alone can reverse the UV-removal of organic solutes suggests that the biological role of the natural organic ligands is generalized adjustment of the chelation poise to within certain favorable limits. This kind of chemical modification of the environment is quite different from the highly specific biochemical warfare[37],[38] and mutual stimulation[39] that certain species of marine phytoplankton engage in.

Further information on the functional role of the UV-labile compounds can be attained by examining the growth response in UV-treated water that was enriched with Mn and Fe in 1 µM concentrations. In Figures 71 and 72 the growth of the "UV, Mn Fe" population was improved over the growth of the "UV, No addition" population. The addition of Mn and Fe, without an organic ligand, reversed to some degree the detrimental effect of UV-treatment. One interpretation of this result is that, by providing Mn and Fe in the form of a freshly-made distilled water solution of the chloride salt,[9] the phyto-plankton are supplied with a new source of the essential metals that they cannot get from the UV-treated seawater. This explanation suggests that the chemical benefit to phytoplankton of naturally occurring organic ligands is that the ligands increase the availability of Mn and Fe by increasing their kinetic mobility. Another explana-tion of growth-enhancement in UV-treated water by either EDTA or metals was pointed out by Stumm.[40] Fe and Mn, when added to seawater, immediately form particulate oxyhydroxides that effectively scavenge other metals from solution.[41] The organic ligands, both synthetic and naturally occurring, can reduce the concentration of a toxic metal ion by forming a stable, nontoxic complex. Both kinds of additions could have the effect of reducing the concentration of a toxic metal ion.

It is impossible to determine which of the two potential chemical benefits is more important on the basis of the information provided by growth experi-ments such as those described in Figures 71 and 72. Growth is a complex process with many potentially rate-regulating processes making it hard to determine whether an observed increase in growth results from acceleration of a process or the removal of a previously-existing inhibition.

SUMMARY

High density inocula of a pure culture of *Chaetoceros socialis* and a mixed population of *Chaetoceros* spp. and *Nitzschia* spp. grew rapidly in nutrient-rich deep ocean water while lower density inocula had a long lag period and grew at a lower rate. When low density inocula were provided with a synthetic organic ligand, EDTA, the lag period and specific growth rate were equal to those of the higher density inocula. Regression analysis of carbon versus chlorophyll *a* indicated that the composition of the EDTA-enhanced phytoplankton was the same as the composition of density-enhanced phytoplankton. The similarity of EDTA and density-enhancement in the aspects studied here suggests that phytoplankton are the source of a metabolic product whose function is analogous to EDTA.

Ultraviolet photooxidation established the existence of UV-labile organic compounds. When the labile compounds were removed from seawater the ability of the seawater to support phytoplankton growth was decreased. The amount of UV-labile, growth-enhancing compounds varied at different locations in the North African upwelling region. The effect of UV removal of organic compounds on phytoplankton growth was completely reversed by the addition of EDTA in μM concentrations offering further evidence for the functional similarity between EDTA and naturally occurring ligands. Mn and Fe in μM concentrations partially reversed the detrimental effect of UV photooxidation, but the reversal was less complete than EDTA reversal. These results suggest that naturally occurring ligands function by increasing the mobility of essential metals such as Mn and Fe, but these findings do not eliminate the idea that natural ligands enhance phytoplankton growth by complexing toxic metals.

ACKNOWLEDGMENTS

The major support for this work was National Science Foundation Grant GA-28742. Other support was provided through the Duke Cooperative Oceanographic Program which is supported by NSF grants GA-27725 and GD-28333. We thank Dr. Ramon Margalef for the opportunity to take part in the Sahara II cruise of the CORNIDE DE SAAVEDRA.

REFERENCES

1. Hood, D. W., ed. *Organic Matter in Natural Waters*. Vol. 1. (University of Alaska, Institute of Marine Science, 1970).

2. Faust, S. J., and J. V. Hunter, eds. *Organic Compounds in Aquatic Environments*. (New York: Marcel Dekker, Inc., 1971).

3. Provasoli, L., J. J. A. McLaughlin, and M. R. Droop. "The Development of Artificial Media for Marine Algae," *Archiv für Mikrobiologie,* Bd. *25,* 392 (1957).

4. Hutner, S. H., and L. Provasoli. "The Phytoflagellates," In *Biochemistry and Physiology of Protozoa*. Lwoff, A., ed. (New York: Academic Press, Inc., 1951).

5. Provasoli, L., and J. F. Howell. "Culture of a Marine *Gyrodinium* in a Synthetic Medium," *Proc. Amer. Soc. Protozool. 3,* 6 (1952).

6. Provasoli, L., and I. J. Pintner. "Ecological Implications of *in vitro* Nutritional Requirements of Algal Flagellates," *Ann. New York Acad. Sci. 56(5),* 839 (1953).

7. Droop, M. R. "Some New Supra-littoral Protista," *J. Mar. Biol. Assoc. U. Kingd. 34,* 233 (1955).

8. Barber, R. T., and J. H. Ryther. "Organic Chelators: Factors Affecting Primary Production in the Cromwell Current Upwelling," *J. Exp. Mar. Biol. Ecol. 3,* 191 (1969).

9. Barber, R. T., R. C. Dugdale, J. J. MacIsaac, and R. L. Smith. "Variations in Phytoplankton Growth Associated with the Source and Conditioning of Upwelling Water," *Inv. Pesq. 35(1),* 171 (1971).

10. Siegel, A. "Metal-Organic Interactions in the Marine Environment," In *Organic Compounds in Aquatic Environments,* Faust, S. J. and J. V. Hunter, eds. (New York: Marcel Dekker, Inc., 1971).

11. Spencer, L. J., R. T. Barber, and R. A. Palmer. "The Detection of Ferric Specific Organic Chelators in Marine Phytoplankton Cultures," In *Food and Drugs from the Sea,* Worthen, N. C., ed. (University of Rhode Island Press, in press).

12. Ryther, J. H. "The Measurement of Primary Production," *Limnol. Oceanog. 1,* 72 (1956).

13. Ryther, J. H., and D. W. Menzel. "Comparison of the ^{14}C-technique with Direct Measurement of Photosynthetic Carbon Fixation," *Limnol. Oceanog. 10,* 490 (1965).

14. Eppley, R. W. "An Incubation Method for Estimating the Carbon Content of Phytoplankton in Natural Samples," *Limnol. Oceanog. 13,* 574 (1968).

15. Eppley, R. W., and J. D. H. Strickland. "Kinetics of Marine Phytoplankton Growth," In *Advances in Microbiol. of the Sea,* Vol. 1, Droop, M. R., and E. J. Ferguson Wood, eds. (New York: Academic Press, 1968).

16. Menzel, D. W. and R. F. Vaccaro. "The Measurement of Dissolved Organic and Particulate Carbon in Seawater," *Limnol. Oceanog. 9*, 138 (1964).

17. Cahn, R. D. "Detergents in Membrane Filters," *Science, 155*, 195 (1967).

18. Lorenzen, C. J. "A Method for the Continuous Measurement of *in vivo* Chlorophyll Concentration," *Deep-Sea Research 13*, 223 (1966).

19. Margalef, R. "Hidrografía de la región de afloramiento del noroeste de Africa. Datos básicos de la campaña (Sahara II) del (Cornide de Saavedra)," *Inv. Pesq. Suplemento 1*, 1 (1972).

20. MacIsaac, J. J., and R. C. Dugdale. "The Kinetics of Nitrate and Ammonia Uptake by Natural Populations of Marine Phytoplankton," *Deep-Sea Research 16*, 45 (1969).

21. MacIsaac, J. J., and R. C. Dugdale. "Interactions of Light and Inorganic Nitrogen in Controlling Nitrogen Uptake in the Sea," *Deep-Sea Research, 19*, 209 (1972).

22. Ryther, J. H., D. W. Menzel, E. M. Hulbert, C. J. Lorenzen, and N. Corwin. "The Production and Utilization of Organic Matter in the Peru Coastal Current," *Inv. Pesq. 35*, 43 (1971).

23. Blasco, Dolores. "Composición y distribución del fito-plancton en la región del afloramiento de las costas peruanas," *Inv. Pesq. 35*, 61 (1971).

24. Hobson, Louis A. "Relationships Between Particulate Organic Carbon and Micro-organisms in Upwelling Areas Off Southwest Africa," *Inv. Pesq. 35*, 195 (1971).

25. Jannasch, Holger W. "Starter Populations as Determined Under Steady State Conditions," *Biotechnology and Bioengineering 7*, 279 (1965).

26. Bunt, J. S. "The Influence of Inoculum Size on the Initiation of Exponential Growth by Marine Diatom," *Zeitschrift fur Allg. Mikrobiologie 8*, 289 (1968).

27. Ketchum, B. H. "The Absorption of Phosphate and Nitrate by Illuminated Cultures of *Nitzschia closterium*," *Am. J. Botany* (June), 26 (1939).

28. Strickland, J. D. H., and T. R. Parsons. "A Manual of Sea Water Analysis," *Bull. Fish. Res. Bd. Can.*, No. 125 (1965).

29. Armstrong, F. A. J., P. M. Williams, and J. D. H. Strickland. "Photooxidation of Organic Matter in Sea-water by Ultraviolet Radiation, Analytical and Other Applications," *Nature 211*, 481 (1966).

30. Armstrong, F. A. J., and S. Tibbitts. "Photochemical Combustion of Organic Matter in Seawater, for Nitrogen, Phosphorus and Carbon Determination," *J. Mar. Biol. Ass. U. K. 48*, 143 (1968).

31. Williams, P. M. "The Determination of Dissolved Organic Carbon in Seawater: A Comparison of Two Methods," *Limnol. Oceanogr. 14*, 297 (1969).

32.   Margalef, R.   "Fitoplancton de la region de afloramiento de Noroeste de Africa.   "igmentos y produccion de la Campana Sahara II del Cornide de Saavedra," *Inv. Pesq. Suplemento 1,* 23 (1972).

33.   Zirino, A., and S. Yamamoto.   "A pH-Dependent Model for the Chemical Speciation of Copper, Zinc, Cadmium, and Lead in Seawater," *Limnol. Oceanogr. 17(5),* 661 (1972).

34.   Zirino, A., and M. L. Healy.   "Inorganic Zinc Complexes in Seawater," *Limnol. Oceanogr. 15,* 956 (1970).

35.   Zirino, A., and M. L. Healy.   "pH-Controlled Differential Voltammetry of Certain Trace Transition Elements in Natural Waters," *Environ. Sci. Technol. 6,* 243 (1972).

36.   Barber, R. T., and S. Huntsman.   Unpublished Results (1972).

37.   Sieburth, J. McN.   "Acrylic Acid, an 'Antibiotic' Principle in Phaeocystis Blooms in Antarctic Waters," *Science 132(3428),* 676 (1960).

38.   Shilo, M.   "Formation and Mode of Action of Algal Toxins," *Bacteriol. Rev. 31(3),* 180 (1967).

39.   Bentley-Mowat, J. A., and S. M. Reid.   "Effect of Gibberellins, Kinetin and Other Factors on the Growth of Unicellular Marine Algae in Culture," *Botanica Marina 12,* 185 (1969).

40.   Stumm, W.   Personal communication (1969).

41.   Langmuir, D. and D. O. Whittemore.   "Variations in the Stability of Precipitated Ferric Oxyhydroxides," In *Nonequilibrium Systems in Natural Water Chemistry,* Advances in Chemistry Series 106, Gould, R. F., ed. (Washington, D.C.: American Chemical Society, 1971).

12. IMPLICATIONS OF METAL-ORGANIC COMPOUNDS IN RED TIDE OUTBREAKS

Dean F. Martin and Barbara B. Martin. Department of Chemistry, University of South Florida, Tampa, Florida

INTRODUCTION

The term "red tide" refers to a massive disturbance of the marine balance of nature caused by a sudden local bloom or growth of organisms, leading to discolored water. Depending upon the organism and conditions, the discoloration may be colors other than red. The general phenomenon is very old; the passage in *Exodus* 7:20-21 describing the plague of the red river probably refers to red tide. It is ubiquitous; the Red Sea and the Gulf of California (formerly the Vermillion Sea) owe their names to similar discolorations, and red tides are observed throughout the world.[1,2]

"Red tide" organisms cause mortalities in three ways. The first cause is the toxin associated with the organism, which either may be given off while the organism is alive or may be contained in the organism and released when it dies or is broken up by wave action. Red tide toxins have a range of pharmacologic properties. For instance, the toxin of the Florida red tide organism, *Gymnodinium breve*, depolarizes membranes, and the toxin from *Prymnesium parvum* causes hemolysis of red blood cells. The toxin of *Gonyaulax catenella* (saxotoxin) is a neurotoxin useful for studying the mechanism of sodium transport across membranes. Some of the major red tide organisms containing toxins are summarized in Table 63. A more extensive listing of nontoxigenic red tides was given by Ryther.[1]

The second cause of mortality may be secondary effects due to deficiency of oxygen, or hydrogen

Table 63

Summary of Some Toxigenic Organisms

Organism	Type of Organism	Location	Reference
Gonyaulax catanella	armored dinoflagellate	Pacific coasts, U.S. and Canada	3
Gonyaulax tamarensis	dinoflagellate	Bay of Fundy, south to New England	4
Gonyaulax acatenella	dinoflagellate	Bay of Fundy	4
Gonyaulax monilata	dinoflagellate	Gulf of Mexico	5
Gymnodium breve	unarmored dinoflagellate	Gulf of Mexico	6
Gymnodinium veneficum	dinoflagellate	English Channel	7
Prymnesium parvum	crysomonad	Mediterranean Sea, Europe and Mid-East fresh water	8

sulfide produced by decay of dead fish, or interruption of the food chain. These may be caused by a variety of red tides, toxic and otherwise.

A third effect, paralytic shellfish poisoning, is caused by ingesting toxic filter-feeding shellfish, such as oysters, mussels and clams. Blooms of *Gonyaulax catenella, G. acatenella*, and *G. tamarensis* (Table 63) are directly associated with poisoning. Over 220 human fatalities throughout the world have been attributed to paralytic shellfish poisoning.[9]

The Florida red tide deserves special attention because its effects are spectacular though similar to others throughout the world and it has received much study. The causative organism is the unarmored dinoflagellate *Gymnodinium breve* Davis.[6] Blooms cause fish kills and airborne broken cells cause respiratory distress to humans on land.[10] Red tide has been observed along the west coast of Florida sporadically since 1844. More recently, seven major outbreaks have occurred off the Florida coast between 1947 and 1960 (Table 64). Most have occurred from Tampa Bay south to just north of the Florida Keys (Figure 73).

Table 64

Seasonal Occurrence of Florida Red Tides[a]

Year	General Area	J	F	M	A	M	J	J	A	S	O	N	D	Total	Intense
						Months[b]								Outbreak Duration (months)	
1946	Boca Grande and South											X̲	X		
1947	Reached Tampa Bay 7/47	X	-	X	X	X	X	X	X	X				11	10
1952	Near Sanibel Island											X̲	X		
1953	Near Sanibel Island	X												3	3
1953	Tampa and South									X̲	X	X	X		
1954	Venice and South	X	X	X	-	-	-	X	X	X	X	X	X		
1955	Venice and South	X	X											18	15
1957	Tampa and South									X̲	X	X	X	4	4
1959	Tampa to Venice									X̲	X	X	-		
1960	Egment Key (near Tampa Bay)	-	-	X	-	-	-	X	X					12	6

[a] From Reference 10.

[b] Initial month of an outbreak is underlined; portions of same outbreak during temporary recession are joined by dashes.

These outbreaks have major implications in terms of ecology, public health, and economy. The organism bloom disturbs the food chain, and the large numbers of dead fish washed on shore pose a tremendous sanitation and disposal problem. During the 1971 outbreak, the city of St. Petersburg paid an estimated cost of $10,000 per day to dispose of the fish. Another economic effect is the loss of tourist trade. It was estimated that businesses of the city of Clearwater lost 3.75 million dollars of tourist business during and after the 1947 outbreak. Complete data for the 1971 outbreak are not available, though it appears the effects are largely confined to business on affected beaches.

Other effects of a red tide outbreak on sport and commercial fishing are hard to assess. The

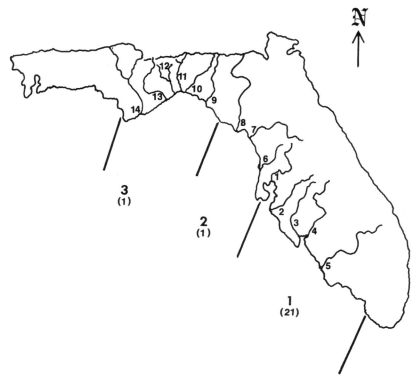

Figure 73. *Red tide outbreak sectors of west Florida coastal*
waters with stream numbers (see Table 67 for key).
The number of outbreaks observed 1878-1970 is
listed in parentheses below sector numbers.

estimated number of fish killed is probably much
lower than the actual, since some of the fish are
very small, some are eaten by larger fish which may
or may not be affected by the toxin, and some just
disappear. A kill of 500 million fish, allowing ten
fish per pound, amounts to 50 million pounds, but
about one billion pounds of menhaden are caught
commercially every year along the north Florida
coast apparently without harming the supply.[10] The
claim that sportfishing is harmed is balanced by
the view that, because trash fish are killed, some
sportfishing is actually improved. In any case,
the extent of mortality is impressive. Steidinger
et al.[11] list 100 species of dead animals found
during red tide outbreaks and includes fish (trash,
commercial, sport), shellfish (shrimp, crabs,

oysters), some birds, porpoises, turtles, barnacles, and invertebrates (sponges).

The origins of the outbreaks are obviously of much interest. The mechanisms involved will be discussed in the next section.

CAUSATIVE MECHANISMS FOR THE ORIGIN OF DINOFLAGELLATE BLOOMS

Three general mechanisms have been proposed: the predator or preceding organism disappearance, concentrating mechanisms, and the special nutrient source mechanism. Steeman Nielsen[12] once noted that "catastrophies like the red tide . . . occur presumably because the planktonic algae for some reason or other grow wild. The normal 'brakes,' ordinarily regulating the growth of the algae, do not function." Many workers now might suggest that blooms, at least microblooms, are a normal event and the problem is to account for the spread (or lack of it) into a macrobloom. For example, Finucane[13] and Wilson[14] suggested resident populations of *G. breve* exist as cysts in local sediments, and encystment-excystment now appears to be a regular stage in the life cycle of many dinoflagellate species, and particularly bloom species.[15] Thus, the significance of the three mechanisms may be more pertinently related to the excystment stage or to the development of the macrobloom from the microbloom.

Predator Organisms

Several examples of predator or inhibiting organisms are available. The importance of zooplankton as regulators of algal production was first considered by Nathansohn,[16] though the significance of his contribution to understanding the spread of red tides has received little attention. Blue-green algae seem to be notorious for suppressing other algae. For example, Jakob[17] found that the freshwater alga, *Nostoc muscorum*, in bacteria-free culture produced a dihydroxyanthroquinone that evidently inhibited the growth of several algae (*Cosmarium, Phormidium,* and *Euglena*). Safferman and Morris[18-20] were able to isolate a virus that seemed to regulate the growth of a blue-green alga.

It also appears likely that competing organisms could effectively eliminate the spread of *G. breve* or other alga by competition, rather than predation. Some workers have also stressed the symbiosis that

exists between bacteria and algae, and the possibility
that the spread of *G. breve* is related to the presence
of bacteria was considered and should be pursued more
avidly.[10]

The possibility that a bloom organism succeeds
other organisms has been considered. *Oscillatoria
erythraea* (or *Trichodesmium thiebautii*), a nitrogen-
fixing blue-green alga, occurs prior to most, if not
all, *G. breve* blooms,[15] and possibly this alga or
associated sulfur bacteria precondition neritic
waters by adding nutrients (nitrogen, trace metal-
organic compounds) or by deleting undesirable
nutrients and/or metabolites. Blooms often occur
in waters depleted of nutrients by preceding
organisms. Brongersma-Sanders[21] noted, "The greatest
outbreaks of red water probably occur toward the end
of a phytoplankton season. . . ." Hornel and Nayudu[22]
found red tide due to peridineans occurred annually
along the Malabar coast after the heavy diatom blooms,
heavy rainfall, and the Southwest monsoon were over.
Menon[23] reported annual blooms of *Gymnodinium sp.*
occurred along the Trivandrum coast of India following
periods of heavy rains and diatom blooms. Rounsefell
and Dragovich[24] found a fairly high *negative* correlati
(-0.48) between abundance of *G. breve* and silicon
levels, though longer study would be needed to
determine whether diatom blooms are involved.

Concentration Mechanisms

These hypothesize a concentration of dinoflagellates at a time when they are dominant or abundant
in the phytoplankton. Several concentrating mechanisms
have been proposed that depend upon the positive
photaxis of some species of dinoflagellates (including
G. breve[25]). These species concentrate at the surface
or depth of optimum light intensity.[1,26,27]

Once the microbloom has spread at the surface,
several mechanisms could account for further concentration, either by a large-scale convergence (as
Chew[28] has claimed) or by smaller-scale mechanisms
proposed by Ryther[1] that include "concentration
against a shore by onshore winds, concentration
between pairs of wind-induced current vortices
(Langmuir cells; cf. Figure 74; Reference 29), and
concentration at convergences where water masses of
different densities are mixing." As Ryther[1] noted,
a concentrating mechanism makes it possible to
explain the high nutrient concentrations found in
dinoflagellate blooms without postulating a special

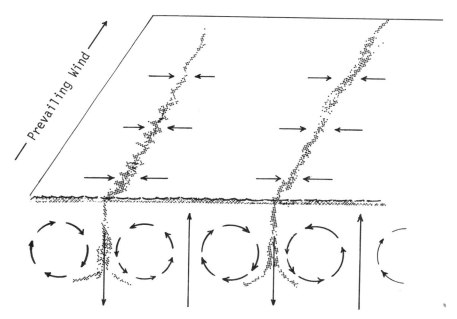

Figure 74. *Schematic representation of wind-driven convection cells (Langmuir circulation) leading to concentration of nutrients and red tide organisms. Dots represent streaks of accumulated floating matter. Arrows indicate direction of net effect.*

nutrient source. As Pomeroy and co-workers[30] noted, however, the two mechanisms are not mutually exclusive, and both might act together in some cases to bring about a bloom.

Special Nutrient Source Mechanisms

Much of Florida is covered by a fine-grained quartz sand of low mineral content, and land drainage may lack the relative abundance and range of metal-organic compounds that characterizes drainage of other areas. Of the trace metals that might be considered, six appear to be promising: the so-called invariable microconstituents of living matter (manganese, copper, cobalt, molybdenum, zinc, and iron[31],[32]). The possibility that one or more of these might be growth-limiting to algae has been considered by various authors.[33] The addition of "vitamins only" to enrich sea water sometimes has a

beneficial action,[34] and some have suggested this
may indicate a deficiency of cobalt, as vitamin B_{12},
from marine waters; though limited data are available,
an unusual red tide (*i.e.*, one found in 1965 north
of the usual areas, in Apalachicola Bay) was studied,
and no correlation was found between levels of B_{12}
and *G. breve* abundance.[35]   Iron has also been considered and a gradient in the iron requirement of
marine plants has been demonstrated;[36]   inshore and
coastal species require greater amounts than oceanic
species.   Thus, spring diatoms that require a greater
amount of iron give way to peridinians that require
or tolerate lower levels.   The significance of
molybdenum, particularly in nitrogen-fixing algae
has been explored,[37] but few studies of the distribution of molybdenum are available; none discuss
west Florida rivers.

SUGGESTIVE EVIDENCE FOR INVOLVEMENT
OF METAL-ORGANIC COMPOUNDS IN
RED TIDE OUTBREAKS

Available evidence falls into two categories:
indirect (enrichment studies) or suggestive (field
studies), but collectively, they provide an impressive body of evidence that indicates the involvement
of metal-organic compounds in red tide outbreaks,
particularly those in Florida coastal waters.

Seasonal Pulses

Blooms of neritic dinoflagellate species often
show seasonal pulses, though the significance of
this is open to several interpretations.   Temperature,
obviously, is a factor, and Ryther[1] suggested that
one nearly universal factor in red tide outbreaks is
the occurrence of high water temperature; in temperate
water, outbreaks appear to be restricted to summer
months.   Rounsefell and Nelson,[10] in analyzing field
data, found that *G. breve* reaches maximum abundance
at temperatures in the range 16-27°C.   The reported
optimum is 26-28°.[38]

Frequently, observations reported in the
literature note that the outbreaks are preceded by
periods of unusually hot, calm weather, and during
smooth seas.   Some outbreaks are noted off the
Peruvian and southwest African coasts when upwelling
is at a minimum.   Possibly, these observations
support the contention that convergence mechanisms
are significant in addition to the importance of
temperature.

Undoubtedly, also, salinity must be a factor, though few reviewers are inclined to attach great significance to it. There is evidence that a salinity "barrier" exists to limit extensive propagation of *Gonyaulax tamerensis*[39] and *Gymnodinium breve*,[40] but the barrier is a low one.

Salinity is a significant factor in the environment of *G. breve*. Rounsefell and Nelson,[10] in an extensive study and computer analysis of field and laboratory data, concluded that the optimum salinity range for *G. breve* growth is about 34-36 ppt (parts per thousand), though the organism survives in salinity between 27 and 37 ppt.

The seasonal pulse may also be related to land drainage that would provide dinoflagellates with necessary organic matter, though the entraining water might not appreciably dilute sea water in the potential outbreak area. The correlation of river runoff with red tide outbreaks has occurred to several workers. Prakash and Rashid[4] noted, for example, that substantial amounts of humic substances enter the Bay of Fundy (where blooms of *G. tamarensis* are prominent) just before the onset of dinoflagellate maximum. Ingle[41] suggested the involvement of organic substances (tannic acids, humic acids) in outbreaks of *G. breve* in Florida coastal waters. These materials were presumed to aid transport of iron from bogs and swamps to potential outbreak areas.

Analysis of the onset of *G. breve* blooms (1844-1960) indicated all outbreaks (for which data seemed reasonably reliable) occurred between August and September, half began in September (the outbreak of June 1971 was clearly exceptional), and during 48 months of toxic blooms, 11 existed during February through July (see Table 64, and Reference 10). This indicates that blooms generally occurred following diatom blooms during periods of nutrient-depleted waters, that *G. breve* can thrive at temperatures from 16-27°C, and that the outbreaks tended to occur after or during heavy seasonal rainfall. Certain rivers, *e.g.*, the Peace River, have increased iron concentrations following periods of heavy rainfall (Figure 75).

Nutrient Depletion and Nitrogen-Phosphorus Requirements

In some respects, the involvement of trace metal or metal-organic compounds looks attractive by default, as it were, when the requirements of nitrogen and phosphorus are considered. The fact that blooms

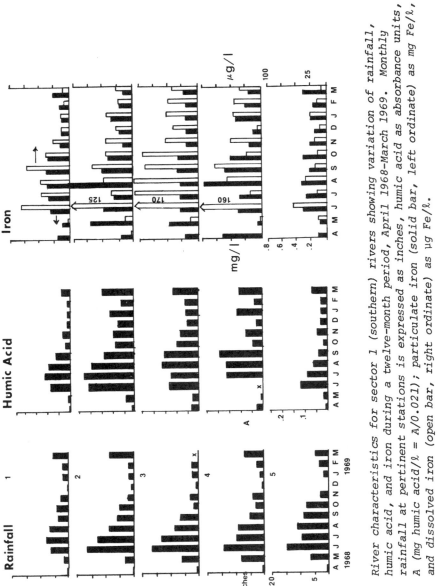

Figure 75.   *River characteristics for sector 1 (southern) rivers showing variation of rainfall, humic acid, and iron during a twelve-month period, April 1968-March 1969. Monthly rainfall at pertinent stations is expressed as inches, humic acid as absorbance units, A (mg humic acid/ℓ = A/0.021); particulate iron (solid bar, left ordinate) as mg Fe/ℓ, and dissolved iron (open bar, right ordinate) as μg Fe/ℓ.*

occurred in nutrient-depleted waters after diatom blooms when major nutrients (inorganic nitrogen, phosphorus, silicon) were depleted suggested that dinoflagellates could bloom in low concentrations of major inorganic nutrients, possibly because of low rates of metabolism.[42]

Also, Rounsefell and Dragovich[24] attempted to correlate *G. breve* abundance during eight years (1954-1961) with nine chemical variables obtained during outbreaks: organic phosphate, total phosphate, nitrate-nitrite nitrogen, copper, alkalinity, silicon, calcium, total organic nitrogen, and ammonia nitrogen. None of these showed significant correlations, though, as noted previously with silicon, -0.48 was the closest to being significant.

In addition, the concentration of phosphorus necessary to sustain a prebloom (3.2 x 10[4] cells/ liter) population of *G. breve* is very low. Using values that were probably high for dinoflagellates in general this value was calculated by Ryther[1] to be 1.5 µg P/liter. Wilson[43] studied serially trans-ferred *G. breve* cultures in an artificial medium and found significant growth required 9.0 µg P/liter. *G. breve* in media with 6.0 µg P/liter or less did not continue to grow. Media with 38 µg/liter or more supported higher population levels. Nitrate-nitrite nitrogen was not required, ammonium ion probably serving as a nitrogen source. Cultures did not grow unless the media contained some form of inorganic nitrogen. It should be noted that only 4% of approximately 6000 samples selected from off shore west coast Florida waters had less than 9 µg P/liter (1955-1960).[43]

Effect of Additives

The history of development of completely artificial "defined" sea water media for dinoflagel-lates indicates the need for organic compounds as additives. Few dinoflagellates can be grown in completely inorganic media, and most marine dino-flagellates seem to require some organic growth factors. Initially, soil extracts were used, and these had growth-promoting actions that were ascribed to the chelating action of the humus component, presumably through a reduction in toxicity or enhancement of the availability of trace elements; but soil extracts should also be a source of vitamins, auxins, growth hormones, etc.[44] In the instance of *G. breve*, for example, soil extracts

were used as media additives,[45] as were river water
samples,[46] to provide growth enhancement, and these
were effectively replaced by vitamins, iron, and a
chelating agent.[47,48]

For example, sea-water samples were assayed
using *G. breve* as an assay organism to evaluate the
suitability of the samples for survival and growth,
and their suitability differed.[47] Samples were also
tested with various additives. Addition of a
chelating agent (EDTA, 50 mg/l) enhanced the suitability of the water samples (in 74 of 95 samples)
and iron (0.02 mg/l) plus EDTA (50 mg/l) enhanced
the suitability (in 81 of 95 samples) to the extent
that *G. breve* either survived or increased in number.

Enrichment experiments, particularly those
involving iron, evidently can suffer from an experimental artifact, that soluble iron ($e.g.$, as
[59]Fe) disappears from natural sea water,[49,50] as
much as 50-80% within the first week of storage in
enclosed containers. A chelating agent, EDTA, was
effective in making iron available in fresh (30
min.) sea water, though not at all in aged (11 days)
sea water, based on a bioassay with six phytoplankton
species.[49] These workers concluded that "experiments
in laboratory containers . . . that show an effect
of EDTA growth rate or photosynthetic rate, may only
be meaningful for the situation within the culture
flask, and one cannot infer from these results that
chelators would have had similar effects in nature."

Though it may not be reasonable to extrapolate
specific results from laboratory to field conditions,
several workers do find similar trends. For example,
Lewin and Chen[49] found that *Coccolithus huxleyi*
showed the greatest facility of six phytoplankton
species in making use of iron. For comparison,
Ryther and Kramer[36] found that only this organism
among nine phytoplankton species studied was able
to completely satisfy iron requirement from the iron
present in off shore sea water (< 5 mg/l Fe) that was
used to prepare the culture media. Other enrichment
experiments indicating that available ($i.e.$ chelated)
iron or other trace metal is a limiting factor for
marine phytoplankton were described by Ryther and
Guillard.[51] Davies[52] described the experimental
criteria for iron-limiting conditions, and the
growth kinetics of the euryhaline flagellate
Dunaliela tertiolecta. Davies also concluded
phytoplankton chlorophyll production may be governed
by the supply of organo-iron complexes.

Lewin and Chen[49] noted that extrapolation of enrichment experiments is always subject to one valid criticism or another, but in this instance several additional problems were not considered. EDTA, though a common chelating agent, is hardly a natural one, is far from representative of naturally occurring chelating agents, and differs in several respects. First, though EDTA may assist assimilation of trace elements,[34] the chelating agents are not metabolized (see Barber, Chapter 11). Second, EDTA is a far superior chelating agent, on the basis of complex formation (equilibria) constants which are 10^{26} for EDTA-ferric complex,[34] $10^{14.3}$ for EDTA-ferrous complex,[53] and 10^2-10^6 for fulvic acid-divalent transition metal complexes.[54] Fulvic acid is a representative naturally-occurring chelator (see Chapter 9). Third, it was assumed that in EDTA-enriched sea water media, the EDTA swamped natural chelators and that the EDTA was added at concentrations (0.15×10^{-3} M) sufficient to react with all trace elements (at 10^{-3} M); in sea water, concentrations of dissolved organic carbon are about $10^{-4.4}-10^{-4}$ M, at the surface though the concentrations are probably much higher in neritic waters,[55] and the chelating tendencies of this dissolved material are probably significant. For example, several workers[56,57] found iron remains in solution in the presence of yellow organic acids which compete with hydroxide ions for coordination sites on aquated iron. Interestingly enough, though most of the iron could be removed from the colored acids by a strong chelating agent (triethanolamine or Versenex 80), a small amount of iron remained bound to the large molecular-weight fractions.[57]

It has been generally assumed that the function of naturally occurring and artificial chelating agents is two-fold: to chelate and detoxify certain metal ions and to assist the transport or availability of other metal ions. Certain experiments suggest other functions may be involved as well. Gibberellic acid (10^{-7} M) a growth hormone, and a weak chelating agent, reportedly assists the growth of *G. breve* in unspecified media; the effect is due to reduction of the lag time from about six days to one, but the lag growth seems unchanged from control to test.[58] Humic substances (*vide infra*) may have intrinsic effect on phytoplankton growth that is not ascribable to chelation processes.[4] These workers found that humic acid (isolated from marine sediments) added to cultures of *Gonyaulax spp.* stimulated the yield growth rate, and ^{14}C-uptake (Table 65). The positive effects on growth were generally independent of nutrient concentration, and therefore the effects were presumably due in part to a process other than chelation. Possibly, the authors concluded, the positive effects were linked to algae cell metabolism Whether the effects could be ascribed to residual iron is uncertain.

Table 65

Relative Responses of G. tamarensis *to Humic Substance Fractions*[a]

	Relative Conc.[b]	*Relative Final Yield*[c]	*Relative[d] Growth constant, k, days^{-1}*
Humic acid	1.0	3.0	0.08
	2.0	3.8	0.10
	4.0	4.2	0.10
Fulvic acid	0.25	1.0	0.02
	1.05	1.6	0.05
	2.1	2.1	0.07
Hymatomelonic acid	1.5	---[e]	--
	3.0	---	--

[a] From Reference 4.

[b] 1.0 = 3.5 mg/l.

[c] 1.0 = 283 cells/ml.

[d] $k = [1/(t_2 - t_1)] \log_2 (N_2/N_1)^a$; where N_2 and N_1 are cell numbers at times t_2 and t_1.

[e] No appreciable growth.

SOURCES OF NATURALLY-OCCURRING METAL ORGANIC COMPOUNDS

In west Florida coastal waters, and probably in any similar environment, three sources or reservoirs of metal-organic compounds are available to micro-blooms. These immediate reservoir sources are terrestrial, sediments, and intramarine. Saxby[59] considered that four major types of metal-organic compounds were available in sediments (and presumably in the other two reservoirs), *viz.*, metal porphyrins, high molecular weight metal porphyrins, metal salts of long-chain organic acids, and humic acids. The last group is of particular interest because more information is available and appears to be pertinent to the area of interest.

Humic substances are formed from biochemical degradation and transformation of plant and animal matter in soil or sediments. On the basis of solubility characteristics, several fractions can be defined operationally, though not precisely (humins and ulmins, humic, fulvic, and hymatomelonic acids). The last three are water-soluble and, therefore, potentially available to phytoplankters. Humic acid is an alkali-soluble, acid-insoluble fraction; fulvic acid is water-soluble and acid-soluble; and hymatomelonic acid is an alcohol-soluble fraction of humic acid. Many workers regard the three acids as closely related compounds and not distinct chemical entities.[60]

Though the humic material in coastal water is probably from the terrestrial reservoir, it is not exclusively so.* Plankton decomposition products ("gelbstoff"[61]) appear to have characteristics similar to those of humic substances. Moreover, though the sediment reservoir of metal-organic compounds, especially humic substances, was originally terrestrial in origin, it could have a significant effect on expansion of the microbloom for two reasons. First, though much of the humate material appears to be precipitated at the fresh-water-marine interface,[62,63] it can sorb metal ions. Naturally and chemically extracted humate can sorb between 1 and 17% by weight of trace metals (cobalt, copper, iron, lead, manganese, molybdenum, nickel, silver, vanadium and zinc[62]). Thus, it seems logical to assume that humate beds, which are extensive (100,000 to 1,000,000 tons in discontinuous beds over an area of 300-500 square miles along the Florida panhandle[62]) could exert a slow, general

---

*Nissenbaum and Kaplan[64] reviewed evidence for the *in situ* origin of marine humic substances and proposed a five-step pathway for marine formation and transformation in sediment: "(1) degraded cellular material, (2) water-soluble complex containing amino acids and carbohydrates, (3) fulvic acids, (4) humic acids, and (5) kerogen."

ion-exchange effect.   The effect may be more direct, however, if a current hypothesis is correct, *i.e.*, microblooms arise from excystment of *G. breve* in contact with sediments.[15]
    A second reservoir that might be considered briefly is the intramarine production of humic substances through the death of organisms that precede *G. breve* blooms.   The possibility has not been explored with *G. breve* though it is reported that studies involving *O. erythraea* are in progress.[15] Other systems, however, have been examined.   For example, Nordli[65] found that aqueous extracts of benthic algae stimulated growth of several dino-flagellate species.   Also, decomposition products of *Nitzschia* stimulated growth of *Gonyaulax sp.* and this growth was concentration dependent, according to Prakash.[4]   Finally, Yentsch and Reichert[66] reported that an increase in oxygen uptake by bacteria was observed when yellow filtrate from algal cultures was added to raw sea water.   This particular obser-vation may be especially significant in view of suspected bacteria-*G. breve* symbiosis.
    The final reservoir, terrestrial, is one that has attracted much study[67] particularly with regard to humic substances.   Two points deserve special attention.   First, what is the effect of humic sub-stances on marine phytoplankters; and second, what is the input of humic substances into the marine environment?
    The first question was considered by Prakash and Rashid,[4] as noted in the preceding section.   These workers isolated and studied two groups of fractions, type and molecular weight.   First, humic acid was found to be more active than fulvic acid, and hymatomelonic acid had no appreciable effect on the growth (Table 65) of *Gonyaulax tamarensis* (cf. Table 63), an alga that is responsible for paralytic shellfish poisoning in the Bay of Fundy.   For humic and fulvic acids, the final yield of cells was dependent on fraction concentration, though the growth constant of humic and hymatomelonic acid showed no dependence on concentration of added fraction.   In a second phase of study (Table 66), molecular weight-fractions of humic acid were separated by Sephadex gel filtration chromatography. The effect of different fractions on *Gonyaulax* species was studied (Table 66), and it appears that the low molecular-weight fractions produced the greatest relative yield.   This is significant because the authors found the molecular-weight distribution

Table 66

Growth Response of *Gonyaulax* spp. to Humic Acid Fractions[a]

Molecular Weight Fraction	Distribution %		Relative Yield[b]		
	River Water	Marine Sediment	G. tamarensis	G. acatenella	G. catenella
Parent	---	---	1.5	5	8
< 700	2.74	6.5	7	6	2
700–1500	7.3	5/0	4	5	6
1500–5000	65.3	5.6	0	0.5	1.5
5000–10000	---	9.9	2	2	4.5
>10000	---	73.0	---	---	---

[a]Reference 4. Fractions were added to media at concentrations of 4 mg/1 from a marine sediment.

[b]Scale, 0-10; Data interpolated from Figure 4, Reference 4.

varied with source; river water had a relatively
high percentage of low molecular-weight material
(<1500 g/mole) and marine sediments a low percentage
of this fraction.

The second question, that of inputs of humic
substances in the marine environment is considered
in the next section.

HUMIC ACID INPUT INTO
WEST FLORIDA COASTAL WATERS

The interest in humic acid input can be placed
in context by considering two Florida red tide
predictive guides, environmental factors and the
iron index.

A multiple curvilinear correlation analysis
performed by Rounsefell and Dragovich[24] associated
61% of the monthly variability of *G. breve* abundance
with certain environmental factors (salinity,
temperature, and onshore winds greater than 7 kts)
plus the *G. breve* abundance of the previous month.
Though these environmental factors might have a fair
predictive value, they suggested factors that might
have better predictive value. The salinity, an
obvious choice for first consideration, should be
governed by two factors: the salinity of gulf
waters and the volume of river discharge. The first
factor is probably fairly constant, though the significance variability in near shore direction of
current and significance to red tide outbreaks needs
to be explored. The second factor, river discharge,
shows considerable variation among 14 major rivers
along the west coast of Florida (Table 67), and it
seemed likely that the contents of the rivers might
be much more significant than the volume.

Pursuing this approach, the iron index was
proposed by Ingle and Martin[68] as a means of predicting occurrence of Florida red-tide outbreaks.
The iron index was defined as the total amount of
iron *potentially* delivered to a potential outbreak
area during a quarterly period. The critical iron
index, the value associated with an actual outbreak
area, would need to be determined for each area.

For example, one prominent outbreak area is
Charlotte Harbor to Sanibel Island, Florida. For
this area the critical index was 235,000 pounds of
iron (as measured at Arcadia, Florida) potentially
delivered by the Peace River, the major river of the
area. Three exceptions were found in accounting
for past outbreaks for the 25-year period 1944-1969:

Table 67

West Coast Florida River Properties[a]

Sector	River Number	River	Mean Annual River Flow cfs	Mean Humic Acid Conc. mg/l	Monthly Conc. µg/l[b] S Fe	P Fe
1	1	Hillsborough	673	1.86	30	185
	2	Manatee	109	5.27	45	303
	3	Myakka	264	4.23	74	208
	4	Peace	1,267	5.47	46	294
	5	Caloosahatchee	4,690	2.02	19	230
2	6	Anclote	86	2.89	52	261
	7	Withlacoochee	1,183	1.92	26	160
3	8	Suwannee	10,740	1.87	18	189
	9	Steinhatchee	336	1.30	17	533
	10	Econfina	136	1.38	13	483
	11	Aucilla	407	0.57	5	589
	12	St. Marks	750	0.77	5	155
	13	Ochlockonee	1,614	2.53	28	35
	14	Apalachicola	25,180	1.45	9	468

[a]From Reference 69.　[b]S, Soluble; P, Particulate.

fall of 1968, late fall of 1960, and perhaps January, 1957 when the prevalent outbreak was not sustained. In all instances, adverse (to *G. breve*) conditions were noted, either turbulence due to hurricanes (first two exceptions) or a record-setting temperature drop. Evident success of the critical iron index would invite inquiry on the basis of the index.

In part, the answer emerged from a study of selected metal-chelate inputs of 14 west coast Florida streams during 1968-1969. Water samples were analyzed for humic acid concentration,[70] as well as concentration of particulate and soluble forms of iron, copper, manganese and zinc. An association between humic, iron concentration and rainfall emerged (see Figure 75).

Details are given elsewhere,[69] and only the salient features are reviewed here.

Temporal and spatial variations of three parameters--iron and humic acid concentrations and rainfall--followed a common pattern for a group of five southwestern rivers (Table 68 rivers 1-5; Hillsborough, Manatee, Myakka, Peace, Caloosahatchee).

Table 68

Correlation Parameters for Iron, Humic Acid, and Rainfall Variation for Selected Florida Streams[a]

River	Variable	Coefficient of Determination[b] r^2_{12}	Coefficient of Multiple Determination[b] $R^2_{1.23}$
1-5	S. Fe	0.504	0.529
	P. Fe	0.152	0.193
1	S. Fe	0.557	0.644
	P. Fe	0.427	0.169
2	S. Fe	0.289	0.431
	P. Fe	0.055	---
3	S. Fe	0.544	0.694
	P. Fe	---	---
4	S. Fe	0.843	0.854
	P. Fe	0.121	0.135
5	S. Fe	0.372	0.432
	P. Fe	0.267	0.342

[a]From Reference 69.

[b]Variable 1, iron; Variable 2, humic acid concentration; Variable 3, rainfall.

The parameter intensity was low in late spring, increased to a maximum in late summer, decreased to a minimum in winter, then increased in spring. (A similar pattern for humic acid was observed in subsequent years.[63] A similar, though less intense, pattern was observed for a group of three west central streams (Anclote, Withlacoochee, Suwannee) and for a group of six northwest streams (Steinhatchee, Econfina, Aucilla, St. Marks, Ochlockonee, and Apalachicola).

The southwestern rivers were characterized by uniformly high concentrations of humic acids, west central by intermediate concentrations and northwestern rivers by uniformly low concentrations. The data in Table 67 indicate that, because of the effect of high flow rates, northwestern streams deliver more humic acid to a potential outbreak. This observation and the apparent colloidal nature of humic acids would account for extensive humate beds and humate-impregnated sands in northwestern Florida. The observations might also suggest that *concentration* not total amount of humic acid and/or selected trace elements should be significant in accounting for red tide outbreaks.

ACKNOWLEDGMENTS

Some of the research described here was supported by the Florida Bureau of Marine Science and Technology, by the National Marine Fisheries Service (NOAA), by the USF Research Council, and by a PHS Research Career Development Award (KO4-GM 42569-04, National Institute of General Medical Sciences).

REFERENCES

1. Ryther, J. H. In *Luminescence of Biological Systems,* Johnson, F. H., ed. (Washington, D.C.: American Association for the Advancement of Science, 1955).
2. Sasner, J. J., Jr. In *Marine Pharmacognosy,* Martin, D. F. and G. M. Padilla, eds. (New York: Academic Press, 1973).
3. Schantz, E. J. In *Properties and Products of Algae,* Zajic, J. E., ed. (New York: Plenum, 1970).
4. Prakash, A. and M. A. Rashid. *Limnol. Oceanogr.* 13, 598 (1968).
5. Howell, J. F. *Trans. Amer. Microscop. Soc.* 72, 153 (1953).
6. Davis, C. C. *Botan. Gaz.* 109, 358 (1948).
7. Ballantine, D. *J. Marine Biol. Assoc. U.K.* 35, 467 (1956).
8. Parnas, I. and B. C. Abbott. *Toxicon 3,* 133 (1965).
9. Halstead, B. W. *Poisonous and Venomous Animals of the World* (Washington, D.C.: U.S. Govt. Printing Office, 1965).

10. Rounsefell, G. A. and W. R. Nelson. *U. S. Fish Wildlife Serv. Spec. Sci. Rep. Fish, No. 535* (1966).

11. Steidinger, K. A., M. A. Burkley, and R. M. Ingle. In *Marine Pharmacognosy*, Martin, D. F. and G. M. Padilla, eds. (New York: Academic Press, 1973).

12. Steeman Nielsen, E. In *The Sea*, Hill, M. N., ed. (New York: Interscience, 1963).

13. Finucane, J. Natl. Mar. Fish. Serv., St. Petersburg Beach, Fla. (1955).

14. Wilson, W. B. *Contrib. Mar. Sci. 12*, 120 (1967).

15. Steidinger, K. A. and R. M. Ingle. *Environ. Letters 3*, 271 (1972).

16. Nathansohn, A. *Intern. Rev. ges. Hydrobiol. Hydrog. 1*, 37 (1910).

17. Jakob, H. *These No. 4485*, Fac. Sciences, Univ. de Paris (1961).

18. Safferman, R. S. and M. E. Morris. *Science 140*, 679 (1963).

19. Safferman, R. S. and M. E. Morris. *J. Amer. Water Works Assn. 56*, 1217 (1964).

20. Safferman, R. S. and M. E. Morris. *J. Bacteriology 88*, 771 (1964).

21. Brongersma-Sanders, M. In *Treatise on Marine Ecology and Paleoecology*, Hedgpeth, J. W., ed. Vol. I. (Baltimore, Md.: Waverly, 1957).

22. Hornell, J. and M. R. Nayudu. *Madras. Fish. Bull. 17*, 129 (1923).

23. Menon, M. A. S. *Proc. Ind. Acad. Sci. B22*, 31 (1945).

24. Rounsefell, G. A. and A. Dragovich. *Bull. Mar. Sci. 16*, 402 (1966).

25. Aldrich, D. V. *Science 137*, 988 (1962).

26. Halse, G. R. *Oikos 2*, 162 (1950).

27. Slobodkin, L. B. *J. Marine Res. 12*, 148 (1953).

28. Chew, F. *Bull. Marine Sci. Gulf and Caribbean 6*, 292 (1956).

29. Langmuir, I. *Science 87*, 119 (1938).

30. Pomeroy, L. R., H. H. Haskin, and R. A. Ragotzkie. *Limnol. Oceanogr. 1*, 54 (1956).

31. Mason, B. *Principles of Geochemistry* (New York: John Wiley and Sons, 1958).

32. Lowman, F. G. In *Radioecology*, Schultz, B. and A. W. Kelement, eds. (London: Chapman and Hall Ltd., 1963).

33. MacKenthum, K. M. and W. M. Ingram. *Algal Growth Aqueous Factors Other than Nitrogen and Phosphorus*, U. S. Department of the Interior, Federal Water Pollution Control Administration (1966).

34. Johnston, R. *J. Mar. Biol. Ass. U.K. 44*, 87 (1964).

35. Florida Board Conservation Marine Lab. *Prof. Papers Serv. No. 8* (1966).

36. Ryther, J. H. and D. D. Kramer. *Ecology 42*, 444 (1961).

37. Naylor, A. W. In *Phylogenesis and Morphogenesis in the Algae*, Fredrick, J. F. and R. M. Klein, eds. (New York: New York Academy of Sciences, 1970).

38. Finucane, J. H. *U. S. Fish Wildl. Serv., Circ. 92*, (1960).

39. Prakash, A. *J. Fish. Res. Bd. Canada 24*, 1589 (1967).

40. Ray, A. M. and D. V. Aldrich. In *Toxic Animals*, Russel, F. E. and P. R. Saunders, eds. (Oxford: Pergamon, 1967).

41. Ingle, R. M. In *Bureau of Commercial Fisheries Symposium on Red Tide*, Sykes, J. E., ed. Spec. Sci. Rep. Fish. No. 521 (1965).

42. Gran, H. H. *Cons. Int. Exp. Mer; Rapport et Proc. Verb. 56 (1926-27)*.

43. Wilson, W. B. *Prof. Papers Ser., No. 7*, Florida Board of Conservation (1966).

44. Provasoli, L., J. J. McLaughlin, and M. R. Droop. *Arch. Mikrobiol. 25*, 392 (1957).

45. Wilson, W. B. In *Special Scientific Report-Fisheries No. 535*, Rounsefell, G. A. and W. R. Nelson, eds. (Washington, D.C.: U.S. Department of the Interior, 1955).

46. Collier, A., W. B. Wilson, and M. Borkowski. *J. Phycol. 5*, 168 (1969).

47. Wilson, W. B. Doctoral Thesis, Texas A & M University (1965).

48. Brydon, G. A., D. F. Martin, and W. K. Olander. *Environ. Letters 1*, 235 (1971).

49. Lewin, J. and C. H. Chen. *Limnol. Oceanogr. 16*, 670 (1971).

50. Burton, J. D. and P. C. Head. *Limnol. Oceanogr. 15*, 164 (1970).

51. Ryther, J. H. and R. R. L. Guillard. *Deep Sea Res. 6*, 65 (1959).

52. Davies, A. G. *J. Mar. Biol. Ass. U.K. 50*, 65 (1970).

53. Bjerrum, J., G. Schwarzenbach, and L. G. Sillén. *Stability Constants*, Part I. (London: The Chemical Society, 1957).

54. Schnitzer, M. and S. I. M. Skinner. *Soil Sci. 103*, 247 (1967).

55. Hood, D. W., ed. *Organic Matter in Natural Waters* (University of Alaska Press, 1970).

56. Shapiro, J. Paper before the Symp. Hungarian Hydrol. Soc., Budapest-Tihany, September 25-28, 1966.

57. Ghassemi, M. and R. F. Christman. *Limnol. Oceanogr. 13*, 583 (1968).

58. Paster, Z. and B. C. Abbott. *Science, 169*, 600 (1970).

59. Saxby, J. D. *Rev. Pure Appl. Chem. 19*, 131 (1969).

60. Felbeck, G. T., Jr. In *Advances in Agronomy* (New York: Academic Press, 1970).

61. Kalle, K. In *Oceanography and Marine Biology Annual Reviews*, Barnes, H., ed. (London: Allen and Unwin, 1966).

62. Swanson, V. E., I. C. Frost, L. F. Rader, Jr., and C.
     Huffman, Jr. U.S. Geol. Survey Prof. Paper (1966).
63. Geotz, C. L. M.S. Thesis, University of South Florida
     (1972).
64. Nissenbaum, A. and I. R. Kaplan. *Limnol. Oceanogr. 17*,
     570 (1972).
65. Nordli, E. *Acta Oecol. Scandinavica 8*, 200 (1957).
66. Yentsch, C. S. and C. A. Reichert. *Botan. Marina 3*, 65
     (1962).
67. Swain, F. M. *Nonmarine Organic Geochemistry* (Cambridge:
     Cambridge University Press, 1970).
68. Ingle, R. M. and D. F. Martin. *Environ. Letters 1*, 69
     (1971).
69. Martin, D. F., M. T. Doig, III, and R. H. Pierce, Jr.
     Professional Papers Series No. 12 (St. Petersburg,
     Florida: Florida Department of Natural Resources, 1971).
70. Martin, D. F. and R. H. Pierce, Jr. *Environ. Letters 1*,
     49 (1971).
71. Ryther, J. H. *Limnol. Oceanogr. 1*, 72 (1956).

13. REMOVAL OF HEAVY METAL POLLUTANTS FROM NATURAL WATERS

Leonard A. Lee* and Howard J. Davis. Celanese Research Company, Summit, New Jersey

The presence of mercury and other toxic metals in our streams and lakes is a matter of increasing concern and one which has stimulated the search for better methods for their removal.[1,2] Previous work on environmental problems and interest in membrane separation systems has led to the study of novel approaches for removing contaminants like those found in the metal plating and finishing industries from water and waste streams. This paper describes a series of experiments having the potential for development into a practical, large-scale process.

The objectives of the study, as symbolized in Figure 76, were twofold: first, selection of a method to capture the heavy metal, and second, once trapped, a means to remove it from the water. After consideration of several approaches, the complexing reaction of the well-known chelating agent, ethylenediaminetetraacetic acid (EDTA) with a variety of heavy metal ions was selected. EDTA forms complexes with many metal ions, including the heavy metals, mercury, lead, cadmium and nickel. Stability constants of metal-EDTA chelates of interest are listed in Table 69 for the association reaction:

$$M^{+n} + Y^{-4} \rightleftharpoons MY^{n-4}$$

$$K = \frac{[MY^{n-4}]}{[M^{+n}][Y^{-4}]}$$

---

*Present Address: Gillette Research Institute, Rockville, Maryland.

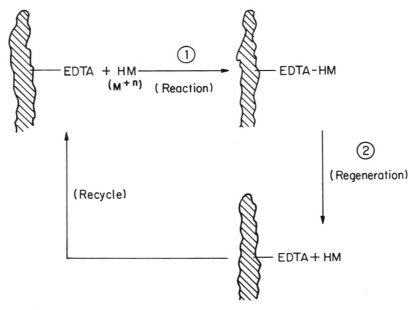

*Figure 76.   Overview of proposed method to remove heavy metal
pollutants.*

Table 69

Stability Constants of Metal Chelates of EDTA [3]

Metal Ion	Log K
Sodium	1.7
Magnesium	8.7
Calcium	10.6
Cadmium (II)	16.5
Lead	18.2
Nickel	18.6
Copper (II)	18.8
Mercury (II)	21.8

Keeping in mind the need for a practical method,
this reaction with heavy metal ions must take place
in the presence of higher concentrations of sodium
salts or hard water salts commonly found in natural
waters. Comparison of the stability constants shows
that the heavy metals, *e.g.*, mercury (II), Hg^{++}, and
cadmium (II), Cd^{++}, will displace sodium, calcium,
and magnesium from the EDTA complex.

EDTA lent itself to a solution of the second
problem, namely, the removal of the metals after
chelation from the water prior to recycling the
chelator. This was made possible by fixing the EDTA
onto an insoluble support or substrate which could
be immersed in water, then lifted out and treated to
displace the heavy metal ions from the complex.
Once restored to its original form, the EDTA-substrate
could then be recycled for further heavy metal pick-up.

EDTA was fixed to a water-insoluble substrate by
combining the magnesium-EDTA chelate with the free
carboxyl groups in cellulose acetate phthalate mixed
ester (CAPh) as shown in Figure 77. The disodium
salt of EDTA was selected in contrast to the free
acid form, which is only very slightly soluble in
water, because of its ready availability and high
water solubility. The magnesium ion is believed to

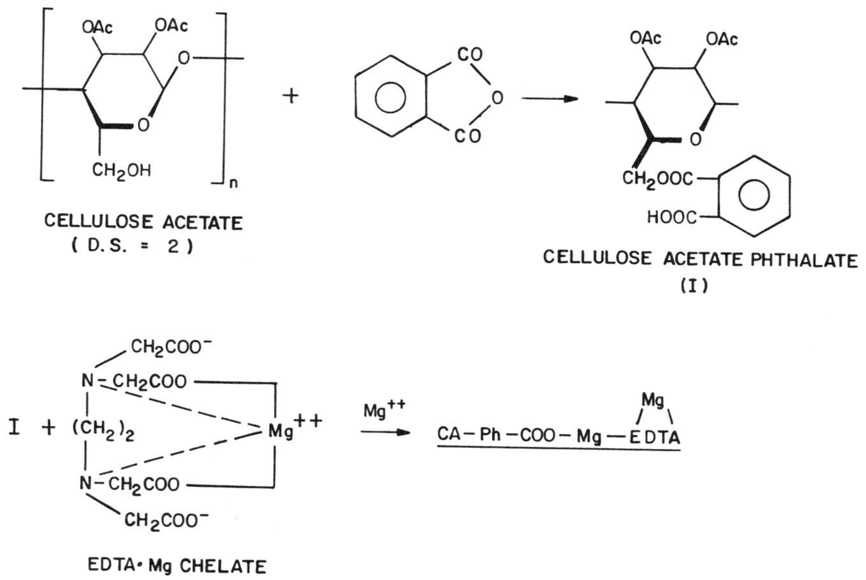

Figure 77. Preparation of polymer EDTA-chelate.

form the insoluble link between the acid-ester and
EDTA. In order to obtain a sufficient number of
carboxyl groups, cellulose acetate with a low degree
of substitution was used in preparing the mixed
ester: the free hydroxyl groups in cellulose acetate
were esterified with phthalic anhydride.[4] The
cellulose acetate phthalate could be used in different physical forms and arrangements, depending
on end-use application. As described later, an
experimentally convenient approach was to use the
EDTA/cellulose acetate phthalate in trapping heavy
metals in a reverse osmosis-like film through which
the simulated contaminated water could be passed to
remove the contaminants.

EXPERIMENTAL

Preparation of Cellulose Acetate
Phthalate (CAPh)

    The mixed ester, cellulose acetate phthalate
(CAPh), was prepared by esterification of the free
hydroxyl groups in a low acetyl value cellulose
acetate (CA), 50.5 A.V., with phthalic anhydride in
the presence of pyridine.[4] The preparation of
cellulose acetate with a low degree of substitution
has been described.[5] In cellulose acetate with a
2.0 degree of substitution (D.S. = 2), two of the
three available hydroxyl positions in the cellulose
basic unit have been esterified to acetate groups,
leaving on the average one hydroxyl group per
anhydroglucose unit for esterification to phthalate
groups. The unit molecular weight of cellulose is
162; that of completely acetylated cellulose, *i.e.*,
cellulose triacetate, with a degree of substitution
equal to 3, is 288. The unit molecular weight of
cellulose acetate with a 2.11 degree of substitution,
acetyl value of 50.5, used in the preparation of
CAPh, is 251. Acetyl value is defined as the grams
of acetic acid liberated per gram of cellulose
acetate, multiplied by 100. About 60% of the
available hydroxyl groups were esterified to
phthalate groups, resulting in a unit molecular
weight of about 340 for the mixed ester. After
precipitation and washing, the mixed ester was
dried and ground to about 20-mesh particle size.
It was soluble in 60/40 acetone/formamide, which is
the solvent used in the preparation of Loeb-type
reverse osmosis membranes from cellulose acetate.[6]

Preparation of Films

Films of 50/50 CA/CAPh were cast from a dope consisting of 20% solids in 60/40 acetone/formamide (by weight) solvent onto glass plates. The cellulose acetate (CA) in the 50/50 mixture was a commercial cellulose acetate with an acetyl value of 55.1; the acetate-phthalate used was prepared as described above. The film was cast by hand on a glass plate, using a hand-casting knife unit, and immediately immersed in water. All films were stored under water. The wet film thickness was about 0.008 cm. Circles 47 mm in diameter were punched out of the film and two circles, one on top of the other, were mounted for use in the pressure filter. The dry weight of one 47-mm circle was 50 $\pm$ 5 milligrams. The $\pm$ 10% weight limits reflect the wide variations encountered in a hand-casting operation.

It was experimentally convenient to mount the film circles in a (Gelman) pressure filter. Effective diameter of a circle in this unit was 35mm. With this unit, it was possible to use accurately measured volumes of feed solutions and to collect effluent at a convenient rate simply by varying the applied pressure. Pressures ranging from 50 to 150 psig were used.

Treatment of CA/CAPh Films
with Chelating Agent

The films in the pressure filter were treated with the magnesium form of disodium-EDTA, $Na_2[Mg(EDTA)]$, by passing about 30 ml of 0.5% $Na_2[Mg(EDTA)]$ through the films at an applied pressure of 50-150 psi. A brief washing with water was needed to remove free $Na_2[Mg(EDTA)]$ and the wash water was tested to insure this was accomplished. The $Na_2[Mg(EDTA)]$-treated films were then ready for testing.

Solutions and Fraction Collections

The following solutions were prepared from analytical grade reagents:

0.00050 M mercuric nitrate
0.00050 M cadmium nitrate
0.00100 M ethylenediaminetetraacetic acid
disodium salt, Na_2EDTA
0.00100 M magnesium sulfate
0.00050 M mercuric nitrate in 0.1% sodium
chloride solution

0.00050 M cadmium nitrate in 0.1% sodium
chloride solution
Indicator: 0.1 g Eriochrome Black T in 100ml
alcohol
Buffer solution, pH10: 67.5 g ammonium chloride
and 570 ml ammonium hydroxide
(sp.gr. 0.88) in one liter.

A pressure filtration funnel, such as Gelman
No. 4280, was used for mounting the film circles.
Exactly 150 ml of mercuric nitrate or cadmium
nitrate solution was added to the pressure filter
in which two film circles were mounted. Solutions
with (0.1%) and without sodium chloride were used.
Treated films, *i.e.*, treated as described above with
Mg[Na₂(EDTA)], and untreated films were tested. A
driving pressure sufficient to collect filtrate at
the rate of about 0.5 ml per minute was used.
Usually 50-150 psi was sufficient. The effluent
was collected in 10 ml fractions. Each 10 ml fraction was analyzed by titration for the presence of
mercury (II), (Hg^{++}) or cadmium (II), (Cd^{++}). Ten
ml of 0.0010 M EDTA, an excess amount, was added to
the fraction and the excess EDTA was then titrated
with 0.0010 M magnesium sulfate with Eriochrome
Black T indicator. The solution was buffered to
pH 10 for the titration. Fractions were collected
until the effluent titration approached the feed
titration. The solution remaining (retentate) in
the pressure filter was also similarly titrated,
using 10.0 ml portions, to insure that the concentration of mercury or cadmium had not changed
(increased) as a result of rejection via a reverse
osmosis mechanism. Any mercury or cadmium removed
was then tied up as the metal-chelate in the CAPh
film.
Along with these test films, plain CA films
were also tested in a similar manner. In addition
alginic acid, a natural polymer built from mannuronic
acid units, a structure resembling cellulose, was
tested in particle form for removal of mercury and
cadmium, as indicated below.

RESULTS AND DISCUSSION

Untreated CA and 50/50 CA/CAPh films were
tested for mercury (II) removal with results that
are plotted in Figure 78. The feed solution concentration, dotted line, was 100 mg Hg^{++} per liter.
It is apparent that CA itself removed very little

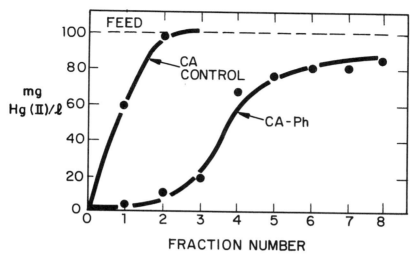

Figure 78.   Removal of mercury by untreated films (no NaCl).

mercuric ion while the CAPh in the 50/50 CA/CAPh
film removed an appreciable amount. Through fraction 3, *i.e.*, 30 ml cumulative volume, practically
all the mercury was removed. Breakthrough occurred
after fraction 3. Each 10 ml of feed contained one
milligram mercury, 0.01 milliequivalent (meq) Hg^{++}.
Through fraction 9, a total of approximately 0.04
meq mercury had been removed by 28 milligrams CAPh
(M.W. = 340) or about 0.082 meq CAPh. This ratio
is about one meq mercury per two meq CAPh. In
Figure 79, the results are shown for the removal of
mercury from a 100 mg/l feed solution, with EDTA-
treated films of (1) CA and (2) CA/CAPh. On
comparing the EDTA-treated CAPh in Figure 79 with
the untreated CAPh in Figure 78, it appears that
the latter is more efficient. This apparent difference is attributed to the variations in film
structural and flow properties, which in turn
affect the residence and contact times of the feed
solution with the film. It is not unusual to observe such differences in cellulosic asymmetric
membranes prepared by the hand-casting technique
employed here. The data show again that CA itself
removed a minor amount of mercury, similar to the
result obtained with untreated CA film in Figure 78,
while the CAPh removed an appreciable amount. In
another experiment it was shown that alginic acid

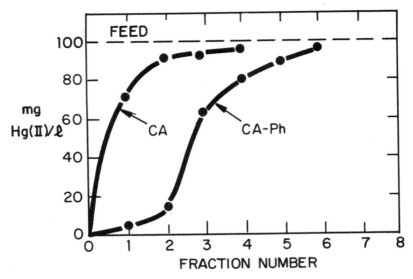

Figure 79.   Removal of mercury by EDTA-treated films (no NaCl).

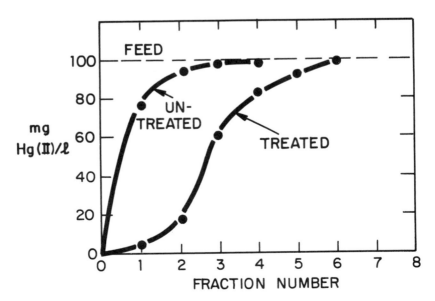

Figure 80.   Removal of mercury by EDTA-treated and untreated films (1000 mg/l NaCl).

also removed mercuric ions, demonstrating the availability of the free carboxyl group for reaction with
mercuric ion.

Figure 80 represents a situation of some practical significance where mercury is selectively
removed by 50/50 CA/CAPh treated with Mg-EDTA from
a feed solution containing a much higher concentration, 1000 mg/l, of sodium chloride while the
untreated CA/CAPh film removed very little. Through
fraction 2, removal was over 90% for the EDTA-treated
film. Total removal through fraction 5, where removal leveled off, amounted to about 0.024 meq
mercury per 0.08 meq CAPh. This ratio is about 1
to 3. Variations in this ratio are expected as a
result of sample or experimental variations, *e.g.*,
high flux-too short contact time. In all cases,
the retentate, the solution remaining in the
pressure filter, was titrated also to confirm that
the concentration of mercury (or cadmium) did not
increase as a result of rejection as in a reverse
osmosis situation. In no case was an increase in
concentration observed, thereby confirming that all
of the mercury (II) (or cadmium II) removed was
removed in the film as the metal-chelate.

Figures 81 and 82 show cadmium removal data
obtained with EDTA-treated and untreated 50/50 CA/
CAPh film. In Figure 81, the feed solution of

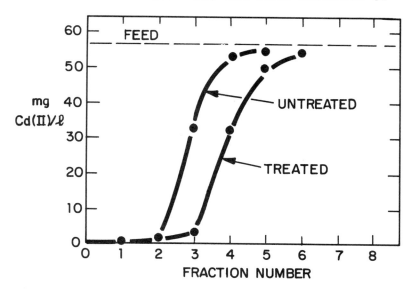

*Figure 81.   Removal of cadmium by EDTA-treated and untreated
films (no NaCl).*

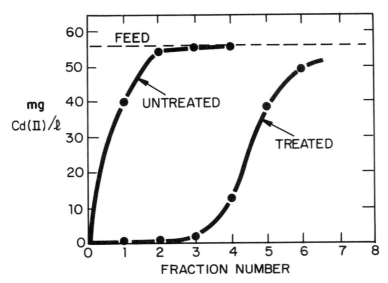

Figure 82.   *Removal of cadmium by EDTA-treated and untreated*
films (1000 mg/l NaCl).

56 mg/l cadmium (1 meq per 10 ml) contained no sodium
chloride. The treated film removed a moderately
larger amount of cadmium than did the untreated film.
Through fraction 3, removal was practically quantitative for the treated film. Through fraction 5,
after which the effluent approached feed concentration, about 0.035 meq cadmium was removed by 0.08
meq CAPh, a ratio of almost 1 to 2 meq cadmium to
CAPh. In Figure 82, the feed solution contained a
large amount, 1000 mg/l of sodium chloride, which
did not affect the performance of the treated film.
Practically all of the cadmium was removed through
fraction 4 by the treated film. A total of 0.042
meq cadmium was removed through fraction 6.

Regeneration of the CAPh-EDTA chelating system,
completing the loop sketched in Figure 76, was
achieved by passing dilute (0.5% potassium thiocyanate or potassium cyanide--the former being
preferable--through the exhausted film. Thiocyanate
and cyanide ions also form very stable complexes
with mercury (II) and cadmium (II). These complexes
are stronger than the corresponding complexes with
EDTA and therefore, it is believed, these ions remove the metal ion from the metal chelate-CAPh
structure. Following regeneration and washing with

water, the capacity of the EDTA-treated CAPh film
to remove mercury (II) and cadmium (II) is restored.

Treating the CAPh first with magnesium sulfate
and then with Na_2EDTA in separate steps did not
prove to be as effective as treating the CAPh with
$Na_2[Mg(EDTA)]$ in the single step treatment described
earlier for removing mercury or cadmium ions. This
was apparently due to the fact that the magnesium
salt of CAPh is water soluble.

CAPh ground to 20-mesh particle size and loosely
packed in a column was equally effective in removing
mercury (II) from solution with and without treatment
with $Na_2[Mg(EDTA)]$. No driving pressure other than
that due to the height of liquid above the column
was needed. A high throughput rate, about 4 ml/
minute, was obtained through a column 3 inches high
and 3/4 inches diameter containing 3 grams of CAPh.
With 1,000 mg NaCl/l present, only the EDTA-treated
powder was effective, as was found with EDTA-treated
film.

The flat film configuration used in these experiments can easily be scaled up and therefore can
be made the basis of a practical large unit design.
Other possible arrangements include a bed or column
of material as described above, a continuous band
of film or fiber tow for uninterrupted recycling,
spiral rolls, and so forth. Probably no one configuration will be best for all applications and
specific trial runs will be needed, before selection
of the best design for a particular situation or
set of circumstances.

This series of experiments on a small laboratory
scale has demonstrated the scope and potential of
using a complexing agent on a water-insoluble substrate to remove heavy metal pollutants from water.
A complexing site of magnesium-EDTA chelate on
water-insoluble cellulose acetate phthalate mixed
ester was employed to preferentially remove mercuric
and cadmium ions from water containing much larger
quantities of sodium chloride. Because cellulose
esters undergo hydrolysis at high (>10) and low (< 3)
pH, feed solutions should be controlled within these
limits, preferably pH 7 $\pm$ 2.

REFERENCES

1.   Katz, A. *Critical Reviews in Environmental Control 2*,
      517 (1972).
2.   Wood, J. M. *Environment 14* (1972).

3. Welcher, F. J. *The Analytical Uses of Ethylenediamine Tetraacetic Acid* (Princeton, N.J.: D. Van Nostrand Co., Inc., 1965).

4. Malm, C. J. and C. R. Fordyce. *Ind. Eng. Chem., 32,* 405 (1940).

5. Model, F. S., H. J. Davis, A. A. Boom, C. Helfgott, and L. A. Lee. *The Influence of Hydroxyl Ratio on the Performance of Reverse Osmosis Desalination Membranes.* Research and Development Report No. 657, U.S. Department of the Interior, Office of Saline Water (Washington, D.C., June, 1971).

6. Manjikian, S., S. Loeb, and J. W. McCutchan. First International Symposium on Water Desalination, Washington, D.C., October, 1965.

SUBJECT INDEX

SUBJECT INDEX

377